Application of Artificial Intelligence in Industrial Process Modelling and Optimization

Application of Artificial Intelligence in Industrial Process Modelling and Optimization

Guest Editors

Pan Yu
Sheng Du
Li Jin
Haipeng Fan

MDPI
Basel • Beijing • Wuhan • Barcelona • Belgrade • Novi Sad • Cluj • Manchester

Guest Editors
Pan Yu
School of Information Science and Technology
Beijing University of Technology
Beijing
China

Sheng Du
School of Automation
China University of Geosciences
Wuhan
China

Li Jin
School of Automation
China University of Geosciences
Wuhan
China

Haipeng Fan
School of Automation
China University of Geosciences
Wuhan
China

Editorial Office
MDPI AG
Grosspeteranlage 5
4052 Basel, Switzerland

This is a reprint of the Special Issue, published open access by the journal *Processes* (ISSN 2227-9717), freely accessible at: https://www.mdpi.com/journal/processes/special_issues/0OX7N1I66C.

For citation purposes, cite each article independently as indicated on the article page online and as indicated below:

Lastname, A.A.; Lastname, B.B. Article Title. *Journal Name* **Year**, *Volume Number*, Page Range.

ISBN 978-3-7258-4801-0 (Hbk)
ISBN 978-3-7258-4802-7 (PDF)
https://doi.org/10.3390/books978-3-7258-4802-7

Contents

About the Editors

Pan Yu

Pan Yu received the B.S. degree in industrial automation and the Ph.D. degree in control engineering from Central South University, China, in 2014 and 2019, respectively. She was a Research Student with the Department of Electrical and Electronic Engineering, Chiba University, Japan, from 2017 to 2019. In 2019, she joined Beijing University of Technology, Beijing, China, where she is currently an Associate Professor with the Faculty of Information Technology. Her research interests include multi-agent systems, disturbance estimation and rejection, time-delay systems, and robust control.

Sheng Du

Sheng Du received the B.S. degree in Measurement & Control Technology and Instrument and the Ph.D. degree in Control Science and Engineering from China University of Geosciences, Wuhan, China, in 2016 and 2021, respectively. He was a joint Ph.D. student with the Department of Electrical and Computer Engineering, University of Alberta, Edmonton, Canada, from 2019 to 2021. He was a Postdoctoral Fellow in Control Science and Engineering, China University of Geosciences, Wuhan, China, from 2021 to 2023. He is currently a Professor with the School of Automation, China University of Geosciences, Wuhan, China. His research interests include process control, intelligent control, intelligent optimization, computational intelligence, and artificial Dr. Du is a Senior Member of the Chinese Association of Automation (CAA) and a winner of the Outstanding Doctoral Thesis Nomination Award of the CAA. He is an Associate Editor of Measurement and Control.

Li Jin

Li Jin received the B.S. degree in Automation and the Ph.D. degree in Control Science and Engineering from China University of Geosciences, Wuhan, China, in 2016 and 2021, respectively. She was a joint Ph.D. student with the Department of Electrical Engineering and Electronics, University of Liverpool, Liverpool, U.K., from 2018 to 2020. She was a Postdoctoral Fellow in Control Science and Engineering, China University of Geosciences, Wuhan, China, from 2022 to 2024. She is currently a Professor in the School of Automation, China University of Geosciences, Wuhan, China. Her current research interests include power system stability analysis and control, time-delay systems, robust theory, and applications.

Haipeng Fan

Haipeng Fan received the B.S. degree in engineering from North China Electric Power University, Baoding, China, in 2015, and received the Ph.D. degree in the School of Automation, China University of Geosciences, Wuhan, China, in 2023. He is currently an associate professor with the School of Automation, China University of Geosciences, Wuhan, China. His research interests include condition monitoring, process control, and performance assessment.

processes

MDPI

Review

A Review of Data-Driven Intelligent Monitoring for Geological Drilling Processes

Sheng Du [1,2,3,4], Cheng Huang [1,3,4], Xian Ma [2,3,4] and Haipeng Fan [2,3,4,*]

1 School of Future Technology, China University of Geosciences, Wuhan 430074, China; dusheng@cug.edu.cn (S.D.); cheng_huang@cug.edu.cn (C.H.)
2 School of Automation, China University of Geosciences, Wuhan 430074, China; xianma@cug.edu.cn
3 Hubei Key Laboratory of Advanced Control and Intelligent Automation for Complex Systems, Wuhan 430074, China
4 Engineering Research Center of Intelligent Technology for Geo-Exploration, Ministry of Education, Wuhan 430074, China
* Correspondence: fanhaipeng@cug.edu.cn

Abstract: The exploration and development of resources and energy are fundamental to human survival and development, and geological drilling is a key method for deep resource and energy exploration. Intelligent monitoring technology can achieve anomaly detection, fault diagnosis, and fault prediction in the drilling process, which is crucial for ensuring production safety and improving drilling efficiency. The drilling process is characterized by complex geological conditions, variable working conditions, and low information value density, which pose a series of difficulties and challenges for intelligent monitoring. This paper reviews the research progress of the data-driven intelligent monitoring of geological drilling processes, focusing on the above difficulties and challenges. It mainly includes multivariate statistics, machine learning, and multi-model fusion. Multivariate statistical methods can effectively handle and analyze complex geological drilling data, while machine learning methods can efficiently extract key patterns and trends from a large amount of geological drilling data. Multi-model fusion methods, by combining the advantages of the first two methods, enhance the ability to handle complex multivariable and nonlinear problems. This review shows that existing research still faces problems such as limited data processing capabilities and insufficient model generalization capabilities. Improving the efficiency of data processing and the generalization capability of models may be the main research directions in the future.

Keywords: geological drilling process; intelligent monitoring; multivariate statistics; machine learning; multi-model fusion

1. Introduction

Geological resources, including petroleum, natural gas, minerals, and water, are indispensable natural resources for human social development. They are not only the raw material basis for industrial production but are also directly related to national energy security and economic independence. With the continuous growth of the global economy, the demand for geological resources is increasing, making the exploration and development of geological resources particularly important.

Since the implementation of the "14th Five-Year Plan", China's coalbed methane exploration and development have entered a higher stage. Breakthrough achievements have been made in exploring new fields and strata such as deep coal seams and thin coal seams, while significant results have been achieved in the enhancement and transformation of old gas fields [1]. Among these, the development of deep coalbed methane is particularly important. China's deep coalbed methane resources are abundant, with deep coalbed methane resources of 29 major basins (groups) estimated to reach 40.71 trillion cubic meters, significantly exceeding the shallow coalbed methane resources within 2000 m [2]. Especially in the depth range of 1500 to 2000 m, the proportion is 31.5%.

China's deep coalbed methane resources are mainly concentrated in the Junggar, Ordos, and Turpan–Hami–Santanghu basins, with these three regions accounting for 37%, 32%, and 27% of the total, respectively [3]. According to the China Petroleum Exploration and Development Research Institute, coalbed methane resources at depths of 2000 to 3000 m in China total about 18.47 trillion cubic meters [4]. Therefore, it is necessary to strengthen the exploration of deep geological resources. Moreover, the decision made at the 2023 National Natural Resources Work Conference also emphasized the importance of a new round of strategic mineral prospecting actions [5]. This policy aims to achieve breakthroughs in mineral prospecting by promoting technological projects and strengthening technical support in the field of resource exploration, thereby ensuring the economic and social development needs of the country.

In this context, the geological drilling process plays a crucial role in the development of deep coalbed methane resources. Precise drilling not only enables the acquisition of critical data on underground coalbed methane reserves but also provides core samples for analyzing their physical and chemical properties, thereby assessing the storage capacity and extractability of coalbed methane. However, current drilling technologies face numerous challenges; complex geological conditions significantly increase risks and costs, particularly in deep environments where the risk of tool wear and breakage escalates [6]. These issues can impact the efficiency of the geological drilling process and even compromise its safety. Therefore, the comprehensive monitoring of the geological drilling process must be implemented to ensure that operations are conducted safely and efficiently.

For an extended period, the detection of deep geological environments has encountered significant challenges due to limitations in technology and equipment. Current sensing technologies for deep geological detection often struggle to accurately capture the complexity and variability of subsurface structures, resulting in low information density during the drilling process [7]. This limitation hampers the comprehensive monitoring capabilities essential for effective geological drilling. To mitigate these challenges, the exploration and application of intelligent monitoring technologies have become crucial for achieving the precise monitoring of inefficiencies and abnormal statuses in drilling. Such intelligent monitoring can analyze drilling parameters in real time, predict and avoid potential risks, and offer considerable scientific and economic benefits in enhancing drilling efficiency, reducing energy consumption and ensuring operational safety [8].

Intelligent monitoring in geological drilling integrates anomaly detection, fault diagnosis, and fault prediction. Anomaly detection serves as the foundation of intelligent monitoring, continuously tracking key indicators such as drilling speed, pressure, and torque to identify deviations from normal operating patterns, thereby facilitating early fault detection. Fault diagnosis entails a thorough analysis of these detected anomalies to pinpoint specific fault types and their locations, which is essential for resolving issues and implementing targeted corrective actions. Finally, fault prediction leverages both historical and real-time data, alongside diagnosed fault types, employing advanced data analysis and machine learning models to forecast the likelihood and timing of future faults. This comprehensive approach enhances safety, efficiency, and reliability in the geological drilling process through real-time monitoring, prompt fault diagnosis, and precise predictions of potential future faults [9].

In summary, scholars have conducted extensive research in the field of data-driven intelligent monitoring, which plays a crucial role in the exploration and development of geological resources. These studies not only reduce the cost and risk of geological drilling but also improve the efficiency of data processing and decision making. This paper will review the academic contributions in the field of intelligent monitoring methods in the geological drilling process, including anomaly detection, fault prediction, and fault diagnosis, while introducing representative methods in these three aspects. By summarizing valuable experiences in the research process, it points out the existing problems and prospects for the future development of intelligent monitoring in the geological drilling process.

2. Descriptions and Analysis of Drilling Process

Geological drilling is a complex engineering process that faces numerous challenges due to uncertain geological conditions and the demanding nature of equipment operation. In response, key technologies like Measurement While Drilling (MWD), Logging While Drilling (LWD), Managed Pressure Drilling (MPD), and Rotary Steerable Systems (RSSs), along with advancements in intelligent monitoring, have played a crucial role in improving both the efficiency and safety of drilling operations [10,11]. These technologies enable the real-time monitoring of critical drilling parameters and fault diagnosis, allowing for more precise control, the prediction of drilling performance, and overall success in the drilling process.

2.1. Characteristic Analysis of Drilling Process

Geological drilling involves a highly complex process, requiring the coordination of multiple components and technologies to drill into the Earth's crust. As shown in Figure 1, the system consists of a traveling block, draw works, a rotary table, and other components essential for controlling the drill pipe's movement. The drill bit, positioned at the bottom of the drill string, penetrates underground formations. The bottom-hole assembly includes various tools for guiding and monitoring the drilling process. Mud is pumped down the drill pipe via the slurry pump and returns through the mud pit, carrying debris back to the surface [12]. This process involves managing various parameters such as drilling speed, weight on the bit, and torque to ensure efficiency and safety. Intelligent monitoring plays a critical role in tracking these factors and optimizing drilling operations in real time.

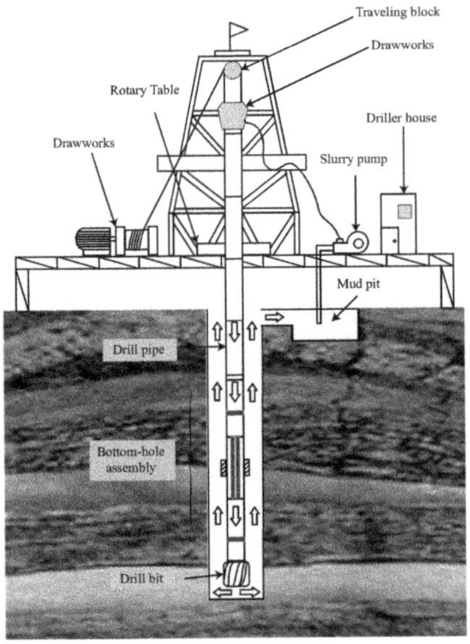

Figure 1. Description of the drilling process.

(1) Multi-condition Characteristics

Multi-condition characteristics refer to the different behaviors and performance features exhibited by systems or equipment under different operating conditions, environmental settings, or working states. The geological drilling process is a highly complex and variable engineering environment, involving various working states and conditions. As drilling progresses, systems or equipment often need to operate under multiple conditions, including different loads, speeds, temperatures, pressures, or environmental conditions. Each condition can uniquely impact the system's performance and behavior, posing significant challenges to the data-driven intelligent monitoring of the drilling process.

Parameter Coupling: One significant challenge faced by the monitoring system in the geological drilling process is the diversity of the drilling environment. Due to the varying geological formations, such as transitioning from sandstone to shale or encountering fractured zones and increasing pressure and temperature with drilling depth, multiple drilling parameters become interdependent. For instance, an increase in the drilling fluid density to control formation pressure can affect the rate of penetration and equivalent circulating density, which in turn influences the risk of wellbore instability. These parameters are not independent but are influenced by other parameters and environmental conditions, interacting closely. This parameter coupling requires an intelligent monitoring system to analyze the interrelationships and linkage effects among multiple parameters, such as how changes in rotary speed and torque affect bit wear and drilling efficiency, rather than simply tracking changes in individual parameters [13].

High-dimensional Data: The diverse conditions involved in geological drilling result in numerous parameter variables that the monitoring system needs to handle. These variables include rock hardness, type, drilling depth, drilling technology, drilling speed, and more, each directly or indirectly affecting drilling efficiency and safety [14]. High-dimensional data not only increase information volume but also introduce challenges in data analysis as data dimensions increase.

In summary, the multi-condition characteristics of the geological drilling process reflect its complexity and dynamic variability. Different conditions may result in completely different data distributions, and condition changes are frequent and complex. This necessitates high adaptability in data-driven intelligent monitoring to accommodate multiple condition switches. Real-time adaptation to changing conditions is essential, capturing different condition changes accurately to ensure drilling process efficiency and safety.

(2) Non-stationary Characteristics

Non-stationary characteristics, a key concept in time series analysis, refer to data not maintaining stability over time, exhibiting trends, seasonality, volatility, and autocorrelation. This means that the data's mean, variance, and correlation change over time, complicating analysis and prediction [15]. In the drilling process, the non-stationary characteristics impact data-driven intelligent monitoring in the following ways:

Temporal Dynamics: As the drill bit penetrates deeper, it encounters diverse geological challenges, leading to changes in statistical properties of monitored parameters. For instance, the drilling pressure may increase due to harder rock formations, while the drilling speed might decrease as the bit encounters more resistance. This variability necessitates that monitoring systems adapt to these fluctuations, ensuring real-time adjustments in analysis models. Continuous changes in physical parameters—such as increased mud flow to maintain borehole stability and variations in temperature and the density of the drilling fluid—highlight the need for a responsive monitoring system to maintain accuracy throughout the drilling cycle [16].

Spatial Dynamics: The spatial variability in geological conditions, such as differences in rock types, fault distributions, and stratigraphic features, directly impacts the selection of drilling parameters and overall drilling performance. For instance, drilling through varying rock types may require adjustments in the weight on the bit and rate of penetration. Intelligent monitoring systems need to account for these spatial variations to accurately detect anomalies. The proper interpretation of spatial data, including real-time measurements of rock hardness and fracture density, is essential for anticipating risks and enhancing safety. This knowledge enables optimized parameter settings and drilling strategies, improving efficiency while reducing hazards [17].

Due to the non-stationary characteristics of the geological drilling process, parameters and conditions continuously change in a nonlinear manner. This means the drilling process exhibits temporal and spatial dynamics, with a high degree of spatiotemporal information coupling. This complexity and unpredictability demand high sensitivity in intelligent monitoring to adapt to these dynamic characteristics.

(3) Low Information Value Density

In the geological drilling process, despite collecting vast amounts of data, the proportion of truly valuable information is relatively low, indicating low information value density. This is mainly due to two reasons:

Transmission delay: The significant distance between surface and downhole monitoring sensors may result in data transmission delays. While these delays are generally short, they can still affect real-time monitoring and decision making in fast-changing drilling environments. Even slight delays in data transmission can reduce the timeliness of information and may occasionally limit the effectiveness of prompt adjustments, potentially impacting decision accuracy [18].

Noise interference: Drilling equipment generates significant mechanical and acoustic noise, which can interfere with various sensors used in Measurement While Drilling and Logging While Drilling systems, such as acoustic, pressure, and vibration sensors. Mechanical noise is produced by the movement and operation of the drilling equipment, while acoustic sensors in Measurement While Drilling systems may struggle to differentiate between background noise and meaningful signals. Pressure sensors in Logging While Drilling systems can be affected by mud flow variations or sudden pressure changes [10,11]. As the drilling depth increases, the higher temperature and pressure present additional challenges, leading to potential signal drift or distortion. Furthermore, equipment wear, such as drill bit damage or irregular drill rod movement, introduces additional vibration noise, complicating the analysis of sensor data [19].

In summary, the low information value density in the geological drilling process is primarily due to transmission delays and noise interference. Delays reduce information timeliness, and noise can lead to inaccurate or distorted data, affecting decision accuracy. Addressing this challenge requires intelligent monitoring to handle large data volumes and quickly extract valuable information to support effective drilling decisions and operations.

2.2. Functions of Intelligent Monitoring in Geological Drilling

Intelligent monitoring plays a crucial role in the geological drilling process, with key functions including anomaly detection, fault diagnosis, and fault prediction. Anomaly detection monitors real-time drilling parameters such as the torque, weight on bit, drilling speed, and mud flow rate to detect deviations from normal patterns, signaling potential issues like bit wear or formation changes [20]. Fault diagnosis then identifies the root causes of these anomalies and recommends corrective actions to maintain operational continuity and efficiency [21]. Fault prediction leverages historical and real-time data to anticipate future faults, enabling proactive adjustments to prevent incidents like drill string failure or wellbore instability [22]. Together, these functions form the foundation of intelligent monitoring, making it indispensable for modern geological drilling operations.

(1) Anomaly Detection

In geological drilling, data-driven anomaly detection techniques play a crucial role. In analyzing drilling process data in real time, these techniques identify behaviors that may cause deviations from normal operations, enhancing safety and efficiency. Real-time data analysis forms the core of data-driven anomaly detection.

To achieve real-time anomaly detection, Reeber et al. proposed a drilling tool wear-monitoring method based on Extreme Gradient Boosting and autoencoders. They utilized Extreme Gradient Boosting to model complex nonlinear relationships within drilling data and employed autoencoders to detect anomalies by identifying deviations between original and reconstructed data. This combination allows for the efficient processing and accurate detection of tool wear in real time [23]. These methods process and analyze drilling data in real time for efficient and accurate anomaly detection. Similarly, Alsaihati et al. proposed an intelligent system using real-time data analysis and machine learning models to predict surface torque during drilling, using the Mahalanobis distance for the anomaly detection of downhole issues like stuck pipes [24].

And Zhong et al. employed a Convolutional Long Short-Term Memory Neural Network model for real-time anomaly detection in drilling data. This model integrates convolutional layers to capture spatial features and utilizes long short-term memory units to model temporal dependencies in sequential data, enhancing the detection of anomalies by learning both spatial and temporal patterns in drilling data [25]. Li et al. proposed an anomaly detection method based on the relationship between input and output signals in the drilling process, developing mathematical models to establish normal operational behavior and detecting deviations from this behavior to identify anomalies, emphasizing the critical role of real-time data analysis in data-driven anomaly detection [26].

These studies demonstrate that data-driven anomaly detection in geological drilling can process vast amounts of data to identify behaviors deviating from normal drilling operations. As the initial part of intelligent monitoring, the goal is to quickly extract potential anomaly information from real-time drilling data, providing immediate optimization guidance for drilling teams.

(2) Fault Diagnosis

Data-driven fault diagnosis involves analyzing specific causes, nature, and solutions for faults after anomaly detection. In geological drilling, fault diagnosis is a key function of intelligent monitoring. It helps drilling teams promptly identify and resolve issues, enhancing overall safety and efficiency. This includes determining fault types, locating sources, and proposing corresponding repair or adjustment measures.

Fault diagnosis in geological drilling has evolved significantly, starting from theoretical model establishment to incorporating data-driven methods, particularly machine learning and deep learning techniques. Reiss provided the theoretical foundation for fault diagnosis in geological drilling [27]. As technology advanced, data-driven methods were introduced. Shen et al. developed a condition monitoring and fault diagnosis system by integrating serial communication protocol bus technology [28]. Zhang et al. advanced the field with an automatic fault diagnosis system based on drilling parameters [29], using Principal Component Analysis and Self-Organizing Maps for accurate fault diagnosis. Additionally, deep learning techniques enable fault classification from sound signals, reducing dependence on expert experience and improving diagnosis accuracy [30]. Vununu et al. combined Principal Component Analysis and Artificial Neural Networks to develop an automatic machine fault diagnosis system based on sound, demonstrating the application value of machine learning in fault diagnosis [31].

These studies show the evolution from theoretical model-based methods to modern automated diagnosis based on data. This progression has enhanced fault diagnosis accuracy and efficiency, providing reliable decision support for drilling operations. Fault diagnosis offers in-depth problem analysis and solutions, minimizing downtime and reducing potential safety risks, making it an indispensable core function in intelligent monitoring.

(3) Fault Prediction

Data-driven fault prediction is crucial in geological drilling, involving various technical applications to predict faults based on anomaly detection and fault diagnosis results. It significantly reduces unplanned downtime and improves operational safety and efficiency, making it a research hotspot in geological drilling. Advanced data analysis, machine learning algorithms, and artificial intelligence extract valuable information from drilling data to identify potential risks and fault signs.

To predict potential faults during drilling, scholars proposed a method for predicting open-hole cable logging faults [32], aiming to reduce costs and time increases caused by faults and improve drilling efficiency. They employed three machine learning techniques—support vector machine, naive Bayes, and decision tree—to predict open-hole cable logging results based on drilling process data. The support vector machine showed an optimal prediction accuracy. This predictive capability is closely linked to the drilling and logging processes. Logging provides essential data about the geological formations encountered, which can influence drilling parameters and decision making. By integrating fault prediction with logging, operators can better anticipate potential issues, ensuring a smoother drilling operation. This connection underscores the importance of fault prediction not only in improving efficiency but also in enhancing safety during drilling activities.

Similarly, Noshi et al. used supervised and unsupervised learning data mining algorithms [33], including logistic regression, hierarchical clustering, and decision tree, to analyze comprehensive data from eighty land wells for predicting casing failure. Zhai et al. developed an intelligent prediction model for drilling complexity based on case-based reasoning, integrating adjacent well data, computer technology, artificial intelligence, and data mining [34] to diagnose and predict potential faults before drilling operations.

3. Intelligent Monitoring in Drilling Process

Currently, data-driven intelligent monitoring research for geological drilling processes can be categorized into multivariate statistical methods, machine learning techniques, and multi-model fusion methods. Multivariate statistical methods can effectively reveal key patterns and trends in data [35], aiding in the detection of abnormal drilling conditions. However, while multivariate statistical methods are effective in data simplification and interpretation, they may struggle with nonlinear complex patterns. In contrast, machine learning methods are highly powerful in pattern recognition [36] and prediction [37], but their opacity [38] presents challenges for result interpretation and validation. Further, combining multivariate statistical methods with machine learning techniques, known as multi-model fusion methods, can leverage their respective strengths, enhancing the accuracy and reliability of intelligent monitoring. This section will analyze current application research status of intelligent monitoring in geological drilling processes from these three directions, exploring their advantages and existing issues.

3.1. Intelligent Monitoring Based on Multivariate Statistics

Multivariate statistical analysis involves using mathematical and statistical methods to analyze and interpret relationships and patterns in multivariable datasets. In geological drilling, this includes analyzing multiple related variables such as the drilling pressure, speed, torque, and mud flow rate [13]. Multivariate statistical analysis is essential in the monitoring of the geological drilling process as it can reveal complex, multidimensional relationships within the data, helping to identify geological features, optimize drilling efficiency, and detect potential risks in real time [39]. This analysis supports intelligent monitoring systems by providing a scientific basis for data interpretation, thus enhancing the accuracy and reliability of prediction models, as shown in Table 1.

Table 1. Overview of methods for geological drilling monitoring.

Considering Issue	Method	Characteristics	Application Scenarios
Multi-condition Characteristics	DB Clustering [39]	Handling data abruptly and slow changes	Local similarity analysis of multi-condition data in drilling process
	MB-SFA, MB-ICA [40]	Considering static, dynamic, and large-scale characteristics	Complex condition monitoring in modern industrial processes
	SFA, BN [41]	Extracting static and dynamic features and clustering analysis	Anomaly detection in multi-mode switching during drilling process
	DIPCA [42]	Extracting dynamic, linear, and nonlinear features	Real-time monitoring of nonlinear dynamic processes
	Multi-step DSFA [43]	Precisely partitioning dynamic conditions, and changing control limits	Full-condition monitoring of dynamic systems

Table 1. *Cont.*

Considering Issue	Method	Characteristics	Application Scenarios
Non-stationary Characteristics	MCC [44]	Capturing complex dynamic characteristics	Multivariate anomaly detection in non-stationary processes
	IPCA [45]	Block processing and handling dynamic characteristics	Process monitoring under non-stationary characteristics
	Deterministic Alg. [46]	Extracting the slowest varying features	Monitoring dynamic changes in time series data
	KPCA, KL Div. [47]	Handling minor shifts	Detecting subtle changes in non-stationary processes
	CA [48]	Extracting non-stationary features from historical data	Predicting fouling in steam generator pipes
	CA [49]	Constructing a stationary feature data set	Dynamic monitoring of data non-stationary characteristics
	ACA [50]	Distinguishing true faults from normal variations	Fault identification in dynamically changing environments

(1) Considering Multi-condition Characteristics

Applying multivariate statistical methods is a key step in analyzing geological drilling data, especially considering the multi-condition characteristics of the drilling process. Geological drilling is influenced by various complex factors, including the physical and chemical properties of geological formations, technical parameters of drilling tools, and operational conditions, collectively forming the multi-condition characteristics. Multivariate statistical analyses, such as principal component analysis (PCA), cluster analysis, and factor analysis, can effectively handle and analyze these complex multivariable data.

Given the correlations among various conditions in the geological drilling process, Zhang et al. used time series feature extraction and density-based (DB) clustering methods to analyze the extracted features, addressing the data fluctuations and slow variations due to multi-condition characteristics, particularly in detecting and diagnosing wellbore instability issues such as lost circulation and kick [39]. Additionally, Huang et al. proposed a dual-layer distributed monitoring structure [40] based on multiblock slow feature analysis (MB-SFA) and multiblock independent component analysis (MB-ICA), to handle complex static, dynamic, and large-scale characteristics in modern industrial processes. Xu et al. proposed multiple subspace slow feature analysis (SFA) [41], suitable for addressing issues arising from multi-condition characteristics. This method uses domain subtraction clustering to divide different modes, further dividing each mode into Gaussian and non-Gaussian subspaces, extracting static and dynamic features, and performing Bayesian network (BN) fusion monitoring for anomaly detection. Guo et al. proposed a multi-feature extraction technique based on principal component analysis [42] for nonlinear dynamic process monitoring, combining dynamic internal principal component analysis (DIPCA), PCA, and kernel principal component analysis (KPCA) in a serial structure to extract dynamic, linear, and nonlinear features. Ma et al. proposed a new multi-step dynamic slow feature analysis (DSFA) algorithm [43], carrying out full-condition monitoring for dynamic systems, precisely dividing dynamic conditions, and adjusting control limits based on condition changes.

(2) Considering Non-stationary Characteristics

Data in the drilling process typically exhibit significant non-stationarity, including trends, periodicity, seasonality, and random noise in time series. These characteristics make traditional statistical methods and models inadequate, necessitating the use of multivariate statistical methods to better understand and analyze these non-stationary characteristics. Messaoud et al. achieved anomaly detection in the drilling process through time series analysis and multivariate control charts (MCCs) [44], capturing complex dynamic characteristics of non-stationary processes. Fan et al. addressed the challenges of non-stationarity by proposing a distributed monitoring method based on integrated probabilistic component analysis (IPCA) and the minimal redundancy–maximum relevance algorithm [45], effectively handling the complex dynamic characteristics due to non-stationarity.

To further address non-stationary characteristics, Zafeiriou extended slow feature analysis (SFA) [46], including a novel deterministic algorithm (Alg.) and an expectation maximization (EM) algorithm to extract the slowest varying features from multiple time-varying data sequences. Cai et al. combined kernel principal component analysis (KPCA) with Kullback–Leibler (KL) divergence [47] to handle changes due to small shifts in non-stationary processes, validated on typical experimental datasets. Kwak et al. used cointegration analysis (CA) to extract non-stationary features accumulated in historical data [48], successfully applied in the fouling prediction of DC steam generator pipes. Wen et al. combined extracted non-stationary features with stationary features to form a new stationary feature dataset, updating monitoring indicators [49].

Given the frequent change patterns in geological drilling monitoring, Zhang et al. proposed an adaptive cointegration analysis (ACA) method [50] to distinguish real faults from normal changes, updating the model with normal samples and adapting to gradual changes in cointegration relationships. Alternatively, Rao et al. proposed a non-stationary process monitoring method based on alternating conditional expectations (ACE) and CA [51], maximizing the linear correlation of transformed variables to handle nonlinear relationships between variables. Zhao et al. proposed a sparse CA-based total variable decomposition and distributed modeling algorithm [52] for non-stationary processes, fully decomposing different cointegration relationships between non-stationary variables and exploring the close linear correlations through local cointegration vectors in each block.

In summary, detection methods based on multivariate statistics are increasingly attracting attention in drilling process monitoring. These techniques allow for the in-depth analysis of complex multivariable data, revealing intrinsic relationships between variables for more precise state monitoring of the geological drilling process. However, implementing these methods often requires determining the types of data to be monitored first and constructing corresponding monitoring models based on data characteristics, leading to multiple modeling processes to address multi-condition characteristics. Future research directions will need to deepen our understanding and application of multivariate statistical methods and innovate algorithms and technologies to meet high-standard monitoring requirements for geological drilling, achieving efficient and accurate monitoring and analysis.

3.2. Intelligent Monitoring Based on Machine Learning

With the rapid development of artificial intelligence technology, machine learning, as a core branch, is increasingly applied in various fields. In geological drilling monitoring, the introduction of machine learning techniques provides new perspectives and methods for traditional geological drilling operations. Machine learning enables computer systems to learn from data and make decisions without explicit programming. In the context of geological drilling process monitoring, machine learning techniques analyze historical geological drilling data, geological information, and real-time monitoring data to learn complex relationships between geological features and drilling process parameters [53]. This enables the accurate detection of anomalies, real-time fault diagnosis, and predictive maintenance, enhancing the overall safety and efficiency of drilling operations, as shown in Table 2.

(1) Anomaly Detection

With the development of machine learning technology, its application in drilling anomaly detection is becoming more mature. Machine learning models can analyze historical geological drilling data and real-time monitoring data, learning the distinctions between normal operations and anomalies, enabling the automatic detection of potential anomalies in the drilling process. Compared to traditional rule-based and experience-based detection methods, machine learning offers higher flexibility and accuracy, effectively reducing the risk of human error. Liao proposed a neural network (NN)-based [53] model, emphasizing the ability to distinguish between normal and abnormal drilling states, optimizing network performance through algorithm improvements. Yang et al. developed a local outlier factor (LOF) anomaly detection algorithm to detect various anomalies [54], validated in NN monitoring models.

Table 2. Machine learning methods applicable to monitoring the drilling process.

Task	Method	Characteristics	Application Scenarios
Anomaly-Detection	NN [53]	Multi-param. fusion, real-time monitoring	Identifying different states in the drilling process
	LOF, NN [54]	Detecting local anomalous data	Anomaly detection in NN monitoring
	RF [55]	Dimensionality reduction, improving efficiency	Extracting effective features for anomaly detection
	DBN, GAN [56]	Reconstructing missing data, feature selection	Anomaly detection in high-dimensional data
	Cascade monitoring, CNN [57]	Analyzing spatial–temporal info, combining sub-models	Comprehensive anomaly detection in industrial processes
	GMM, stacked denoising AE [58]	Initial mode identification, extracting deep nonlinear features	Robust monitoring under steady-state modes
Fault-Diagnosis	AC-GAN, Bayesian algo. [59]	Mitigating data scarcity issues	Automatic diagnosis of downhole drilling accidents
	CNN [60]	Incremental learning, including new samples	Dynamically updating fault diagnosis
	Multi-task learning, CNN [61]	Simultaneous anomaly localization and fault classification	Fault diagnosis in complex processes
	CNN [62]	High-precision classification	High-precision fault diagnosis
	DT [63]	Clear rules, easy to interpret	Fault diagnosis and alarm design in industrial processes
	Optimal ELM, Bernoulli transform coyote opt. [64]	Improving classifier performance	Enhancing fault diagnosis accuracy
Fault-Prediction	BN, LSTM [65]	Combining time series prediction	Early fault warning for steam turbines
	MLP, ANN, BN [66]	Combining multiple models, enhancing prediction accuracy	Fault prediction in production processes
	BN [67]	Handling uncertainty	Early warning of wellbore loss and influx accidents
	BN [68]	Flexible modeling, handling complex relationships	Monitoring operational parameters in oil wells

Addressing low information value density, Li et al. proposed a feature simplification random forest (RF) algorithm [55] to extract effective features, reducing dimensionality and improving anomaly detection efficiency. Tian et al. proposed a feature-based deep belief network (DBN) method [56], using generative adversarial networks (GANs) to reconstruct random and non-random missing data, selecting feature variables using Spearman's rank correlation coefficients from high-dimensional data, and successfully employing a DBN for deep abstraction, learning, and tuning in anomaly detection.

Additionally, Yu et al. proposed a cascade monitoring network to simultaneously analyze spatiotemporal information for detecting industrial process anomalies [57]. This method uses convolutional neural networks (CNNs) to extract spatiotemporal information from each variable, combining multiple sub-models into a final monitoring model. Gao et al. used Gaussian mixture models (GMMs) for preliminary mode identification [58], employing stacked denoising autoencoders (AEs) to extract deep nonlinear features embedded in process variables, establishing robust monitoring models for each steady-state mode.

(2) Fault Diagnosis

Machine learning techniques learn complex relationships between normal and abnormal states in geological drilling by analyzing historical and real-time data, enabling automatic fault identification and classification. Wang et al. proposed a downhole drilling accident diagnosis method [59] using an auxiliary classifier generative adversarial network (AC-GAN) to expand the dataset and a Bayesian algorithm for the diagnosis model, addressing low information value density. Yu et al. proposed an incremental-learning general CNN [60], updating itself with newly collected abnormal samples and fault categories for fault diagnosis. Zhao et al. proposed a multi-task learning CNN model [61] for simultaneous abnormal variable localization and fault classification, applicable in geological drilling fault diagnosis.

Further, Glaeser et al. successfully achieved high-precision fault diagnosis using advanced CNN classifiers [62]. Dorgo et al. proposed a decision tree (DT) classifier-based alarm information design method [63] for industrial process fault diagnosis, applicable in geological drilling fault diagnosis. Hu et al. proposed a new fault diagnosis method based on optimal extreme learning machine (ELM) [64], using a Bernoulli transform–coyote optimization algorithm to optimize the kernel ELM classifier, improving fault diagnosis accuracy. These methods, employing typical machine learning techniques, vary in strategies but are suitable for geological drilling fault diagnosis, enhancing fault diagnosis precision.

(3) Fault Prediction

Timely and accurate fault prediction is crucial in geological drilling for ensuring operational safety, reducing costs, and improving efficiency. Machine learning techniques analyze historical and real-time monitoring data, learning complex relationships between normal and abnormal states, enabling the early identification and prediction of potential faults. Bayesian networks (BNs), which are a typical machine learning algorithm, are often used for fault prediction.

For instance, Zhang et al. proposed a method for predicting wellbore loss and influx accidents in drilling processes by constructing a BN-based prediction model [67]. They selected critical drilling parameters, such as mud weight, formation pressure, and drilling rate, that represent accident characteristics and considered the uncertainty of parameter changes during accidents. Their model effectively predicted potential wellbore instability incidents, providing valuable guidance for drilling operations.

Similarly, Mamudu et al. utilized Bayesian networks to develop fault prediction models for monitoring operational parameters in oil wells [68]. By incorporating various operational data like pressure, temperature, and flow rates into the BN model, they could predict potential faults in real time, enhancing the reliability and safety of oil well operations.

Liu et al. proposed an early fault warning method based on a combination of Bayesian networks and long short-term memory (LSTM) neural networks [65]. They developed an LSTM prediction model that addresses data uncertainty and considers complex equipment operations. Tested with real steam turbine data, their method provided accurate early warnings during fault creep stages. Although applied to steam turbines, this approach is applicable to geological drilling, where equipment complexity and data uncertainty are significant challenges.

In another study, Mamudu et al. also proposed a method combining multilayer perceptrons (MLPs) and artificial neural networks (ANNs) with BN techniques for effective production fault warning [66]. By integrating MLP and ANN models with a BN, they improved the fault prediction accuracy in production systems. This method is applicable to geological drilling fault prediction due to similar operational complexities and the need for accurate fault forecasting.

These methods employ machine learning, especially Bayesian network techniques, differing in technical approaches and focuses but are applicable in geological drilling fault prediction. They demonstrate the effectiveness of advanced machine learning models in handling complex relationships between drilling parameters and predicting potential faults.

In summary, with the rapid advancement of artificial intelligence, especially machine learning, its application in geological drilling monitoring is becoming increasingly widespread. Through deeply analyzing historical geological drilling data, geological information, and real-time monitoring data, machine learning techniques can identify complex relationships between drilling parameters, achieving effective state monitoring. Despite significant potential in enhancing safety, efficiency, and accuracy, challenges remain in data quality and availability, model generalization and adaptability, real-time computational efficiency, and model interpretability. Future research must explore data processing, model optimization, and technological innovation to address these challenges, further advancing geological drilling monitoring technology.

3.3. Intelligent Monitoring Based on Multi-Model Fusion

As shown in Table 3, multi-model fusion involves organically combining multivariate statistical analysis and machine learning methods, providing comprehensive and precise monitoring for geological drilling. This approach integrates various data analysis techniques, retaining the interpretability of multivariate statistical methods while leveraging machine learning's strength in handling nonlinear features, better managing complex data, and offering robust fault prediction and precise fault diagnosis. Consequently, the effectiveness of multi-model fusion methods in intelligent monitoring applications in geological drilling has been widely researched.

Table 3. Multi-model fusion for the drilling process.

Task	Method	Characteristics	Application Scenarios
Anomaly Detection	Hybrid PCA, multivariate CNN-LSTM [69]	Enhancing anomaly detection performance	Anomaly detection and optimization
	Ensemble learning, KCVA, Bayesian inference [70]	Improving monitoring performance	Monitoring complex industrial processes
Fault Diagnosis	Enhanced RF, SFA [71]	Analyzing static and dynamic nodes	Dynamic fault classification
	Ensemble learning, ICA [72]	Enhancing model generalization	Monitoring non-Gaussian processes
	EWC, PCA [73]	Continuous learning, preventing forgetting	Monitoring complex and variable industrial processes
Fault Prediction	Machine learning models, multivariate statistics [74]	Integrating multiple techniques	Predicting potential faults in the drilling process

To validate the effectiveness of multi-model fusion methods, Islamov used machine learning models combined with multivariate statistical methods to predict potential faults in geological drilling [74]. The study compared various machine learning algorithms, including logistic regression, naive Bayes classifier, K-nearest neighbors, decision trees, support vector machines, RF, gradient boosting, and NNs, to identify and classify abnormal states. Multivariate statistical methods were used to evaluate different machine learning algorithms' performance, including accuracy, recall, and F-score, to determine the most suitable fault prediction model for geological drilling. Barbosa emphasized machine learning's potential in predicting and optimizing drilling rates, highlighting multivariate statistical methods' importance in model performance evaluation and feature selection [75].

While machine learning techniques may outperform traditional models in fault prediction accuracy, multivariate statistical analysis remains crucial in feature selection and model evaluation, demonstrating the effective fusion of both methods. Chai proposed an enhanced RF [71], analyzing static and dynamic nodes simultaneously, classifying faults, and using a modified SFA method to design new slow indices for supervised fault classification, reflecting the fusion of machine learning and multivariate statistics in logic and standard design, applicable in drilling fault classification.

Further, more in-depth multi-model fusion methods combine machine learning and multivariate statistics. Tariq proposed using hybrid probabilistic PCA combined with multivariate convolutional long short-term memory (CNN-LSTM) models [69], integrating neural networks and probabilistic clustering to enhance anomaly detection performance. Li and Yan proposed an independent component analysis (ICA) method based on ensemble learning [72] for non-Gaussian process monitoring, improving model generalization by integrating ensemble learning logic. Wang and Wu proposed a similar method, introducing ensemble learning and kernel canonical variable analysis (KCVA) to develop a novel ensemble kernel canonical variable analysis method [70], combining multiple KCVA models using Bayesian inference to improve process monitoring performance.

The fusion of machine learning and multivariate statistical modeling addresses their respective shortcomings. For instance, Zhang used elastic weight consolidation (EWC) to solve catastrophic forgetting in PCA models [73], achieving continuous learning, applicable in complex and variable industrial process monitoring, including intelligent geological drilling monitoring.

In summary, multivariate staistical analysis and machine learning each have their own strengths in geological drilling monitoring. Multivariate statistical analysis excels at identifying relationships between multiple variables, providing clear insights into trends and correlations in drilling data. Machine learning, on the other hand, is powerful in processing large datasets and detecting complex patterns, allowing for real-time anomaly detection and predictive fault diagnosis. The fusion of these two methods combines their strengths, resulting in a more robust approach to monitoring, where data interpretation and fault prediction are both enhanced. Together, these three approaches play a critical role in improving the accuracy, reliability, and efficiency of geological drilling monitoring, as shown in Figure 2.

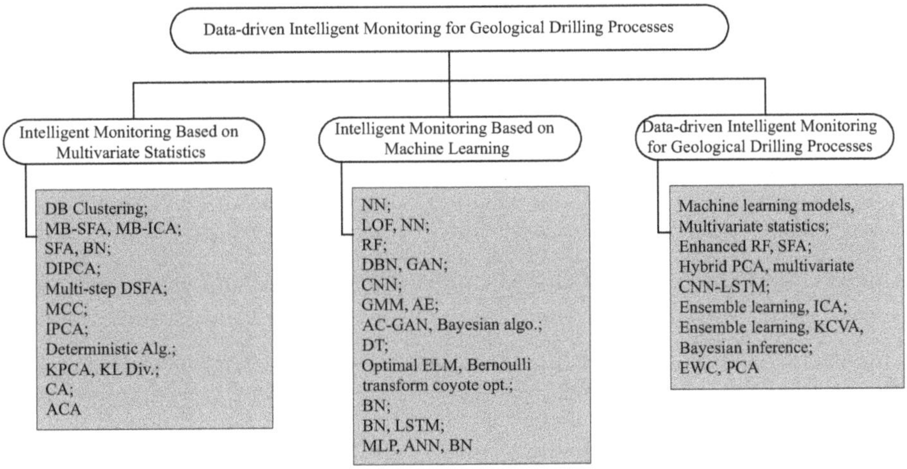

Figure 2. Data-driven intelligent monitoring for geological drilling processes.

4. Challenges and Prospects

In recent years, significant progress has been made in the application of multivariate statistical methods and machine learning techniques in the field of geological drilling data processing and monitoring. These technologies leverage their unique advantages in handling complex data, providing new opportunities to enhance the intelligent monitoring of geological drilling processes. Moreover, their integration has the potential to facilitate real-time decision making and improve operational efficiency. Despite achieving a range of results, some challenges remain in practical applications. Future research directions will focus on exploring more effective solutions to address these challenges and promote further advancements in this field.

4.1. Challenges

Multivariate statistical methods play a crucial role in the processing of geological drilling data, effectively revealing multidimensional relationships within the data and providing a scientific explanation for intelligent monitoring. These methods demonstrate unique advantages in addressing the multifactorial and non-stationary characteristics of the drilling process. Concurrently, the introduction of machine learning techniques has significantly enhanced the level of intelligence in monitoring, allowing for more accurate predictions of drilling performance and potential risks through the analysis of historical and real-time data, thereby facilitating effective anomaly detection, fault diagnosis, and prediction.

The integration of multivariate statistical analysis and machine learning techniques through a multi-model fusion approach offers a more comprehensive and precise solution for geological drilling monitoring. This innovative method not only improves data processing capabilities but also enhances the accuracy of fault warnings and the adaptability of diagnostic models. Nevertheless, current intelligent monitoring technologies still face several limitations, which constrain their effectiveness and further development, primarily manifesting as the following:

(1) Lack of Comprehensive Consideration of Global and Local Features

Current drilling process-monitoring technologies face two main issues when addressing multi-condition and non-stationary characteristics. On one hand, focusing on the global features of the drilling process often overlooks the importance of local features, which may have a decisive impact on drilling efficiency and safety under specific conditions. On the other hand, when concentrating on local features, the relationships between these features and their collective impact on the overall drilling process might be missed. In geological drilling monitoring, maintaining a dynamic balance between global and local features is crucial. Global features provide an overall trend of the drilling process, while local features reveal subtle, short-term changes, which are often key to predicting anomalies and avoiding potential risks.

(2) Scarcity and Low Information Value Density of Drilling Data

Existing data-driven intelligent monitoring methods can somewhat mitigate the challenges of data scarcity and low information value density in geological drilling. Nevertheless, these methods still face challenges in processing complex geological drilling process data. The uncertainty and variability of the drilling environment require intelligent monitoring systems to handle large volumes of data and possess high generalizability. Drilling data may become scarce due to equipment limitations, costs, and external environmental factors, and the valuable information density within the data might be low. This necessitates more effective identification and utilization of potential value in sparse data during intelligent geological drilling monitoring.

(3) Lack of Spatiotemporal Information Coordination

Although recent data-driven intelligent monitoring methods for geological drilling processes have made significant technological advancements, particularly in addressing the temporal characteristics and interrelations of local variables, they still have limitations. These methods often focus on either temporal analysis or spatial feature analysis, failing to effectively combine these two critical dimensions. However, geological drilling is a highly complex and dynamic process involving temporal evolution and multiple spatial variables, such as drilling equipment and formation characteristics. These spatiotemporal interactions collectively determine the efficiency and safety of the drilling process. Thus, relying solely on single-dimensional analysis makes it challenging to fully capture and understand the complex phenomena in the drilling process, limiting the accuracy and applicability of monitoring methods.

4.2. Future Directions and Solutions

To address these challenges, future research should focus on several key areas. First, enhancing the integration of global and local features is essential for effectively capturing their interactions. Second, the development of intelligent monitoring technologies capable of managing large datasets with high generalizability will enhance the analysis of the drilling performance. Lastly, incorporating robust spatiotemporal analyses will facilitate a deeper understanding of the complexities inherent in the drilling process. These directions will significantly contribute to advancing intelligent monitoring capabilities in geological drilling and may outline future development paths. The following aspects will be the focus of ongoing research:

(1) Intelligent Monitoring Based on Multi-scale Information Granulation

To address the lack of comprehensive consideration of global and local features in geological drilling process monitoring, multi-scale information granulation methods can be adopted. Through analyzing data at different granularity levels, this approach can effectively capture both the overall trends and local detail changes in the drilling process, providing a comprehensive understanding of drilling data. This method can identify subtle anomalies that might be overlooked in conventional data analysis and explore their relationships with the overall drilling process, deepening the understanding of anomaly causes and their potential impacts on drilling efficiency and safety.

(2) Intelligent Monitoring Based on Sample Augmentation and Transfer Learning

To tackle the issues of data scarcity and low information value density in drilling processes, the combination of sample augmentation and transfer learning offers an effective method to overcome traditional data limitations. Since drilling data are often scarce and have low information value density, transfer learning methods can significantly enhance the monitoring model performance. This approach leverages pre-trained deep learning models from other domains or related tasks to achieve knowledge transfer, reducing the dependency on large labeled datasets and enhancing model generalization under limited data conditions. Sample augmentation techniques further supplement this by artificially expanding the training set, improving model generalization.

(3) Intelligent Monitoring Based on Spatiotemporal Correlation Analysis

To address the lack of spatiotemporal information coordination in geological drilling process monitoring, intelligent monitoring methods based on spatiotemporal correlation analysis can be employed. Through analyzing the relationships between temporal sequence data and spatial distribution data, this approach can deeply capture the dynamic changes and interactions in the drilling process across time and space. Detailed spatiotemporal analysis not only helps to understand the changing characteristics of the drilling process more precisely but also identifies potential spatiotemporal anomalies that might be overlooked. This provides richer and more accurate data support for the drilling process, significantly improving efficiency and safety.

Author Contributions: Conceptualization, S.D., C.H. and X.M.; methodology, S.D.; software, S.D.; validation, S.D., C.H. and X.M.; formal analysis, S.D.; investigation, S.D.; resources, S.D.; data curation, S.D.; writing—original draft preparation, S.D.; writing—review and editing, S.D.; visualization, S.D.; supervision, S.D.; project administration, S.D.; funding acquisition, H.F. All authors have read and agreed to the published version of the manuscript.

Funding: This work was supported in part by the China Postdoctoral Science Foundation under Grant No. 2023M733306, in part by the Hubei Provincial Natural Science Foundation of China under Grant No. 2022CFB582, in part by the 111 Project under Grant No. B17040, in part by the Fundamental Research Funds for the Central Universities, China University of Geosciences, under Grant No. 2021237, and in part by the Natural Science Foundation of Wuhan under Grant No. 2024040801020280.

Conflicts of Interest: The authors declare no conflicts of interest.

References

1. Wu, Y.; Men, X.; Lou, Y. New progress and prospect of coalbed methane exploration and development in China during the 14th Five-Year Plan period. *China Pet. Explor.* **2024**, *29*, 1.
2. Zheng, M.; Li, J.; Wu, X.; Wang, S.; Guo, Q.; Chen, X.; Yu, J. Potential of oil and natural gas resources of main hydrocarbon-bearing basins and key exploration fields in China. *Earth Sci.* **2019**, *44*, 833–847.
3. Qin, Y.; Shen, J.; Shi, R. Strategic value and choice on construction of large CMG industry in China. *J. Coal Sci. Eng.* **2022**, *47*, 371–387.
4. Chen, G. Deep Low-Rank Coalbed Methane System and Reservoiring Mechanism in the Case of the Cainan Block in Junggar Basin. Ph.D. Thesis, China University of Mining and Technology, Xuzhou, China, 2014.
5. Ministry of Natural Resources of the People's Republic of China. The 2023 National Natural Resources Work Conference Held [EB/OL]. 2023. Available online: https://www.mnr.gov.cn/dt/ywbb/202301/t20230-202_2772713.html (accessed on 1 November 2024).

6. Chunxu, Y.; Laiju, H.; Yuhuan, B. New development and future direction of modern vertical drilling technology. *Pet. Drill. Tech.* **2007**, *35*, 16.
7. Wang, H.; Huang, H.; Bi, W.; Ji, G.; Zhou, B.; Zhuo, L. Deep and ultra-deep oil and gas well drilling technologies: Progress and prospect. *Nat. Gas Ind. B* **2022**, *9*, 141–157. [CrossRef]
8. Li, G.; Song, X.; Tian, S.; Zhu, Z. Intelligent drilling and completion: A review. *Engineering* **2022**, *18*, 33–48. [CrossRef]
9. D'Almeida, A.L.; Bergiante, N.C.R.; de Souza Ferreira, G.; Leta, F.R.; de Campos Lima, C.B.; Lima, G.B.A. Digital transformation: A review on artificial intelligence techniques in drilling and production applications. *Int. J. Adv. Manuf. Technol.* **2022**, *119*, 5553–5582. [CrossRef]
10. Zhong, R.; Johnson, R.L.; Chen, Z. Using machine learning methods to identify coal pay zones from drilling and logging-while-drilling (LWD) data. *Spe J.* **2020**, *25*, 1241–1258. [CrossRef]
11. Liu, N.; Zhang, D.; Gao, H.; Hu, Y.; Duan, L. Real-time measurement of drilling fluid rheological properties: A review. *Sensors* **2021**, *21*, 3592. [CrossRef]
12. Guo, B.; Gao, D. New development of theories in gas drilling. *Pet. Sci.* **2013**, *10*, 507–514. [CrossRef]
13. Su, Q.; He, S.; Hu, X.; He, F. Study and application of rock drillability and bit selection for difficult-to-drill formations in Shuangyu Stone structure, western Sichuan. *Drill. Prod. Technol.* **2019**, *42*, 124.
14. Tan, X.; Wang, J.; Guo, X.; Duan, L. Application of PDM drilling technology in Well-GR1 drilling in hot dry rock. *Drill. Eng.* **2021**, *48*, 49–53.
15. Zhang, S.; Qi, L. *A Concise Introduction to Time Series Analysis*; Tsinghua University Press: Beijing, China, 2003.
16. Frantziskonis, G.; Denis, A. Complementary entropy and wavelet analysis of drilling-ability data. *Math. Geol.* **2003**, *35*, 89–103. [CrossRef]
17. Sun, Q.; Tang, Y.; Yang Lu, W.; Ji, Y. Feature extraction with discrete wavelet transform for drill wear monitoring. *J. Vib. Control* **2005**, *11*, 1375–1396. [CrossRef]
18. Xia, W.; Meng, Y.; Li, W. Study on multipath channels model of microwave propagation in a drill pipe. *J. Electromagn. Waves Appl.* **2018**, *32*, 129–137. [CrossRef]
19. Yang, Q.; Xu, B.; Zuo, X.; Jiang, H. An unscented Kalman filter method for attitude measurement of rotary steerable drilling assembly. *Acta Pet. Sin.* **2013**, *34*, 1168.
20. Brophy, B.; Kelly, K.; Byrne, G. AI-based condition monitoring of the drilling process. *J. Mater. Process. Technol.* **2002**, *124*, 305–310. [CrossRef]
21. Chen, G.; Wu, Y.; Fu, L.; Bai, N. Fault diagnosis of full-hydraulic drilling rig based on RS–SVM data fusion method. *J. Braz. Soc. Mech. Sci. Eng.* **2018**, *40*, 1–11. [CrossRef]
22. Sabah, M.; Talebkeikhah, M.; Wood, D.A.; Khosravanian, R.; Anemangely, M.; Younesi, A. A machine learning approach to predict drilling rate using petrophysical and mud logging data. *Earth Sci. Inform.* **2019**, *12*, 319–339. [CrossRef]
23. Reeber, T.; Henninger, J.; Weingarz, N.; Simon, P.M.; Berndt, M.; Glatt, M.; Kirsch, B.; Eisseler, R.; Aurich, J.C.; Möhring, H.C. Tool condition monitoring in drilling processes using anomaly detection approaches based on control internal data. *Procedia CIRP* **2024**, *121*, 216–221. [CrossRef]
24. Alsaihati, A.; Elkatatny, S.; Mahmoud, A.A.; Abdulraheem, A. Use of machine learning and data analytics to detect downhole abnormalities while drilling horizontal wells, with real case study. *J. Energy Resour. Technol.* **2021**, *143*, 043201. [CrossRef]
25. Zhong, Z.; Sun, A.Y.; Yang, Q.; Ouyang, Q. A deep learning approach to anomaly detection in geological carbon sequestration sites using pressure measurements. *J. Hydrol.* **2019**, *573*, 885–894. [CrossRef]
26. Li, Y.; Cao, W.; Gopaluni, R.B.; Hu, W.; Cao, L.; Wu, M. False alarm reduction in drilling process monitoring using virtual sample generation and qualitative trend analysis. *Control Eng. Pract.* **2023**, *133*, 105457. [CrossRef]
27. Reiß, T. Model based fault diagnosis and supervision of the drilling process. *IFAC Proc. Vol.* **1991**, *24*, 211–216. [CrossRef]
28. Shen, Z.; Dong, H.; Yao, N.; Li, X. Condition monitoring and fault diagnosis system of fully hydraulic drilling in coal mine. In Proceedings of the 3rd International Conference on Mechanical, Industrial, and Manufacturing Engineering (MIME 2016), Los Angeles, CA, USA, 30–31 January 2016; pp. 167–170.
29. Zhang, N. Research on automatic fault diagnosis system of coal mine drilling rigs based on drilling parameters. In Proceedings of the IEEE 4th Advanced Information Technology, Electronic and Automation Control Conference (IAEAC), Chengdu, China, 20–22 December 2019; pp. 2373–2377.
30. Tran, T.; Lundgren, J. Drill fault diagnosis based on the scalogram and mel spectrogram of sound signals using artificial intelligence. *IEEE Access* **2020**, *8*, 203655–203666. [CrossRef]
31. Vununu, C.; Moon, K.S.; Lee, S.H.; Kwon, K.R. Sound based machine fault diagnosis system using pattern recognition techniques. *J. Korea Multimed. Soc.* **2017**, *20*, 134–143. [CrossRef]
32. Pootisirakorn, M.; Chongstitvatana, P. Failure Prediction in Open-hole Wireline Logging of Oil and Gas Drilling Operation. In Proceedings of the 23rd International Computer Science and Engineering Conference (ICSEC), Phuket, Thailand, 30 October–1 November 2019; pp. 203–208.
33. Noshi, C.; Noynaert, S.; Schubert, J. Casing failure data analytics: A novel data mining approach in predicting casing failures for improved drilling performance and production optimization. In Proceedings of the SPE Annual Technical Conference and Exhibition, Dallas, TX, USA, 24–26 September 2018; p. D011S001R003.

34. Zhai, H.; Liu, B.; Chen, Y.; Lv, C. Construct a Drilling Complexity Intelligent Prediction Model Based on the Case-Based Reasoning. In Proceedings of the International Field Exploration and Development Conference, Wuhan, China, 19–21 September 2023; Springer: Singapore, 2023; pp. 346–352.
35. Wen, C.; Lu, F.; Bao, Z.; Liu, M. A review of data-driven-based incipient fault diagnosis. *Acta Autom. Sin.* **2016**, *42*, 1285–1299.
36. Bhamare, D.; Suryawanshi, P. Review on reliable pattern recognition with machine learning techniques. *Fuzzy Inf. Eng.* **2018**, *10*, 362–377. [CrossRef]
37. Pei, H.; Hu, C.; Si, X.; Zhang, J.; Pang, Z.; Zhang, P. Review of machine learning based remaining useful life prediction methods for equipment. *J. Mech. Eng.* **2019**, *55*, 1–13. [CrossRef]
38. Burrell, J. How the machine 'thinks': Understanding opacity in machine learning algorithms. *Big Data Soc.* **2016**, *3*, 2053951715622512. [CrossRef]
39. Zhang, Z.; Lai, X.; Wu, M.; Chen, L.; Lu, C.; Du, S. Fault diagnosis based on feature clustering of time series data for loss and kick of drilling process. *J. Process Control* **2021**, *102*, 24–33. [CrossRef]
40. Huang, J.; Yang, X.; Peng, K. Double-layer distributed monitoring based on sequential correlation information for large-scale industrial processes in dynamic and static states. *IEEE Trans. Ind. Inform.* **2020**, *17*, 6419–6428. [CrossRef]
41. Xu, H.; Yu, H. Anomaly detection method for multimode complex industrial process based on multiple subspaces slow feature analysis. *IEEE Access* **2021**, *9*, 119722–119734. [CrossRef]
42. Guo, L.; Wu, P.; Lou, S.; Gao, J.; Liu, Y. A multi-feature extraction technique based on principal component analysis for nonlinear dynamic process monitoring. *J. Process. Control* **2020**, *85*, 159–172. [CrossRef]
43. Ma, X.; Si, Y.; Yuan, Z.; Qin, Y.; Wang, Y. Multistep dynamic slow feature analysis for industrial process monitoring. *IEEE Trans. Instrum. Meas.* **2020**, *69*, 9535–9548. [CrossRef]
44. Messaoud, A.; Weihs, C.; Hering, F. Detection of chatter vibration in a drilling process using multivariate control charts. *Comput. Stat. Data Anal.* **2008**, *52*, 3208–3219. [CrossRef]
45. Fan, H.; Lai, X.; Du, S.; Yu, W.; Lu, C.; Wu, M. Distributed monitoring with integrated probability PCA and mRMR for drilling processes. *IEEE Trans. Instrum. Meas.* **2022**, *71*, 1–13. [CrossRef]
46. Zafeiriou, L.; Nicolaou, M.A.; Zafeiriou, S.; Nikitidis, S.; Pantic, M. Probabilistic slow features for behavior analysis. *IEEE Trans. Neural Netw. Learn. Syst.* **2015**, *27*, 1034–1048. [CrossRef]
47. Cai, P.; Deng, X. Incipient fault detection for nonlinear processes based on dynamic multi-block probability related kernel principal component analysis. *ISA Trans.* **2020**, *105*, 210–220. [CrossRef]
48. Kwak, S.; Ma, Y.; Huang, B. Extracting nonstationary features for process data analytics and application in fouling detection. *Comput. Chem. Eng.* **2020**, *135*, 106762. [CrossRef]
49. Wen, J.; Li, Y.; Wang, J.; Sun, W. Nonstationary process monitoring based on cointegration theory and multiple order moments. *Processes* **2022**, *10*, 169. [CrossRef]
50. Zhang, J.; Zhou, D.; Chen, M. Adaptive cointegration analysis and modified RPCA with continual learning ability for monitoring multimode nonstationary processes. *IEEE Trans. Cybern.* **2023**, *53*, 4841–4854. [CrossRef] [PubMed]
51. Rao, J.; Ji, C.; Wen, J.; Wang, J.; Sun, W. Nonstationary process monitoring based on alternating conditional expectation and cointegration analysis. *Processes* **2022**, *10*, 2003. [CrossRef]
52. Zhao, C.; Sun, H.; Tian, F. Total variable decomposition based on sparse cointegration analysis for distributed monitoring of nonstationary industrial processes. *IEEE Trans. Control Syst. Technol.* **2019**, *28*, 1542–1549. [CrossRef]
53. Liao, M. Drilling state monitoring and fault diagnosis based on multi-parameter fusion by neural network. *J. China Univ. Pet.* **2007**, *31*, 149–152.
54. Yang, A.; Wu, M.; Hu, J.; Chen, L.; Lu, C.; Cao, W. Discrimination and correction of abnormal data for condition monitoring of drilling process. *Neurocomputing* **2021**, *433*, 275–286. [CrossRef]
55. Li, G.; Wang, C.; Zhang, D.; Yang, G. An improved feature selection method based on random forest algorithm for wind turbine condition monitoring. *Sensors* **2021**, *21*, 5654. [CrossRef]
56. Tian, W.; Liu, Z.; Li, L.; Zhang, S.; Li, C. Identification of abnormal conditions in high-dimensional chemical process based on feature selection and deep learning. *Chin. J. Chem. Eng.* **2020**, *28*, 1875–1883. [CrossRef]
57. Yu, W.; Zhao, C.; Huang, B. MoniNet with concurrent analytics of temporal and spatial information for fault detection in industrial processes. *IEEE Trans. Cybern.* **2021**, *52*, 8340–8351. [CrossRef]
58. Gao, H.; Wei, C.; Huang, W.; Gao, X. Multimode process monitoring based on hierarchical mode identification and stacked denoising autoencoder. *Chem. Eng. Sci.* **2022**, *253*, 117556. [CrossRef]
59. Wang, C.; Ma, J.; Jin, H.; Wang, G.; Chen, C.; Xia, Y.; Gou, J. ACGAN and BN based method for downhole incident diagnosis during the drilling process with small sample data size. *Ocean. Eng.* **2022**, *256*, 111516. [CrossRef]
60. Yu, W.; Zhao, C. Broad convolutional neural network based industrial process fault diagnosis with incremental learning capability. *IEEE Trans. Ind. Electron.* **2019**, *67*, 5081–5091. [CrossRef]
61. Zhao, W.; Li, J.; Li, H. A multi-task learning approach for chemical process abnormity locations and fault classifications. *Chemom. Intell. Lab. Syst.* **2023**, *233*, 104719. [CrossRef]
62. Glaeser, A.; Selvaraj, V.; Lee, S.; Hwang, Y.; Lee, K.; Lee, N.; Lee, S.; Min, S. Applications of deep learning for fault detection in industrial cold forging. *Int. J. Prod. Res.* **2021**, *59*, 4826–4835. [CrossRef]

63. Dorgo, G.; Palazoglu, A.; Abonyi, J. Decision trees for informative process alarm definition and alarm-based fault classification. *Process. Saf. Environ. Prot.* **2021**, *149*, 312–324. [CrossRef]
64. Hu, X.; Hu, M.; Yang, X. A novel fault diagnosis method for TE process based on optimal extreme learning machine. *Appl. Sci.* **2022**, *12*, 3388. [CrossRef]
65. Liu, G.; Gu, H.; Shen, X.; You, D. Bayesian long short-term memory model for fault early warning of nuclear power turbine. *IEEE Access* **2020**, *8*, 50801–50813. [CrossRef]
66. Mamudu, A.; Khan, F.; Zendehboudi, S.; Adedigba, S. Dynamic risk modeling of complex hydrocarbon production systems. *Process. Saf. Environ. Prot.* **2021**, *151*, 71–84. [CrossRef]
67. Zhang, Z.; Lai, X.; Lu, C.; Chen, L.; Cao, W.; Wu, M. Lost circulation and kick accidents warning based on Bayesian network for the drilling process. *Drill. Eng.* **2020**, *4*, 114–121. 144. [CrossRef]
68. Mamudu, A.; Khan, F.; Zendehboudi, S.; Adedigba, S. Logic-based data-driven operational risk model for augmented downhole petroleum production systems. *Comput. Chem. Eng.* **2022**, *165*, 107914. [CrossRef]
69. Tariq, S.; Lee, S.; Shin, Y.; Lee, M.S.; Jung, O.; Chung, D.; Woo, S.S. Detecting anomalies in space using multivariate convolutional LSTM with mixtures of probabilistic PCA. In Proceedings of the 25th ACM SIGKDD International Conference on Knowledge Discovery & Data Mining, Anchorage, AK, USA, 4–8 August 2019; pp. 2123–2133.
70. Wang, X.; Wu, P. Nonlinear dynamic process monitoring based on ensemble kernel canonical variate analysis and bayesian inference. *ACS Omega* **2022**, *7*, 18904–18921. [CrossRef] [PubMed]
71. Chai, Z.; Zhao, C. Enhanced random forest with concurrent analysis of static and dynamic nodes for industrial fault classification. *IEEE Trans. Ind. Inform.* **2019**, *16*, 54–66. [CrossRef]
72. Li, Z.; Yan, X. Fault-relevant optimal ensemble ICA model for non-Gaussian process monitoring. *IEEE Trans. Control Syst. Technol.* **2019**, *28*, 2581–2590. [CrossRef]
73. Zhang, J.; Zhou, D.; Chen, M. Monitoring multimode processes: A modified PCA algorithm with continual learning ability. *J. Process. Control* **2021**, *103*, 76–86. [CrossRef]
74. Islamov, S.; Grigoriev, A.; Beloglazov, I.; Savchenkov, S.; Gudmestad, O.T. Research risk factors in monitoring well drilling—A case study using machine learning methods. *Symmetry* **2021**, *13*, 1293. [CrossRef]
75. Barbosa, L.F.F.; Nascimento, A.; Mathias, M.H.; de Carvalho, J.A., Jr. Machine learning methods applied to drilling rate of penetration prediction and optimization—A review. *J. Pet. Sci. Eng.* **2019**, *183*, 106332. [CrossRef]

processes

MDPI

Article

Decision Support System (DSS) for Improving Production Ergonomics in the Construction Sector

Laura Sardinha [1], Joana Valente Baleiras [1], Sofia Sousa [1], Tânia M. Lima [1,2] and Pedro D. Gaspar [1,2,*]

1 Department of Electromechanical Engineering, University of Beira Interior, 6201-001 Covilhã, Portugal; laura.sardinha@ubi.pt (L.S.); joana.valente.baleiras@ubi.pt (J.V.B.); sofia.s.sousa@ubi.pt (S.S.); tmlima@ubi.pt (T.M.L.)

2 C-MAST—Center for Mechanical and Aerospace Science and Technologies, 6201-001 Covilhã, Portugal

* Correspondence: dinis@ubi.pt

Abstract: Ergonomics is essential to improving workplace safety and efficiency by reducing the risks associated with physical tasks. This study presents a decision support system (DSS) aimed at enhancing production ergonomics in the construction sector through an analysis of high-risk postures. Using the Ovako Work Posture Analysis System (OWAS), the Revised NIOSH Lifting Equation (NIOSH equation) and Rapid Entire Body Assessment (REBA), the DSS identifies ergonomic risks by assessing body postures across common construction tasks. Three specific postures—X, Y and Z—were selected to represent typical construction activities, including lifting, squatting and repetitive tool use. Posture X, involving a forward-leaning stance with arms above the shoulders and a 25 kg load, was identified as critical, yielding the highest OWAS and NIOSH values, thus indicating an immediate need for corrective action to mitigate risks of musculoskeletal injuries. The DSS provides recommendations for workplace adjustments and posture improvements, demonstrating a robust framework that can be adapted to other postures and industries. Future developments may include application to other postures and sectors, as well as the use of artificial intelligence to support ongoing ergonomic assessments, offering a promising solution to enhance Occupational Safety and Health policies.

Keywords: ergonomics; OWAS; NIOSH equation; REBA; musculoskeletal injuries; decision support system; OSH

1. Introduction

The construction sector is characterised by a high dependency on human labour, with frequent reports of injuries and fatalities due to improper postures and heavy loads [1]. An alarming number of workers in this sector are diagnosed with musculoskeletal disorders (MSDs), a risk amplified by factors such as prolonged awkward postures and the repetitive handling of loads [2].

Ergonomics is the science that focuses on improving the development of the physical and mental health of human beings, providing them with a safe, comfortable and healthy environment, and, in turn improving the efficiency of their work. Ergonomics is associated with other sciences such as psychology, sociology, physiology and anatomy, among others. In industrial terms, ergonomics has been used to prevent work injuries and increase worker safety [3]. However, it should be considered that changes made to improve operators' ergonomics should not cause them any kind of inconvenience, such as stress related to interacting with systems and robots [4].

Several factors can negatively influence workers' ergonomics, some of which are physical, demographic and psychosocial [5]. Regarding physical factors, these are characterised by vibrations transmitted to the whole body, incorrect postures, repetitive work movements, constant application of loads and constant force [5]. Moving on to demographic factors, these also have a major influence on the quality of the work carried out. Age, Body Mass Index (BMI), gender, eating habits, previous medical diagnoses, or addictions (such as smoking and alcoholism) have a significant impact on an individual's personal and professional life [5,6]. Factors such as stress, anxiety, burnout, or depression have also been identified as adversities in conducting the work required. These factors are just a few examples of psychosocial factors that can affect workers [5].

Jobs with an important level of human intervention require a high level of ergonomic risk monitoring for operators. One such case is the construction sector, which is directly dependent on human effort and is therefore a highly dangerous industry with a high number of reports identifying deaths and injuries [1,7]. Due to the hard tasks that construction workers perform and the uncomfortable postures in which they work for long hours, they are highly likely to develop musculoskeletal injuries and/or other health problems in the future [1].

Some of the more traditional methods of analysing movements used fixed characteristics, which limited their use and made them ineffective for analysing repetitive movements [8]. Therefore, more effective approaches were developed that included counting repetitions and could be applied to the various movements used in the construction sector [8].

In 2020, according to Eurostat, the construction industry in the European Union was responsible for around a fifth of deaths at work and, according to the European Agency for Safety and Health at Work (EU-OSHA), approximately 50% of European construction workers were diagnosed with musculoskeletal injuries [1].

Analysing the data available on PORDATA, it is possible to see that between 2011 and 2021 the apparent productivity of labour in the construction industry, i.e., the wealth created on average, increased by EUR 3562.85 (apparent labour productivity in the construction industry in 2011 and 2021 was EUR 21,018.81 and EUR 24,581.66, respectively), reaching its maximum in 2020, with a total of EUR 25,000.96 [9]. Regarding full-time employment generated by this sector, between 2011 and 2021 there was a decrease of 52,030 jobs (the full-time jobs created by the construction sector in 2011 and 2021 were 402,700 and 350,670, respectively). In 2014, the minimum value for this time interval was reached (272.49 thousand), with this figure gradually increasing until 2021 [10]. It is also important to note that, despite the current property crisis, the Portuguese National Statistics Institute (INE) reported that construction output grew by 4.7 percent last February, 0.2 percentage points down on the previous month, and that the year-on-year change (which measures the change in an indicator compared to its value in the same period of the previous year [1]) in the wage index saw an increase of 11 cent percent, 0.9 percentage points up on January [11].

Several studies assess workers' postures while performing their activities in the workplace. According to Rajendran et al. [12], musculoskeletal injuries, mainly in the lumbar region, accompanied by the adoption of inadequate postures and repetitive movements are the main cause among construction workers.

In addition, analysing a study conducted by Shaikh et al. [5] helped to identify the main factors and provide a comprehensive view of their impact on workers' health. The study conducted by Ogedengbe et al. [13] also proves that this information allows us to understand the implications of a poor work environment resulting from the physical and mental health of workers.

2. Case Study Definition

This case study aims to assess ergonomic risk in the workplace, specifically in the construction sector. This assessment will be based on using some ergonomic tools that assess workers' body postures, such as the Revised NIOSH Lifting Equation (NIOSH equation), Rapid Entire Body Assessment (REBA) and Ovako Work Posture Analysis System (OWAS), specified in the methodology.

The area covered by the case study is the construction sector. According to the statistics provided by ACT, the Authority for Working Conditions, on the number of investigations and serious accidents at work, the construction industry is the sector with the highest rate, with around 684 investigations. All these statistics are broken down by category, state, whether the inquiries have been concluded or are being investigated, month and day of the week, district, age group and sector of activity, among others. To choose the sector under study more intuitively, statistics relating to the sector of activity from 2020 to 2024 were used [14].

The risk of musculoskeletal injuries currently affects most workers in the construction sector and is considered one of the biggest occupational health problems, according to the European Agency for Safety and Health at Work (EU-OSHA) [1]. This type of injury can affect the muscles, ligaments, tendons and nerves of the human body, and can be intensified by workers' efforts and inappropriate postures [2].

Although we used some data from ACT, which also presents surveys showing the number of accidents to the different parts of the body affected during the execution of tasks in this sector, we used the scientific database Science Direct and others, to define which parts of the body would be analysed in this case study.

Most of the injuries found in industrial workers, particularly in the construction industry, are related to lifting and transporting loads, repetitive movements and inadequate handling of tools from an ergonomic point of view. These activities cause lower back injuries, affecting between 50–70% of workers, and this type of back injury affects more than a quarter of workers in manual labour environments [7,13,15]. All these injuries can lead not only to future health problems for workers but also to lower productivity and competitiveness for organisations [13,16]. MSDs are considered a problem of an individual, social and organisational nature [16].

Since MSDs are a key factor in the fight against occupational diseases, working conditions must be optimal, by adopting more appropriate postures when conducting tasks and improving workers' health, safety and quality of life [16].

This case study will therefore be based on three postures frequently adopted by workers in the construction sector, using ergonomic risk assessment tools. Throughout its development, certain parts of the body will be assessed (trunk, neck, legs, arm, forearm and wrist) so that all the proposed indicators can be calculated.

3. Methodology

Given the relevance of the case study presented above and the negative consequences of adopting an incorrect posture when conducting construction tasks, this chapter will present some previously weighted indicators—the OWAS, NIOSH and REBA—to assess the ergonomic risk associated with these postures.

This weighting arises from the evaluation of the set of tools that are commonly used to assess ergonomic risk (NIOSH [17], Rapid Upper Limb Assessment (RULA) [12], OWAS [12], REBA [12] and Washington Industrial Safety and Health Act (WISHA) [18]) in various work tasks and the assessment of different parts of the human body. However, ergonomic risk assessment is not straightforward, as different people can make different interpretations of the same position and the danger it represents [19]. Furthermore, not all tools can include all parts of the body, limiting the assessment of other risk factors present in work environments, and they are often expensive and difficult to implement [19]. Additionally, the assessments that are usually conducted do not always take into account the body dimensions of the individual performing the task, their physical abilities, or even something as basic as their age [19].

This chapter also presents a decision support system to help assess production ergonomics, in this case using data from the construction sector.

3.1. OWAS

In addition to the fatigue associated with hard construction work, the constant and repetitive incorrect postures that workers adopt while conducting the activities associated with their work can lead to the development of MSDs in the long term. Therefore, from an ergonomic point of view, these postures need to be identified and signalled so that they can subsequently be reduced [20].

There are various tools used to assess incorrect postures, which include the OWAS, RULA and REBA methods. However, when compared to other tools, the OWAS proves to be more effective in assessing postures in complex and unclear workplaces, as is the case in the construction sector. Consequently, this study uses the OWAS as one of the tools to quantify the risks associated with the movements made by operators while conducting their work [20].

The OWAS was designed to assess body positions during working hours and score them according to the strain identified [12]. In this way, the tool is used to carry out a total ergonomic assessment of body postures, taking into account the movements of the trunk, arms and legs, combined with the weight of the load carried, these four points being defined by a code with four variables, respectively [20,21]. Each of these variables has several steps associated with it, which represent the postures adopted while conducting the tasks and the intensity of the operation [20,21]:

- Trunk/back: 1 (neutral), 2 (leaning forwards), 3 (twisted) and 4 (bent/twisted);
- Arms: 1 (both arms below shoulders), 2 (one arm above shoulders) and 3 (both arms above shoulders);
- Legs: 1 (sitting), 2 (standing with both legs stretched out), 3 (standing with one leg stretched out), 4 (standing with one knee bent), 5 (standing with both knees bent), 6 (kneeling/squatting) and 7 (walking);
- Load: 1 (less than 10 kg), 2 (between 10 and 20 kg) and 3 (more than 20 kg).

Subsequently, a total score is calculated, using worldwide consensus tables, called S, which classify the total risk into four levels: (i) S = 1, the postures have no particular ergonomic risk; (ii) S = 2, the postures have a slight risk; (iii) S = 3, the postures have a harmful effect and (iv) S = 4, the postures have an extremely harmful risk [20]. By way of example, if a worker, while conducting their task, is bent over, with both arms below their shoulders, kneeling and unladen, their level of risk will be the maximum level, i.e., level 4.

In the end, the OWAS Index (*OI*) can be calculated using (1), shown below, where a indicates the percentage of observations with a risk assessment of 1 and b, c and d correspond to the observations with a risk assessment of 2, 3 and 4, respectively [21].

$$OI = (a \times 1 + b \times 2 + c \times 3 + d \times 4) \times 100 \quad (1)$$

The minimum value for the *OI* is 100, which corresponds to an activity with no ergonomic risks, and the maximum value is 400, which indicates an activity with a fairly high risk. Louhevaara and Suurnäkki [21] considered that an activity is not ergonomically critical if its *OI* is below 200.

3.2. NIOSH Equation

The NIOSH equation is used to assess ergonomic risk in object lifting and lowering operations, focusing mainly on identifying the risk of the task during its performance. The parameters used in the NIOSH equation are the recommended weight limit (RWL) and the Lifting Index (LI) [15].

The RWL, established as the maximum safe weight to be lifted/lowered by the worker in a given time in a repetitive manner, is the main part of the NIOSH calculation and is defined by a set of specific information about the task to be performed, such as the load that the healthiest workers (those who have no previously diagnosed health problems that could increase the risk of MSDs) could handle over an extended period, such as 8 h of work a day, without causing a long-term injury. Equation (2) shows how *RWL* is calculated [13,15].

$$RWL = LC \times HM \times VM \times DM \times AM \times FM \times CM \quad (2)$$

In the formula, *LC* (load constant) represents the load constant, usually 23 kg; *HM* (horizontal multiplier) and *HV* (vertical multiplier) translate, respectively, the horizontal and vertical location of the object; *DM* (distance multiplier) indicates the distance the object has moved; *AM* (Asymmetric Multiplier) corresponds to the asymmetry factor; *FM* (frequency multiplier) symbolises the frequency of movement and *CM* (coupling multiplier) represents coupling [12,13].

The HM considers the horizontal location (H), which is measured from the midpoint of the line joining the inner bones of the ankle to a point projected on the ground directly below the midpoint of the hands (centre of load).

The HV is defined considering the vertical location (V), defined as the vertical height of the hands above the floor. To determine the MD, the vertical travel distance (D) is used, characterised as the vertical travel distance of the hands between the origin and destination of the lift.

AM considers that asymmetry refers to a lift that starts or ends outside the midsagittal plane. FM is defined by the number of lifts per minute, the time dedicated to the lifting activity and the vertical height from the floor. The lifting frequency (F) refers to the average number of lifts performed per minute [13].

The *LI* provides an estimate of the level of physical stress associated with lifting work. This estimate is calculated using Equation (3), which relates the weight of the load lifted (*L*) to the recommended weight limit (*RWL*) [13].

$$LI = \frac{Load\ Weight}{Recommended\ Weight\ Limit} = \frac{L}{RWL} \quad (3)$$

Thus, based on the values obtained with Equation (3), the risk of a particular activity can be determined. In this way, if the *LI* is less than 1.0, the posture is considered to have a very low ergonomic risk, if the *LI* is between 1 and 1.50 the ergonomic risk is considered low, if the *LI* is between 1.50 and 2 the level of risk is moderate, if the *LI* is between 2 and 3 the ergonomic risk is high and if the *LI* is greater than 3 the level of risk is considered very high [13,15].

It should be borne in mind that the NIOSH equation considers the operator's origin and destination, analysing their entire movement.

3.3. REBA

The REBA ergonomic assessment tool is a rapid whole-body assessment method that is commonly used to analyse the postures adopted by workers when conducting their tasks and the ergonomic risks associated with them [19].

As this method enables a full-body assessment to be carried out, taking into account heavy lifting, repetitive hand movements, and neck and trunk flexion, it is the most widely used indicator in a wide variety of sectors, such as transport, manufacturing, education, agriculture, health and construction [19].

However, calculating the REBA is not as intuitive as other indicators, and it is necessary to follow predetermined steps and consult the scores assigned to each element in existing tables [19]. In other words, initially, it is essential to define what you want to assess to obtain the REBA scores, which depend directly on the angles of each part of the body (neck, trunk/spine, legs, arms and wrist) made while performing the tasks [19,22].

For calculation purposes, the REBA method divides the body parts into two groups: group A, consisting of the neck, trunk and legs, and group B, consisting of the upper arm, lower arm and wrist [19]. Figure 1 shows groups A and B with their respective elements, as well as the angles corresponding to the movements performed. It should be noted that in Figure 1 each angle (or range of angles) has a score assigned depending on the type of movement performed. This will be the score used to conduct the rest of the REBA method calculations [19,22].

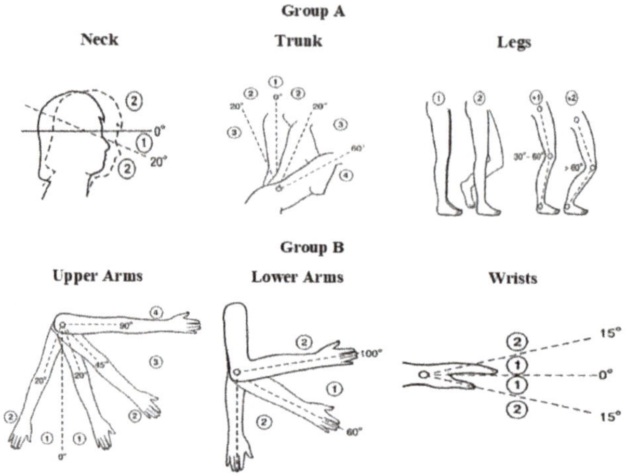

Figure 1. Body elements belonging to groups A and B and their corresponding movement angles [19].

Taking the position of the operator's body parts and their respective scores as a starting point, it is possible to define the next steps and how they will be carried out to obtain a final ergonomic risk score using the REBA method [19].

Group A's scheme of action begins by defining the positions of the constituent elements (neck, trunk and legs), to define a score for each of the assumed angles. Then, with these data, it is necessary to use existing tables to find the score for that posture, thus discovering the score for posture A. The same happens for group B, obtaining the score for posture B [19].

However, to calculate the score attributed to groups A and B, it is necessary to consider the force exerted when acting group A (if applicable) and, in the case of group B, the coupling of objects to the hand. Thus, score A and score B give rise to a new score, determined using a final table relating the two scores. By adding the activity score to the resulting value, the REBA score for the posture in question is calculated and, finally, it is possible to realize the ergonomic risk it represents for the operator [22,23].

Figure 2 shows the calculation diagram that will be used to calculate the REBA score, using all the indicators.

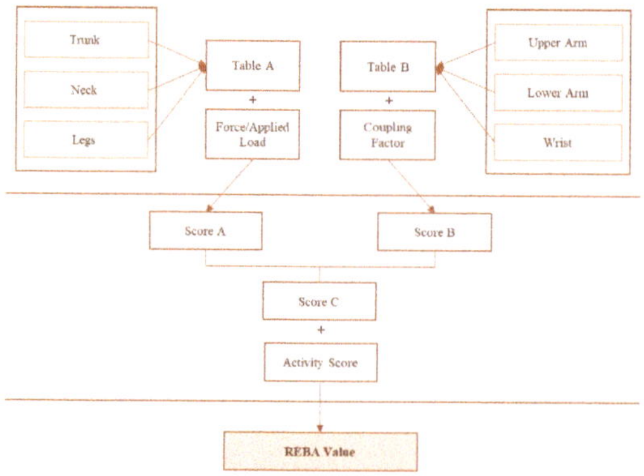

Figure 2. REBA calculation diagram (adapted from [19,23]).

4. Formulation

To put into practice the indicators presented above (OWAS, NIOSH and REBA) for calculating the underlying ergonomic risk of certain postures frequently adopted by workers in the construction sector, a production ergonomics decision support system was built.

This decision support system (DSS) consists of calculating and creating awareness of the ergonomic risk by entering weighted data on the postures that are adopted. It was developed in Microsoft Office Excel to make its implementation in the industry more intuitive and clearer, given that it is one of the most widely used programmes in an industrial environment.

4.1. OWAS

The OWAS values were determined using the table defined by Louhevaara and Suurnäkki [21]. The first step was to identify the variables to be assessed and their respective scores, namely the position of the trunk, arms and legs and the load carried by the worker. The values for the trunk, arms, legs and load are filled in in Table 1, according to the posture the worker performs during their work.

Table 1. OWAS determination table.

Assessment	
Trunk	0
Arms	0
Legs	0
Load	0
OWAS	

With this information, the decision support system determines the level of risk associated with the posture in question and identifies its severity using a colour system. Table 2 shows the association between the possible OWAS values and the level of ergonomic risk, with the corresponding colour coding.

Table 2. OWAS risk levels.

Associated Risk	
OWAS Value	**Risk Level**
1	low
2	medium
3	high
4	very high

4.2. NIOSH

The NIOSH values were calculated using Equations (2) and (3), as previously defined and explained. Thus, in the decision support system developed, it is only necessary to enter the values associated with the calculation, both for the origin and the destination, taken from the tables defined by Waters, Putz-Anderson and Garg [13].

In other words, these values are the load constant, which is 23 kg, the horizontal and vertical multiplier, the distance multiplier, the asymmetry multiplier, the coupling multiplier and the actual weight of the load being transported [13,15].

In this way, the system determines the Lifting Index, which must be less than 1.0. When the LI is within the established limits, the cell in which it was calculated is filled in green. However, when the LI is higher than 1.0, the cell is filled in red, indicating that the posture performed by the worker has a high ergonomic risk. Figure 3 shows the layout of the decision support system, where the values of the variables should be entered.

	LC		HM		VM		DM		AM		FM		CM	
	Origin	Destination	Origin	Destination	Origin	Destination	Origin	Destination	Origin	Destination	Origin	Destination	Origin	Destination
Posture X														
Posture Y														
Posture Z														
Posture VAL														

Figure 3. Decision support system for calculating NIOSH.

With this information, the decision support system will determine the level of risk associated with the posture in question and identify its severity using a colour system. Table 3 shows the association of the possible NIOSH values with the level of ergonomic risk, with the respective colour code, and the recommended actions for each level of risk.

Table 3. NIOSH risk levels and recommended actions.

LI Value	Risk Level	Recommended Actions
$LI \leq 1.00$	Very Low	The load is acceptable for most people and no action is required for the healthy population.
$1.00 < LI \leq 1.50$	Low	The load should be assessed, and medium-term changes introduced. Pay special attention to low frequency/high load conditions and extreme and static postures.
$1.50 < LI \leq 2.00$	Moderate	Reformulate tasks and workplaces according to priorities to reduce LI, analysing the results to confirm the effectiveness of the changes.
$2.00 < LI \leq 3.00$	High	Short-term changes should be introduced to reduce the risk of MSDs. Tasks with this assessment should be redesigned or assigned only to selected workers who will be rigorously monitored.
$LI > 3.00$	Very high	The load presents a risk to most people and action should be taken immediately. This type of task is unacceptable from an ergonomic point of view and should be modified.

4.3. REBA

The decision support system used to calculate the REBA values was developed based on the data and tables suggested by Hignett and McAtamney [24] when developing this indicator, divided into three tables with the addition of some key elements related to the positions chosen. Figure 4 shows the layout of the decision support system for the REBA method, described in the previous chapter, where all the calculation formulas have been entered to obtain the final value for the indicator in question.

The values for the trunk, neck, legs, upper arm, lower arm, wrist, load factor, coupling factor and activity factor are entered manually by the user by assessing the posture adopted by the worker, since different postures can result in different ergonomic risk values.

In addition, score A was obtained by adding the value resulting from table A with the load factor, score B by adding the value resulting from table B with the coupling factor and score C from table C, using the two previous scores. All these values resulting from looking at the tables are obtained automatically by the system, according to the values entered for each body component. So, the operator does not need to spend time analysing the tables, just the postures.

Once the values have been obtained, the system calculates the final REBA value (C score + activity) and issues a visual signal depending on the ergonomic risk presented and the type of intervention required.

The colours relating to the visual signal issued can be found in Table 4, along with the respective ergonomic level and the action that needs to be taken.

The same procedure was conducted for the three postures chosen (X, Y and Z postures).

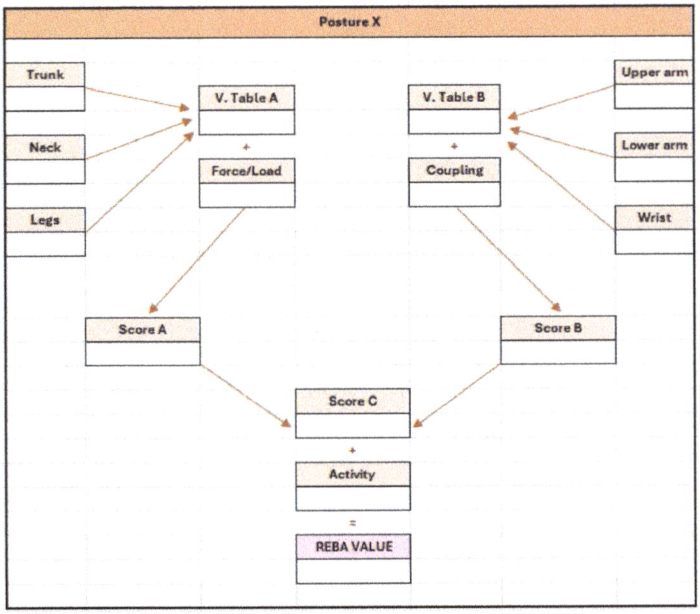

Figure 4. Decision support system for calculating REBA.

Table 4. REBA action levels.

REBA Action Levels			
REBA Value	***Risk Level***	***Action Level***	***Action (After Complementary Action)***
1	null	0	not necessary
2 a 3	low	1	may be necessary
4 a 7	medium	2	necessary
8 a 10	high	3	needed very soon
11 a 15	very high	4	necessary now

4.4. *Posture Explanation*

To be able to apply the ergonomic risk assessment tools defined, three postures typically performed by workers in the construction industry were selected. A brief description of each posture will be given below to assess their ergonomic risk.

4.4.1. Posture X

The first posture involves lifting a load above the shoulders. Thus, the worker has a slight inclination of the trunk, of around 20°. In addition, the worker's lower arms are both being used above the shoulders, making an angle of more than 100°, and the arm makes an angle of 90° with the trunk. In this case, the worker's wrist can be in a straight position. To make it easier to understand what has been described above, a representative sketch of posture X has been drawn up, shown in Figure 5.

4.4.2. Posture Y

In posture Y, we tried to represent the worker squatting. In this sense, the worker has a slight inclination of the trunk, about 10°, with the pelvic area below the usual axis when the body is standing and with both knees bent. The thigh makes a 90° angle with the trunk and the knee is bent more than 60°. To make the description easier to understand, a sketch representing the Y posture was made, illustrated in Figure 5.

Figure 5. Graphic representation of postures: (**a**) posture X; (**b**) posture Y; (**c**) posture Z.

4.4.3. Posture Z

The third and final posture depicts the right hand attaching a tool followed by a rotational movement of the arm to place mortar on a wall. The left lower arm is at a 90° angle to the arm, holding a container with the mortar. In addition, in this posture, the worker is standing with a neutral trunk, without any inclination, and with one knee bent at an angle of between 30° and 60°. As with the other postures, a sketch of posture Z was drawn up, as shown in Figure 5.

5. Results and Discussion

Considering the description of the decision support system developed, this chapter will present and discuss the results obtained for each indicator, by entering the data relating to positions X, Y and Z into the system, as described above.

5.1. Posture X

5.1.1. OWAS

As mentioned in the description of posture X, the worker is leaning with his trunk forward, both arms above his shoulders, standing with both legs stretched out and carrying a 25 kg load, the equivalent of a sack of cement. Thus, the scores used to determine the OWAS were as follows: trunk—2 (leaning forward), arms—3 (both arms above the shoulders), legs—2 (standing with both legs straight) and load—3 (over 20 kg). The level of risk generated by the OWAS is shown in Table 5.

Table 5. OWAS result for posture X.

Assessment	
Trunk	2
Arms	3
Legs	2
Load	3
OWAS	3

Considering the value obtained (3), it can be said that posture X has a high level of risk. This value is essentially due to the position of the arms and the load carried, with these two variables obtaining the maximum score.

5.1.2. NIOSH

To determine the NIOSH values, it was considered that before the worker reaches posture X, the bag of cement is on the floor and that the worker must bend down to get the bag onto his back, so it can be said that the type of coupling he performs is of the bad type, being uncomfortable and entailing some ergonomic risks for the operator. So, for the origin, the calculation variables have the following values: HM—1.00 (H $\leq$ 25 cm), VM—0.78 (V = 150 cm), DM—0.85 (D = 145 cm), AM = 1.00 (A = 0°), FM = 0.94 ($\leq$ 1 h, V $\geq$ 75 cm and F = 1), CM = 0.90 (V $\geq$ 75 cm and Type = Bad) and LW = 25 kg. Regarding destination, the calculation variables have the following values: HM—1.00 (H $\leq$ 25 cm), VM—0.81 (V = 140 cm), DM—0.85 (D = 145 cm), AM = 1.00 (A = 0°), FM = 0.94 ($\leq$ 1 h, V $\geq$ 75 cm and F = 1) and CM = 0.90 (V $\geq$ 75 cm and Type = Bad). The result of the Lifting Index calculation is shown in Table 6.

Table 6. NIOSH result for posture X.

Posture X	
RWL Origin	12.90065
RWL Destination	13.39683
LI Origin	1.937886
LI Destination	1.866113

By calculating the Lifting Index, we can see that its value is 1 at both the origin and destination, which means that, according to this indicator, posture X poses a moderate ergonomic risk to the worker. The task should therefore be reformulated to reduce the LI and the results analysed to confirm the effectiveness of the changes.

5.1.3. REBA

Based on the above description of posture X, in which the worker shows obvious tension in the trunk and arms due to the weight being carried, the following values were assigned to the body elements: trunk—2 (flexion of 20°), neck—1 (flexion between 0° and 20°), legs—1 (no change), upper arm—3 (90° with the trunk), lower arm—2 (flexion >100°) and wrist—1 (no change).

Considering that the operator is carrying a sack of cement weighing around 25 kg, the load factor was given 2 points and the coupling an equal score, because the grip, although possible, is not acceptable. Finally, the activity was scored with 2 points because parts of the body were stationary for more than a minute and because the weight of the load caused an unstable base. Figure 6 shows the final REBA value for this posture, calculated using the decision support system developed.

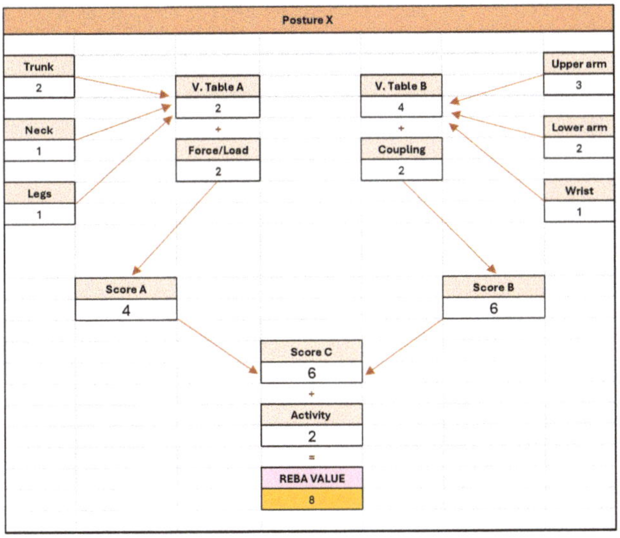

Figure 6. REBA result for posture X.

Looking at Figure 6, this posture has a REBA value of 8, meaning that the ergonomic risk level of this posture is high, which will imply a third level of action, i.e., corrective action, is needed very soon.

In this posture, the factor that has the most implications will be the trunk, given the weight of 25 kg that is exerted on it and the poor positioning of the arms, together with the handle, which is not acceptable and is very harmful to the operator's health.

5.2. Posture Y

5.2.1. OWAS

In posture Y, you can see that the worker has both knees bent but is not standing. In addition, his trunk is slightly inclined, both arms are below his shoulders and the load he is carrying is 5 kg. The decision variables therefore take on the following values: trunk—2 (leaning forwards), arms—1 (both arms below the shoulders), legs—6 (kneeling/squatting) and load—1. The level of risk determined by the decision support system is shown in Table 7.

Table 7. OWAS result for posture Y.

Assessment	
Trunk	2
Arms	1
Legs	6
Load	1
OWAS	2

5.2.2. NIOSH

Before the worker reached posture Y, he had to pick up the bucket from the floor and stoop down. It can be said that the type of coupling he performs is of the regular type, since this movement does not overload the back. Thus, the variables used to calculate the NIOSH for the origin have the following values: HM—0.57 (H = 44 cm), VM—0.93 (V = 50 cm), DM—0.90 (D = 55 cm), AM = 1.00 (A = 0°), FM = 0.91 ($\leq$ 1 h, V $\leq$ 75 cm and F = 2), CM = 0.95 (V < 75 cm and Type = Regular) and LW = 5.00 kg. However, for the destination, the values used are as follows: HM—0.57 (H = 44 cm), VM—0.84 (V = 130 cm), DM—0.86 (D = 130), AM = 1.00 (A = 0°), FM = 0.91 ($\leq$ 1 h, V $\leq$ 75 cm and F = 2), CM = 1.00 (V $\geq$ 75 cm and Type = Regular) and LW = 5.00 kg. The calculation of the Lifting Index is shown in Table 8.

Since the Lifting Index value for both the origin and destination is less than 1.0, posture Y has a very low ergonomic risk for the worker, where the load is acceptable, and no improvements are needed.

Table 8. NIOSH result for posture Y.

Posture Y	
RWL Origin	9.486219
RWL Destination	8.618304
LI Origin	0.52708
LI Destination	0.580161

5.2.3. REBA

The same analysis made of posture X was applied to posture Y, in which the operator is in an unstable position, since all the support is provided by his legs in a very unstable position. The following scores were given to the body parts: trunk—2 (10° flexion), neck—1 (no change), legs—4 (2 unstable posture + 2 flexion > 60°), upper arm—1 (straight with the trunk), lower arm—1 (90° flexion) and wrist—1 (no change).

Given that the operator only supports a weight in both hands of around 5 kg and that the grip is adequate, scores of 1 and 0 were assigned to the load and coupling factor, respectively.

Finally, the activity score reached a value of 3, since one or more parts of the body are stationary for more than a minute, short-range actions are performed with the feet and the posture is highly unstable.

Figure 7 shows the REBA value assigned to posture Y by the decision support system.

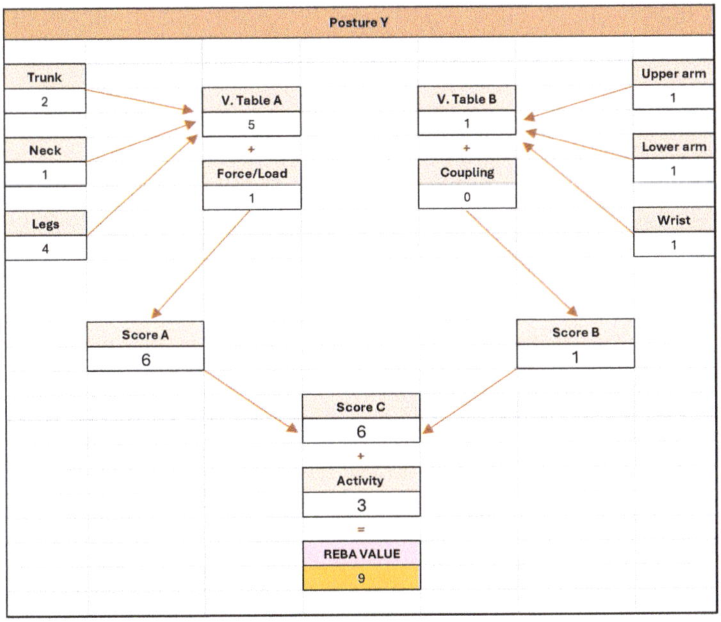

Figure 7. REBA result for posture Y.

Looking at Figure 7, it can be seen that the final REBA value assigned to posture Y was 9, indicating that the risk level is high, and that corrective action is needed very soon, with action level 3.

Comparing the operators' postures and the REBA values for postures X and Y, posture Y has a higher REBA, even if the grip is adequate and there is not as much tension in the trunk and arms as there is in posture X. However, in posture Y, the operator exerts a lot of pressure on the lower limbs, with the weight of the body supported by the legs in an extremely unstable position. As a result, even though the REBA values are relatively similar, the Y posture is more harmful than the X posture.

5.3. *Posture Z*

5.3.1. OWAS

Considering the description given above, it can be said that the worker's trunk is in an upright position, with no inclination, that both arms are below the shoulders and that the load the worker is carrying is less than 5 kg. However, the worker is standing with one knee bent. Thus, the values associated with the variables are trunk—1 (neutral), arms—1 (both arms below the shoulders), legs—4 and load—1. Thus, the value assigned by the system to posture Z is shown in Table 9.

Although most of the variables have been assigned the minimum value, the risk level of this posture is medium. This value is because the worker is standing with one knee flexed, which causes the leg score to be 4, instantly increasing the OWAS value.

Table 9. OWAS result for posture Z.

Assessment	
Trunk	1
Arms	1
Legs	4
Load	1
OWAS	2

5.3.2. NIOSH

Since the Z posture simulates the worker putting cement on the wall, he had to bend down to pick up the necessary tools and then stand up again, without excessive ergonomic effort. Thus, the NIOSH calculation variables, at source, take on the following values: HM—0.57 (H = 44 cm), VM—0.90 (V = 40 cm), DM—0.93 (D = 40 cm), AM = 1.00 (A = 0°), FM = 0.94 ($\leq$ 1 h, V $\geq$ 75 cm and F = 1), CM = 1 (V > 75 cm and Type = Good) and LW = 2.5 kg. However, the NIOSH calculation variables at the destination assume the following values: HM—0.57 (H = 44 cm), VM—0.93 (V = 100 cm), DM—0.87 (D = 100 cm), AM = 1.00 (A = 0°), FM = 0.94 ($\leq$ 1 h, V $\geq$ 75 cm and F = 1), CM = 1 (V > 75 cm and Type = Good) and LW = 2.5 kg. The calculation of the Lifting Index is shown in Table 10.

Table 10. NIOSH result for posture Z.

Posture Z	
RWL Origin	10.3146858
RWL Destination	9.97086294
LI Origin	0.24237287
LI Destination	0.250730555

Since the Lifting Index value for both the origin and destination is less than 1.0, posture Z has a very low ergonomic risk for the worker, where the load is acceptable, and no improvements are needed.

5.3.3. REBA

Finally, of all the postures proposed for analysis, only the Z posture is missing, which represents movements like the operator placing mortar on a wall. In this case, the following values were assigned to the body elements: upper body—1 (as it is straight), neck—1 (no changes), legs—3 (2 unilateral+1 flexion between 30° and 60°), upper arm—2 (one arm aligned with the upper body, but the other rotates), lower arm—1 (flexion between 60° and 100°) and wrist—2 (as well as being straight, the wrist rotates when placing the mortar).

The load and coupling weighting factors were considered satisfactory, given that the tool with which the operator places the mortar is much less than 5 kg, including the "plate" on which it is placed before being applied to the wall. In addition, the grip on these objects is quite acceptable with an equally satisfactory average power.

The activity, on the other hand, was given a score of 2, as there are parts of the body that are stationary for more than 1 min and, in addition, the short-range actions are carried out repetitively.

Figure 8 shows the REBA result for posture Z, obtained by the decision support system developed.

From the result obtained by the decision support system, shown in Figure 8, the REBA value assigned to posture Z was 5, which is considered to present a medium level of ergonomic risk, where corrective action will be required in the future.

Compared to the previous postures, this one present less ergonomic risk because the operator has the correct trunk and neck postures, although the movement of the arms is not the most appropriate, particularly when applying the mortar. The way the operator moves his wrist and arm creates tension in these elements, which could lead to future musculoskeletal injuries.

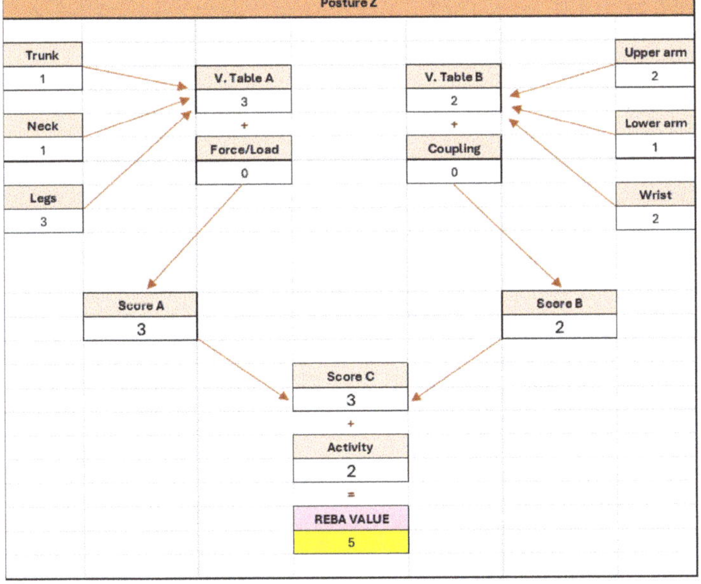

Figure 8. REBA result for posture Z.

Although posture Z exhibited a lower overall risk, the movement involved in applying mortar during the construction tasks (posture Z) revealed the potential for cumulative strain injuries in the arms and wrists. This underscores the importance of regularly rotating workers between tasks to prevent overuse injuries.

5.4. Results Validation

To prove that the decision support system developed was well structured and that all the results presented above are valid, this section presents the application of the system in question to three positions studied in scientific articles.

5.4.1. OWAS Validation Posture

To confirm that the decision support system developed actually works, the skidding posture studied by Enez and Nalbantoğlu [25] will be used. In this posture, the worker is standing with the trunk leaning forwards, both arms below the shoulders and no load. Thus, the values assigned to the OWAS determination variables are as follows: trunk—2 (leaning forward), arms—1 (both arms below the shoulders), legs—2 (standing with both legs straight) and load—1 (less than 10 kg). The level of risk associated with this posture is shown in Table 11.

Table 11. OWAS result for validation posture.

Posture VAL Assessment	
Trunk	2
Arms	1
Legs	2
Load	1
OWAS	2

As can be seen in Table 11, the level of risk attributed to this posture is medium. However, as assigning values to each variable is a subjective activity, the values assigned by the authors of the article to each of the variables is different and are as follows: trunk—1 (neutral), arms—2 (one arm above the shoulders), legs 2—(standing, with both legs stretched out) and load—1 (less than 10 kg). In this way, and with the difference in classification between variables, the risk level for this assessment would be low.

5.4.2. NIOSH Validation Posture

As with the REBA and OWAS validation, the NIOSH validation will use a previously studied posture to validate the decision support system developed. Rajendran et al. [15] carried out an ergonomic assessment of workers during the manual handling of materials. Thus, in this section, the posture used to lift a sealed bag from the bottom shelf of a shelf will be used, where the variables used to calculate the NIOSH, at source, take on the following values: HM—0.50 (H = 50 cm), VM—0.84 (V = 130 cm), DM—0.85 (D = 145 cm), AM = 1.0 (A = 0°), FM = 0.88 (V < 75 cm, > 1 but $\leq$ 2 h, F = 1), CM = 0.95 (V < 75 cm and Type = Regular) and LW = 8 kg. However, at the destination, the variables take on the following values: HM—0.50 (H = 50 cm), VM—0.81 (V = 140 cm), DM—0.8 (D = 145 cm), AM = 0.71 (A = 90°), FM = 0.88 (V < 75 cm, > 1 but $\leq$ 2 h, F = 1), CM = 1.0 (V $\geq$ 75 cm and Type = Regular) and LW = 8 kg. The value for NIOSH calculated by the decision support system is illustrated in Table 12.

Table 12. NIOSH result for validation posture.

Posture VAL	
RWL Origin	6.864396
RWL Destination	4.94701
LI Origin	1.165434
LI Destination	1.617138

Comparing the result obtained at the destination using the system developed (LI = 1.617) with the result achieved by the authors of the initial study (LI = 1.562) [15], it is safe to say that the system developed fulfils all the necessary requirements for the efficient calculation of the LI. Thus, the posture of lifting a sealed bag from a bottom shelf, analysing the result provided by both systems, is a posture that entails a moderate ergonomic risk for the worker, where the task should be reformulated to reduce the LI.

5.4.3. REBA Validation Posture

Taking as an example the third posture studied by Enez and Nalbantoğlu [25] for the purposes of validating the REBA calculation, where the operator is removing/moving tree trunks weighing more than 20 kg in a flatbed truck, the following scores were given to the body elements: trunk—5, neck—2, legs—3, upper arms—3, lower arms—2 and wrist—2.

In this case, only the load factor, derived from the trunks being moved, was considered and given a score of 2, as it was a weight of more than 20 kg. The other factors were not considered in the REBA calculation [25].

So, applying these data to the decision support system developed, Figure 9 shows the REBA result obtained for the posture in question.

As can be seen from Figure 9, the REBA result obtained by the decision support system developed is equal to the value resulting from the application of the scientific article. In fact, this type of posture is extremely harmful to workers' health, as the level of ergonomic risk is very high, requiring immediate action at action level 4.

It is therefore possible to conclude that the decision support system used to calculate REBA fulfils the established requirements and can produce valid and credible final REBA values.

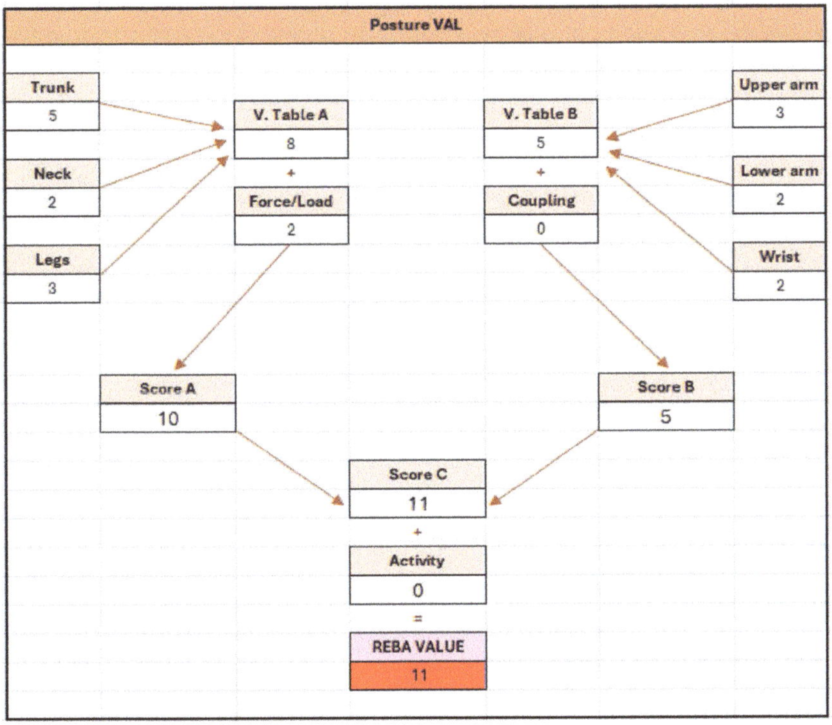

Figure 9. REBA result for validation posture.

5.5. Summary of Results Obtained

After applying the production ergonomics decision support system developed, by calculating the indicators proposed for the postures envisaged, it was possible to compile all the results to compare them.

Table 13 shows all the results obtained for each of the postures, including the validation postures used to gauge the reliability of the system developed.

Looking at Table 13 and excluding the validation postures, it is possible to see that the most critical posture is posture X, which represents the greatest ergonomic risk, as it has the highest OWAS and NIOSH values. However, although the REBA value is not the highest, it does have more weight compared to posture Y, where the OWAS and NIOSH values are lower.

Table 13. Summary of the results obtained for the different indicators for different postures.

	OWAS	NIOSH		REBA
		LI Origin	LI Destination	
Posture X	3	1.93788625	1.866112685	8
Posture Y	2	0.527080388	0.580160535	9
Posture Z	2	0.24237287	0.250730555	5
Posture VAL (OWAS)	2			
Posture VAL (NIOSH)		1.165433929	1.617138368	
Posture VAL (REBA)				11

Therefore, some concern is raised and commitment is needed to correct this type of posture so that it does not have such negative implications for the health of workers in the construction sector, such as the early onset of musculoskeletal injuries, as well as the occurrence of accidents at work that have serious consequences (i.e., serious injuries and even fatal accidents). Analysing these results, there are some strategies that can be adopted by workers. One suggestion for reducing the number of injuries to operators, given the improvement in their body postures, is staff rotation. By implementing this measure, companies are benefiting their workers, as they will not always be in the same position or performing the same tasks. Thus, by rotating operators, i.e., having them change positions, they would enjoy, for example, a few minutes of rest, allowing the upper or lower limbs to stretch, reducing the early onset of injuries during the working period.

Analysing the data relating to the calculation of the OWAS, NIOSH and REBA values, it was concluded that if there were another decision support system, which could be integrated into the one developed, or not, this could automatically evaluate and assign values relating to the body elements of each posture. For this reason, one suggestion for improving the efficiency of this tool would be to use a monitoring device, such as a wristband or other non-invasive mobile device, which would not interfere with the worker's activity.

To support this suggestion, there are already case studies proposing solutions to prevent the risk of musculoskeletal injuries in workers using smart personal protective equipment (PPE) and other monitoring systems [26–29]. This smart PPE emerged from the interaction between Industry 4.0 and International Data Corporation (IoD) technologies [26]. The use of smart PPE enables communication with the environment, as it combines traditional PPE with electronic components and sensors, extracting information about workers and thus reducing the rate of accidents and occupational illnesses [26–29].

The use of these devices is intended to help organisations plan for the long term, with a view to improving Occupational Safety and Health (OSH) policies, using artificial intelligence (AI). In this way, using artificial intelligence algorithms, companies can identify working conditions that are susceptible to accidents. In this way, organisations can maintain safer working environments, with the aim of improving the health and safety of their operators [26–28].

Although this suggestion is a viable option, there are some drawbacks, as these systems can be costly, particularly for small and medium-sized companies. Additionally, another direction for future work could be exploring how this methodology, originally developed for the construction sector, could be adapted and applied to other industries, thus broadening its potential applications beyond construction.

6. Conclusions

The case study presented was aimed at assessing ergonomic risks in the workplace. In this case, the sector in which the case study focused was the construction sector, since it was the area with the highest incidence of serious accidents at work in the last 4 years (from 2020 to 2024).

As presented during the development of this article, some ergonomic indicators that assess workers' postures were used, namely, the OWAS, NIOSH and REBA. To put these indicators into practice, a production ergonomics decision support system was developed by formulating all the data relating to each indicator in Microsoft Office EXCEL.

To validate the system, three pre-defined positions were evaluated: the X, Y and Z postures. In this way, and through scientific validation, it can be concluded that the system developed achieved the objectives set.

To improve working conditions in the construction sector, future studies should focus on using AI techniques to analyse data that can identify which phases of construction pose the greatest risks to workers. The success of this initiative depends on several factors, with the key challenges being the variability of risks and the acceptance of operators.

Author Contributions: Conceptualiszation, L.S., J.V.B. and S.S.; methodology, L.S., J.V.B. and S.S.; software, L.S., J.V.B. and S.S.; validation, T.M.L. and P.D.G.; formal analysis, T.M.L. and P.D.G.; investigation, L.S., J.V.B. and S.S.; resources, L.S., J.V.B. and S.S.; data curation, L.S., J.V.B. and S.S.;

writing—original draft preparation, L.S., J.V.B. and S.S.; writing—review and editing, T.M.L. and P.D.G.; visualisation, T.M.L. and P.D.G.; supervision, T.M.L. and P.D.G. All authors have read and agreed to the published version of the manuscript.

Funding: The authors would like to express their gratitude to FCT—Fundação para a Ciência e a Tecnologia, I.P. and the Centre for Mechanical and Aerospace Science and Technologies (C-MAST) for their support in the form of funding, under the project UIDB/00151/2020 (https://doi.org/10.54499/UIDB/00151/2020; https://doi.org/10.54499/UIDP/00151/2020) (accessed on 7 November 2024).

Data Availability Statement: The data in this study are available from the corresponding author upon request.

Conflicts of Interest: The authors declare no conflicts of interest.

References

1. Zong, H.; Yi, W.; Antwi-Afari, M.F.; Yu, Y. Fatigue in Construction Workers: A Systematic Review of Causes, Evaluation Methods, and Interventions. *Saf. Sci.* **2024**, *176*, 106529. [CrossRef]
2. Çakıt, E.; Karwowski, W. Soft Computing Applications in the Field of Human Factors and Ergonomics: A Review of the Past Decade of Research. *Appl. Ergon.* **2024**, *114*, 104132. [CrossRef] [PubMed]
3. Zhang, M.; Li, H.; Tian, S. Visual Analysis of Machine Learning Methods in the Field of Ergonomics—Based on Cite Space V. *Int. J. Ind. Ergon.* **2023**, *93*, 103395. [CrossRef]
4. Gualtieri, L.; Rauch, E.; Vidoni, R. Emerging Research Fields in Safety and Ergonomics in Industrial Collaborative Robotics: A Systematic Literature Review. *Robot. Comput.-Integr. Manuf.* **2021**, *67*, 101998. [CrossRef]
5. Murtoja Shaikh, A.; Bhusan Mandal, B.; Mangani Mangalavalli, S. Causative and Risk Factors of Musculoskeletal Disorders among Mine Workers: A Systematic Review and Meta-Analysis. *Saf. Sci.* **2022**, *155*, 105868. [CrossRef]
6. Mostafa, N.A. Human Factors and Ergonomics for Intelligent Manufacturing in the Era of Industry 4.0. In Proceedings of the 2023 4th International Conference on Artificial Intelligence, Robotics and Control (AIRC), Cairo, Egypt, 9–11 May 2023; pp. 88–93.
7. Oyekan, J.; Chen, Y.; Turner, C.; Tiwari, A. Applying a Fusion of Wearable Sensors and a Cognitive Inspired Architecture to Real-Time Ergonomics Analysis of Manual Assembly Tasks. *J. Manuf. Syst.* **2021**, *61*, 391–405. [CrossRef]
8. Chen, X.; Yu, Y. Automatic Repetitive Action Counting for Construction Worker Ergonomic Assessment. *Autom. Constr.* **2024**, *167*, 105726. [CrossRef]
9. PORDATA—Estatísticas Sobre Portugal e Europa Produtividade Aparente do Trabalho: Total e por Ramo de Atividade. Available online: https://www.pordata.pt/pt/estatisticas/economia/setores-de-atividade/produtividade-do-trabalho-por-ramo-de-atividade (accessed on 6 May 2024).
10. PORDATA—Estatísticas Sobre Portugal e Europa Emprego: Total e por Ramo de Atividade, Equivalente a Tempo Completo. Available online: https://www.pordata.pt/pt/estatisticas/economia/setores-de-atividade/emprego-por-ramo-de-atividade (accessed on 6 May 2024).
11. INE—Instituto Nacional de Estatística Índices de Produção, Emprego e Remunerações na Construção. Available online: https://www.ine.pt/xportal/xmain?xpid=INE&xpgid=ine_destaques&DESTAQUESdest_boui=643890203&DESTAQUEStema=55534&DESTAQUESmodo=2 (accessed on 6 May 2024).
12. Rajendran, M.; Sajeev, A.; Shanmugavel, R.; Rajpradeesh, T. Ergonomic Evaluation of Workers during Manual Material Handling. *Mater. Today Proc.* **2021**, *46*, 7770–7776. [CrossRef]
13. Ogedengbe, T.S.; Abiola, O.A.; Ikumapayi, O.M.; Afolalu, S.A.; Musa, A.I.; Ajayeoba, A.O.; Adeyi, T.A. Ergonomics Postural Risk Assessment and Observational Techniques in the 21st Century. *Procedia Comput. Sci.* **2023**, *217*, 1335–1344. [CrossRef]
14. ACT—Autoridade para as Condições de Trabalho. Número de Inquéritos de Acidentes de Trabalho Graves. Available online: https://portal.act.gov.pt/Pages/acidentes_de_trabalho_graves.aspx (accessed on 13 May 2024).
15. Kodle, N.R.; Bhosle, S.P.; Pansare, V.B. Ergonomic Risk Assessment of Tasks Performed by Workers in Granite and Marble Units Using Ergonomics Tool's REBA. *Mater. Today Proc.* **2023**, *72*, 1903–1916. [CrossRef]
16. Rodrigues, B.S.; Freitas, M.; Tomé, D.; Neto, H.V. Avaliação de Fadiga Laboral e Lesões Músculo-Esqueléticas Relacionadas com o Trabalho numa Secção de Mistura de Cortiça. *CESQUA* **2020**, *1*, 149–177.
17. Lowe, B.D.; Dempsey, P.G.; Jones, E.M. Ergonomics Assessment Methods Used by Ergonomics Professionals. *Appl. Ergon.* **2019**, *81*, 102882. [CrossRef] [PubMed]
18. Meregalli Falerni, M.; Pomponi, V.; Karimi, H.R.; Lavit Nicora, M.; Dao, L.A.; Malosio, M.; Roveda, L. A Framework for Human–Robot Collaboration Enhanced by Preference Learning and Ergonomics. *Robot. Comput.-Integr. Manuf.* **2024**, *89*, 102781. [CrossRef]
19. Yalcin Kavus, B.; Gulum Tas, P.; Taskin, A. A Comparative Neural Networks and Neuro-Fuzzy Based REBA Methodology in Ergonomic Risk Assessment: An Application for Service Workers. *Eng. Appl. Artif. Intell.* **2023**, *123*, 106373. [CrossRef]
20. Zhang, H.; Lin, Y. Modeling and Evaluation of Ergonomic Risks and Controlling Plans through Discrete-Event Simulation. *Autom. Constr.* **2023**, *152*, 104920. [CrossRef]

21. Calzavara, M.; Glock, C.H.; Grosse, E.H.; Persona, A.; Sgarbossa, F. Models for an Ergonomic Evaluation of Order Picking from Different Rack Layouts. *IFAC-Pap.* **2016**, *49*, 1715–1720. [CrossRef]
22. Joshi, M.; Deshpande, V. Investigative Study and Sensitivity Analysis of Rapid Entire Body Assessment (REBA). *Int. J. Ind. Ergon.* **2020**, *79*, 103004. [CrossRef]
23. Stanton, N.; Hedge, A.; Brookhuis, K.; Salas, E.; Hendrick, H. *Handbook of Human Factors and Ergonomics Methods*; CRC Press LLC: Boca Raton, FL, USA, 2005; ISBN 0-415-28700-6.
24. Hignett, S.; McAtamney, L. Rapid Entire Body Assessment (REBA). *Appl. Ergon.* **2000**, *31*, 201–205. [CrossRef]
25. Enez, K.; Nalbantoğlu, S.S. Comparison of Ergonomic Risk Assessment Outputs from OWAS and REBA in Forestry Timber Harvesting. *Int. J. Ind. Ergon.* **2019**, *70*, 51–57. [CrossRef]
26. Lemos, J.; Gaspar, P.D.; Lima, T.M. Individual Environmental Risk Assessment and Management in Industry 4.0: An IoT-Based Model. *Appl. Syst. Innov.* **2022**, *5*, 88. [CrossRef]
27. Lemos, J.; Gaspar, P.D.; Lima, T.M. Environmental Risk Assessment and Management in Industry 4.0: A Review of Technologies and Trends. *Machines* **2022**, *10*, 702. [CrossRef]
28. Lemos, J.; de Souza, V.B.; Falcetta, F.S.; de Almeida, F.K.; Lima, T.M.; Gaspar, P.D. Enhancing Workplace Safety through Personalized Environmental Risk Assessment: An AI-Driven Approach in Industry 5.0. *Computers* **2024**, *13*, 120. [CrossRef]
29. Márquez-Sánchez, S.; Campero-Jurado, I.; Herrera-Santos, J.; Rodríguez, S.; Corchado, J.M. Intelligent Platform Based on Smart PPE for Safety in Workplaces. *Sensors* **2021**, *21*, 4652. [CrossRef]

processes

MDPI

Article

Priority/Demand-Based Resource Management with Intelligent O-RAN for Energy-Aware Industrial Internet of Things

Seyha Ros [1], Seungwoo Kang [1], Inseok Song [1], Geonho Cha [1], Prohim Tam [2] and Seokhoon Kim [1,3,*]

[1] Department of Software Convergence, Soonchunhyang University, Asan 31538, Republic of Korea; rosseyha003@gmail.com (S.R.); oooksw12@sch.ac.kr (S.K.); sis5041@sch.ac.kr (I.S.); takflue@gmail.com (G.C.)
[2] School of Digital Technologies, American University of Phnom Penh, Phnom Penh 12106, Cambodia; prohimtam@gmail.com
[3] Department of Computer Software Engineering, Soonchunhyang University, Asan 31538, Republic of Korea
* Correspondence: seokhoon@sch.ac.kr

Abstract: The last decade has witnessed the explosive growth of the internet of things (IoT), demonstrating the utilization of ubiquitous sensing and computation services. Hence, the industrial IoT (IIoT) is integrated into IoT devices. IIoT is concerned with the limitation of computation and battery life. Therefore, mobile edge computing (MEC) is a paradigm that enables the proliferation of resource computing and reduces network communication latency to realize the IIoT perspective. Furthermore, an open radio access network (O-RAN) is a new architecture that adopts a MEC server to offer a provisioning framework to address energy efficiency and reduce the congestion window of IIoT. However, dynamic resource computation and continuity of task generation by IIoT lead to challenges in management and orchestration (MANO) and energy efficiency. In this article, we aim to investigate the dynamic and priority of resource management on demand. Additionally, to minimize the long-term average delay and computation resource-intensive tasks, the Markov decision problem (MDP) is conducted to solve this problem. Hence, deep reinforcement learning (DRL) is conducted to address the optimal handling policy for MEC-enabled O-RAN architectures. In this study, MDP-assisted deep q-network-based priority/demanding resource management, namely DQG-PD, has been investigated in optimizing resource management. The DQG-PD algorithm aims to solve resource management and energy efficiency in IIoT devices, which demonstrates that exploiting the deep Q-network (DQN) jointly optimizes computation and resource utilization of energy for each service request. Hence, DQN is divided into online and target networks to better adapt to a dynamic IIoT environment. Finally, our experiment shows that our work can outperform reference schemes in terms of resources, cost, energy, reliability, and average service completion ratio.

Keywords: energy efficient; network functions virtualization; open radio access network; software-defined network; industry internet of things

1. Introduction

With the continued advancement of the 5th generation and the industry internet of things (IIoT), their applications have gained increasing attention. The world is expected to reach 15.9 billion in 2023 to more than 32.1 billion IoT devices in 2030 [1]. Hence, as one of the three scenarios of 5G, massive machine type of communication (mMTC) can handle a strong supply of the development of the IIoT devices [2], and efficient use of networking with IIoT could be achieved, thereby managing the resources for upcoming new service-driven industries [3] economically. It is considered one of the vital driving factors that contribute to achieving Industry 5.0 [4,5]. Hence, IIoT devices are exposed to networking and state-of-the-art features that increase the volume of highly demanding resources. On the other hand, IIoT devices (e.g., sensors and actuators) [5] are accessed and connected to network servers in the core via a fronthaul physical network [6]. However,

IIoT needs a complementary relationship between network computing and communication in terms of satisfaction with quality of service (QoS) and quality of experience (QoE). IIoT may continue to generate different types of tasks, with a large increase in processing time, computation resources, battery lifetime, and efficiency [7].

The core of utility resources, leveraging mobile edge computing (MEC), is typically demonstrated to offer resource sufficiency to affordable computing tasks of IIoT devices [8]. The open radio access network (O-RAN) revolutionizes the convention monoline structure, disaggregating base stations functionalized into distinct [9,10]. O-RAN offers a new architecture and standard to support the multi-vendor forms factor that allows the use of multiple technologies integrated into 5G antennas and next-generation. Meanwhile, MEC is deployed into O-RAN [9–12], which allows the closest user experience and addresses the energy-efficient demands of modern telecommunication and network systems. O-RAN is significantly orchestrated to MEC architecture, which ensures delivery service computation and allows the deployment of network connectivity, an effective way to break bottlenecks and latency packet delivery ratios. Within MEC, illustrating resource provisioning is the crucial technique for adjusting computing, which is one of the challenges. Software-defined network (SDN) and network function virtualization (NFV) are accomplished to archive the network target by providing virtualization and softwarization [13–15]. When MEC-enabled SDN and NFV are the accommodation of new experiences, they enhance the potential computing service for IIoT devices, providing low-latency computation capabilities and ensuring the network is compatible with scalability and reliability [16,17]. MEC improves bandwidth utilization to enhance the user experience and overall network performance. However, tasks generated by IIoT devices incur an additional transmission to the controller that conflicts with delay (queuing) and energy consumption. Additionally, properly prioritizing and demanding resource management at the MEC server will influence task execution.

Therefore, studies have been investigating MEC in terms of managing communication and computing resource capabilities. MEC can be ostensibly affordable in terms of resource utilization and computing capabilities, which are suitable for offering computing tasks in terms of offloading decisions, resource allocation, and resource prioritization. Despite management resource flexibility and dynamic adjustments, there have been crucial impacts in increasing computation time and energy efficiency. Thereby, artificial intelligence-driven (AI-driven) IIoT devices show opportunities to customize intelligent edge computing with heterogeneous hardware, which has good energy efficiency in processing specific AI-based tasks in network performance.

In this paper, we leverage deep reinforcement learning (DRL) to enhance network performance, eliminate long-term cumulative rewards, and design an efficient, jointly optimal approach to address resource priority and demanding resource management. Hence, the system utilizes the deep q-network (DQN) algorithm. This approach tackles the challenge of managing system states and actions. The DQN selects optimal actions within a continuous space, making it perfect for this dynamic environment. The prioritized tasks by learning from a reward system are considered successful task completion and efficient resource usage. As demands change, DQN constantly re-evaluates the situation, adapting resource allocation to meet immediate needs while keeping long-term priorities. DQN bridges priority and demand management, ensuring critical tasks are completed while optimizing resource utilization for IIoT tasks. This paper's contributions are as follows:

- We study efficiency resource management, which enables O-RAN to provide gratitude support for multi-vendor and scalable deployment of MEC servers and improve resources based on the task demands and service priority.
- Then, the problem of resource and energy minimization is conducted to transform into a Markov decision process (MDP). After, we design a novelty distributed DRL-driven resource management policy in the proposed model, which jointly optimal resource and priority/demand based on IIoT criteria usage.

- Our proposed DQG-PD algorithm improves resource management efficiency and reduces task processing time and latency to enhance efficient resource awareness of IIoT applications.
- We enhance network energy efficiency optimization based on the DQN approach. Leveraging the DQN approach, which decouples two stages (e.g., online network and target network) to respond to the network performance by stabilizing long-term learning while enabling rapid adaptation to immediate demands.
- Lastly, we conduct experiments to evaluate and show the witness that our network scenario outperforms reference schemes.

In the rest of the paper, Section 2 gathers the previous studies and motivation to address our work. Section 3 is considered a problematic formulation for optimal resource efficiency while computing and communicating IIoT device tasks. Moreover, we consider the characteristics of different rewards and three setting network scenarios. Simulation results and further analysis are demonstrated in Section 4. Section 5 summarizes the conclusion.

2. Related Work

The industrial manufacturing system integrated with IoT, the IIoT is increasingly complex [18]. In recent years, the proper management and orchestration of various resources in IIoT devices and the optimization of network performance have become the focus of research. In fact, Figure 1 demonstrates the use of the MEC-assisted O-RAN to enhance resource computation and provide accommodations for IIoT devices. In addition, IIoT devices include more network types than IoT [19–24]. The major network types of IoT are WLAN and cellular networks, adopting Wi-Fi, Bluetooth, and 5G/B5G technologies. IIoT device networks further include low-powered wide area network (LPWAN) and WPN [25,26]. Once MEC is enabled, the SDN/NFV controller will ensure the network's hierarchical alignment of resource dynamics and flexibility for the next generation.

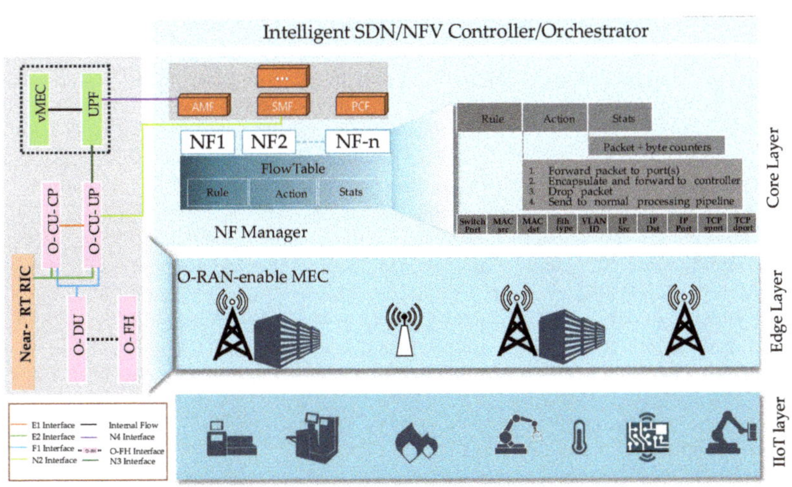

Figure 1. An intelligent SDN/NFV controller-based MEC server is deployed in the O-RAN.

2.1. *Energy Efficiency for IIoT*

Energy is properly utilized in processing to compute and transfer to ensure the packets of network delivery are reachable. The existing studies are conveying their approach to prove that concept and provide the technique to concisely deploy in network systems [27–30]. Energy consumption has three significant impacts: communication, computation, and sensing [31,32]. Sensing is referred to as physical hardware designed to support, which is determined for suspended functionality, and we prioritize computation and communication overhead.

2.2. Optimization Approaches Based on MEC for Virtualization in Energy Utilization

Leveraging an algorithm to enhance energy efficiency in the networking environment, which enables edge cooperation, has been investigated. In [33], the focus is on resource management in end-to-end wireless-powered MEC networks for devices. They addressed the optimization challenges posed by dynamic task arrivals and battery power fluctuations. Employing Lyapunov optimization and virtual queues transformed the long-term optimization problem into a deterministic time slot drift-plus-penalty subproblem, making it more tractable and ensuring optimal performance in the long run. Moreover, in [34], they investigated the trade-off between energy efficiency and delay in multi-user wireless power systems. By incorporating wireless energy transfer into the MEC system, MEC enhances computing capabilities to prolong device battery life. Using Lyapunov optimization theory, they optimized network energy efficiency while ensuring network stability, available communication resources, and meeting energy causality constraints. In [35], green content caching is based on user association mechanisms that aim to minimize content requests by enabling small cell networks while ensuring the QoS of mobile terminals.

2.3. DRL for MEC-Integrated O-RAN Resource Management

Both radio and computing resources play a crucial role in the performance of task offloading. Radio resources influence the data rate and energy consumption during the transmission process, while computing resources limit the computing time and energy consumption of tasks that offload to the MEC server. Hence, ref. [36] proposes an architecture utilizing xApp, powered by machine learning algorithms that quickly identify the network traffic and intelligently allocate resources within O-RAN architecture. Such a structure addresses the defects of static resource allocation in O-RAN architecture by automatically adapting PRBs to traffic load and QoS requirements, leading to better performance and more economical satisfaction for end users. On the other hand, in ref. [37], the reinforcement learning (RL) algorithm is adopted to allocate and efficiently manage PRBs by utilizing modulation and coding schemes based on traffic flow and channel quality key performance indicator (KPI) requirements. The authors of [38] investigated a mMTC scenario in 5G to troubleshoot IIoT services that ensure efficient resource allocation methods to enhance dynamic and complex environments by conducting intelligent end-to-end self-organizing resource allocation IIoT with the asynchronous actor and critic-driven DRL algorithm.

However, the above works did not consider the user offloading to the MEC server in computation by serving as energy efficient in resource utilization. In fact, this complex selection strategy cannot be ignored due to the IIoT devices transferring to the MEC server at the same time and different resource demands. Moreover, the problem of minimizing energy consumption and energy efficiency in O-RAN is notably distinct and challenging from traditional RAN. To overcome encounters related to energy efficiency for diverse IIoT resource types in the MEC server, we leveraged O-RAN-enabled MEC to ensure resource awareness and priority on demands.

3. Problem Formulation and Objectives

We first formulate the problem of resource management IIoT model and slice types of processing in the MEC server and then describe the resource management based on priority and demanding capability to ensure efficient energy consumption.

3.1. IIoT Model and Slice Types

This section uses QoS class identifiers (QCIs) to determine the characteristics and requirements of different setting types that ensure the network controller can be handled by dividing prioritized data flows based on three slice types. QCIs from the 3GPP TS 23.203 V12.2.0 [39] specification can be employed to represent pertinent example services. Table 1 outlines the QCI index, resource type, priority level, packet delay budget (PDB), and packet error loss rate (PELR) for various smart industry scenarios. Regarding resource types, guaranteed bit rate (GBR) ensures that end users receive a minimum bandwidth, even in times of network congestion. This is critical for applications requiring consistent performance. Conversely, non-GBR offers optimal service under normal conditions; however, it does not guarantee that the requested bandwidth will be available during high network usage periods. This type of service is suitable for applications where occasional delays are acceptable, such as video streaming or data uploads from environmental sensors. PDB and PELR serve as upper-bound thresholds, defining the maximum tolerable delays and packet loss rates between IIoT and the policy charging enforcement function. These parameters are crucial for maintaining the QoS in smart industrial applications, ensuring efficient energy utilization and resource demand capability in flexible and dynamic environments. Here, we dive through to describe three slice types of IIoT application use cases as follows:

- QCI 3 ensures that the communication infrastructure supports the reliability and timely exchange of data critical for the automation of industrial processes and real-time monitoring applications. Hence, data is critical for automation in controlling the network environment's charge policy.
- QCI 70 ensures that mission-critical data in IIoT environments receives the highest level of service quality, characterized by ultra-reliability, low latency, high priority, enhanced security, and dedicated bandwidth.
- QCI 82 provides the resource capabilities for defining discrete automation, which involves controlling and monitoring manufacturing processes that handle individual parts or units, generally in environments such as assembly lines or robotics, where precision and real-time performance are crucial.

Table 1. QCI specification is defined to set different types of industry application use cases.

QCI-Index	Resource Types	Priority Level	PDB	PELR	Industry Application Use Case
QCI-3 Process Automation and Monitoring	GBR	30	50 ms	10^{-3}	Robotic monitoring
QCI-70 Mission Critical Data	Non-GBR	55	200 ms	10^{-6}	Safety systems
QCI-82 Discrete Automation	Delay critical GBR	19	10 ms	10^{-4}	Automate quality control

3.2. Designing and Formulating Network Resource Management

Network resource management and priority use cases are addressed from the standardization framework to support the implementation of industry applications. Table 1 demonstrates the importance of setting each class's network sensitivity in the initial usage. Thus, it gives samples for enhancing the network performance and setting network vector management.

3.3. Communication Model

We consider an offloading resource over edge computing paradigms. The MEC is deployed in O-RAN and provides capability via wireless access points in cellular networks through a wired connection. Accordingly, the clustering of the network consists of communication and computation in edge computing. Table 2 shows a system model that was used to formulate MEC resources and IIoT tasks. The number of IIoT tasks denoted as $I = (1,2,3,\ldots,i)$, $\forall\, i \in I$ participate at time slot $T = (1,2,3,\ldots,t)$, $\forall t \in T$. Next, $N = (1,2,3,\ldots,n)$, $\forall n \in N$, where denotes referring to the number of devices and $M = (1,2,3,\ldots,m)$, $\forall m \in M$, where denotes the number of MEC servers used in our network and deployed in O-RAN. We assume that in our network scenario, in which O-RAN is deployed at the base station and coordinated, one or more O-RAN can cooperate with each other for downlink and uplink transmissions to the end devices. Furthermore, downlink transmission is ignored in our work.

Table 2. The notation of network system model.

Symbol	Description
I	Set IIoT tasks $I = (1,2,3,\ldots,i),\ \forall i \in I$
N	Set IIoT devices $N = (1,2,3,\ldots,n),\ \forall n \in N$
T	Set time slots $T = (1,2,3,\ldots,t),\ \forall t \in T$
M	Set MEC server $M = (1,2,3,\ldots,m),\ \forall m \in M$
S^i	Offloading decision from IIoT device to MEC, whether 1 or otherwise
S_n	Data size of the computation task n-th
$D_{n,m}$	Data transmission rate from IIoT n-th to MEC server m-th
$B_{n,m}$	Transmission power device-n to MEC server m-th
$H_{n,m}$	Channel bandwidth
ψ	Ground interference power consumption
P_v	Processing power required by VNF v-th
U_v	Utilization of VNF v-th
B_m	Total bandwidth of MEC server m-th
L_{max}	Satisfaction of latency
C_{max}	Upper bound of total resource usage of the capacity of each MEC server.
$X_{mec}^{n,m}$	Execution at MEC server with task n-th
L_i^n	Time accepted

Decision variables:

$$S^i = \begin{cases} 1, & \text{Offloading decision of IIoT to MEC server} \\ 0, & \text{Otherwise} \end{cases}$$

Regarding Simmons' law, the transmission data rate $D_{n,m}$ of IIoT device can be calculated as:

$$D_{n,m} = H_{n,m}\log_2\left(1 + \frac{B_n G_n}{\psi_n + \sum_{m=1}^{N} B_m G_m}\right), \quad (1)$$

where $H_{n,m}$ is the channel bandwidth, B_n is the transmission power of n-th devices. ψ is the background interference power consumption that includes wireless transmission from other devices ψ_i^0 and noise power consumption can express: $\psi_n = \psi_i^0 + \psi_i^1$. G_m is the channel gains, and as in Formula (1), if multiple devices simultaneously perform calculations and offload via the wireless access channel, significant interference and a reduction in the data transmission rate will occur. Thus, the constraint conditions that the size of $z_{n,m}(t)$ should follow is:

$$z_{n,m}(t) \leq D_{n,m}(t), \forall n \in N \quad (2)$$

The transmission energy consumed by offloading tasks for IIoT i is given as follows:

$$\Delta_{T,i}(t) = B_n(t)\frac{z_{n,m}(t)}{D_{n,m}(t)} \quad (3)$$

Based on the communication model, it consists of many devices simultaneously offloading tasks to the MEC via the wireless channel, which will inevitably result in reduced data transmission rates. When the data rate for the IIoT device decreases, the backhaul link will incur higher energy consumption and longer transmission times for offloading tasks. Utilizing edge servers for computing tasks is an enhancement as it can avoid the long transmission delays associated with cloud computing.

3.4. Offloading Model

In the task offloading model, the task data shall first be transferred to MEC, which is the process of transferring computational tasks and transmission time of the task $n_i \in N$. S_n is the data size of computation task *n-th* to MEC server for computation to address the offloading time of task *n-th* from the IIoT devices can be expressed:

$$CT_{n,m}^{off} = \frac{S_n}{D_{n,m}} \quad (4)$$

The energy consumption is generated by computation tasks. IIoT devices of CPU computation rate in the time slot $F_n(t)$, and $\mathcal{F}$ is the number of cycles required for the CPU. $CB_n(t)$ is the computing task size of device *n-th* in time *slot-t*.

$$CB_n(t) = \frac{F_n(t)}{\mathcal{F}} \quad (5)$$

3.5. Computation Model

In this primary consideration of the computation, our network system addresses MEC computation, a computing task generated by an IIoT device, and utilizes a MEC server to tackle the computation and leverage it to offer resource capabilities. Hence, the total computation time for task *n-th* can be calculated based on what is processed to offload to the MEC server.

- MEC server execution:

$$X_{mec}^{n,m} = \frac{F_n(t)}{\mathcal{F}_m} \quad (6)$$

- Total complete time:

The total time required to complete the task includes both transmission and execution time on the MEC server.

$$CT_{mec}^{n,m} = \frac{S_n(t)}{D_{n,m}} + X_{mec}^{n,m} \quad (7)$$

3.6. Objective Model on Resource Management

In our proposed approach, MEC is leveraged to conduct assessments of resource management and orchestration for the priority of IIoT tasks. MEC provides resource capability and indicates stringent requirements. We aim to minimize energy consumption by leveraging MEC to provide accommodation for NFV to instantiate virtual machines (VM) and adjust and reconfigure the resource capacity of utilizations. In addition, VM minimizes resource utilization over the MEC server, ensuring the efficiency of VNF deployments and enhancing computation. Where R_n denotes resource requirements of task-n, P_v denotes processing power required by VNF in the MEC server and U_v is the utilization of VNF-v.

$$\min_{[H_{n,m},S_i]} \sum_{m\in M} \left(\frac{\sum_{i\in I}\sum_{n\in N} S_n^i \cdot R_n}{C_{mec}} + \frac{\sum_{v\in V} P_v \cdot U_v}{C_{mec}} \right) \quad (8)$$

Subject to

$$\sum_{i\in I}\sum_{n\in N} S_n^i \cdot R_n + \sum_{v\in V} P_v \cdot U_v < X_{mec},\ \forall m \in M \quad (9)$$

$$\sum_{i\in I}\sum_{n\in N} S_n^i \cdot D_n < B_m,\ \forall m \in M \quad (10)$$

$$\sum_{m\in M} S_n^i \cdot C_{mec}^{n,m} < L_{max},\ \forall n \in N \quad (11)$$

$$\sum_{v\in V} P_v \cdot U_v < C_{max},\ \forall m \in M \quad (12)$$

$$E_{local}^i \cdot L_n^i + \sum_{m\in M} E_{off}^{n,m} \cdot S_n^i \le E_{max},\ \forall i \in I \quad (13)$$

4. DQN-Based Priority/Demanding Resource Management

4.1. Markov Decision Process Elements

We study the MDP algorithm to solve the problem, which is a tuple of $(\mathcal{S}, \mathcal{A}, \mathcal{R}, \mathcal{S}')$, where $\mathcal{S}$ denotes the set of the possible states, $\mathcal{A}$ is denoted the action. At *timeslot-t*, the agent observes the state s_t of the time slot and selects action. In our proposed system, we leverage deep q-network-based-priority/demanding resource management (DQG-PD). We have developed the DQG-PD to enhance the network policy in terms of resource management and efficiency. Our approach obtains the optimal network policy for efficient resource offloading in MEC servers that demonstrate resource capacity in terms of maximizing the energy efficiency in O-RAN-enabled MEC. However, to utilize the DRL approach for reinforcement learning (RL) agents, we address the steps of states-space, action-space, and rewards function as:

(1) State-space: in each time slot, each communication link and computation in MEC observe the network state from the environment. Let S denote the state space. The current environment state includes $D_{n,m}(t)$ measurement of the data transmission rate from the IIoT device and MEC server, the status of all resources in the IIoT device $S^i = 1$ is supposed to offload the resource to the MEC server, 0 otherwise. $CB_n(t)$ computation task model of *n-th*. As a result, state S is defined by the following parameters:

$$S(t) = \left[D_{1,1}(t), D_{1,2}(t), \ldots D_{n,m}(t), CB_1(t), CB_2(t), \ldots CB_n(t), B_{1,1}(t), B_{1,2}(t), \ldots B_{n,m}(t), S^i_{n'}, \right]^T \quad (14)$$

(2) Action-space: we utilize agents to make decisions based on gathering the current state of the environment. The goal of the agent is to make the optimal decision based on maximizing the resource utilization in terms of bandwidth, computation resource utilization, and minimizing the overall average service delay with minimal task execution. Action $a(t) \in \mathcal{A}$ at each time step t can be defined as the action in our network system, which considers offloading the *t-th* task $(1 < t \leq N)$ and allocating the resource (bandwidth and computation resource) to the task for execution on the MEC server. Action can be defined as:

$$a(t) = \{H_{n,m},\ S_i, \Psi\} \quad (15)$$

where $H_{n,m}$ is a representation of the channel bandwidth, S_i selection task offloading for task size with Ψ belonging to $\{1, 2, \ldots, Y\}$ MEC server when $S_i = 1$, and $S_i = 0$ in local. The agent will take actions based on tasks in each time step and get the reward from the environment. Note that in each decision epoch, an action also affects the next state s' in next time *slot-t*.

(3) Reward: RL aims to maximize the reward from good actions. Our reward function is to design and optimize to reflect the enhancement of the priority of resource management and efficient energy. The reward function r_t can be defined as:

$$r_t = CT^{n,m}_{mec} = \frac{S_n(t)}{D_{n,m}} + X^{n,m}_{mec} \quad (16)$$

4.2. DQN-Based Solutions

Our work aims to provide flow execution by using the DQN algorithm to handle resource demand and prioritization on the MEC server. In fact, DQN is one of the most powerful tools for assisting network controls and adaptation of resource estimates. Moreover, DQN offers two stages of architecture (e.g., target network and online network). Figure 2 indicates the network conditions, which consist of several stages of the IIoT cluster that are divided into three categories of network determination resource tasks. The proposed method is conducted using DQN with an SDN/NFV controller for interaction and abstracts the resource utilization from MEC resources in computing the IIoT tasks. A strategy chooses the action obtained through long-term optimization, and DQN maximizes the reward value through the selection of the optimal value. Controller gathers the state as in Equation (14) and selects action $a(t) = \pi(s(t))$ to obtain the current reward r_t, while $s(t)$ is transferred to the next state. π is the specific policy. The interaction with the environment proceeds based on the updated state, aiming to maximize the reward value by maintaining the ongoing process. The cumulative discounted reward over the time interval is calculated using $Q((s(t), a(t))$ as in the following equation:

$$Q(s(t), a(t)) = E\left[\sum_{t=1}^{T} \gamma^t r(t) | s(t), a(t)\right], \gamma \in (0,1) \quad (17)$$

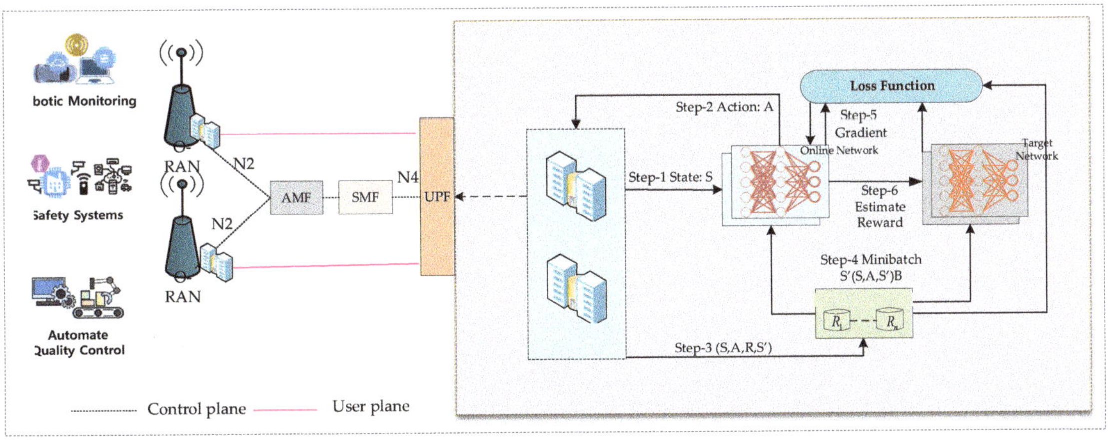

Figure 2. DQN-based MEC for priority/demanding resource utilization.

From the equation, $\gamma \in (0,1)$ is the discount factor. If $\gamma = 1$, the agent only considers the current rewards, whereas if $\gamma \leq 1$, the agent considers the later rewards.

$$Q(s(t), a(t)) = Q(s(t), a(t)) + \alpha(r(t) + \gamma max Q(s(t+1),\ a(t+1)) - Q(s(t), a(t))) \quad (18)$$

When this value is used to indicate iteratively estimate the optimal Q-value for all the state and action pairs s(t), a(t), and the optimal policy π^*:

$$\pi^* = \underset{\pi}{argmax} Q(s(t),\ a(t)\,) \quad (19)$$

α is the learning rate, while the deep neural network (DNN) technique is leveraged. DQN is constructed into two structures, with one of the DNNs being the online network, which fits the value function Q, and another DNN being the target network, which is used to gain the target Q value. $\varnothing$ and $\varnothing^-$ are the weight of online and target on Q-network, respectively.

$$Q^* = r\big(s(t), a(t) + \gamma max Q(s(t+1), a(t+1)); \varnothing^-\big) \quad (20)$$

In addition, in the learning rate phase, the reward and state updates obtained from the iteration with an environment that is stored in the experience buffer in forms $s(t), a(t), r(t), s(t+1)$. The parameters of DNN are updated by interactively minimizing the loss function as follows:

$$LOSS\ FUNCTION = E\Big[(Q^* - Q(s(t), a(t); \varnothing))^2\Big] \quad (21)$$

Algorithm 1 depicts the hierarchical structure proposed for handling the resource priority and demanding satisfaction in terms of energy efficiency in the MEC server for controlling the IIoT tasks. On the other hand, the DQG-PD algorithm is leveraged for priority and demand resource allocation in the MEC server and employs DQN to determine the optimal actions for each time slot. Initially, the algorithm sets up two Q-networks, an online network and a target network with random weights and initializes a replay memory to store previous experiences. In each episode, an initial state proceeds through multiple time steps, deciding between exploration and exploitation. During exploration, an action is selected randomly, while in exploitation, the action that maximizes the Q-value (predicted future reward) is chosen based on the online Q-network. The selected action is then executed, leading to a new state and reward stored in the replay memory. It periodically samples a mini-batch of experiences from this memory to update the online Q-network by performing gradient descent on the loss between predicted and target Q-values. The target Q-network is periodically synchronized with the online network to stabilize learning. This process is repeated for multiple episodes, allowing the MEC server to learn and refine its resource allocation strategies, ultimately returning the optimal action for each time slot based on the learned policy.

Algorithm 1: DQG-PD algorithm for priority/demand resources in the MEC server

Input Discount factor, learning rate, exploration factor greedy, replay memory size
Output Optimal action of each slot $a^{*}(t)$
Initialize Online and target-Q network parameters with random weight $\emptyset$ and $\emptyset^{-}$, respectively. ($\emptyset^{-} = \emptyset$), replay memory B
for episode = 1, 2, …, k $\rightarrow K$ **do**
    Choses an initial state S;
    for t = 1→ T **do**
    if
        Randomly select number $a(t)$ ∇ *Exploration*
        Apply action $a(t)$
        Observe the state $(s+1)$ and reward $r(t)$
        Store transition experience $\{s(t), a(t), r(t), s(t+1)\}$ in replay memory B
        Randomly sample experience from replay memory
    else ∇ *Exploitation*
        Selection action $a(t) = \underset{a}{argmax}\, Q(s(t), a(t); \emptyset)$
        Update weight $\emptyset$ by performing a gradient descent process on $[Q^{*} - Q(s(t), a(t); \emptyset)]^{2}$
        Reset $\emptyset^{-} = \emptyset$ after each C steps
    end
    Return Optimal action $a^{*}(t) = \underset{a}{argmax}\, Q(s(t), a(t); \emptyset)$
end

5. Simulation and Discussions

In this study, we conducted the experiment by setting the network topology integrating the traffic between IIoT local hosts, access points, and SDN/NFV controller environment [40], instructing the policy by our approach agent decisions. In the following sub-section, we described the parameter settings and the performance of the experiment.

5.1. Parameter Settings

In this scenario, the setting needs to illustrate the three clusters of IIoT applications to conduct our experiments, as shown in Table 3. Specifically, we consider the assembly line in industrial manufacturing. The base station is assumed to be the O-RAN-enabled MEC server, which is under the umbrella of resource utilization in vertical resource alignment to the IIoT device required. MEC servers are powerful computers located close to the IIoT devices, which handle the computation tasks offloaded by the IIoT devices. The placement of four MEC servers assists in a balance between processing power and energy consumption. More servers could handle more tasks; however, the infrastructure would consume more energy. IIoT devices are in industrial setups that generate data and require computation. The numbers [50, 100, 150] indicate different scenarios with varying numbers of devices. The size of the tasks generated can range from 5 MB to 30 MB. Bandwidth is the maximum rate of data transfer across the network. A bandwidth of 20 MHz ensures that data can move quickly between the IIoT devices and MEC servers, reducing the time and energy needed for communication. CPU frequency determines how fast the MEC servers can process data. Depending on the workload, a range of 5 GHz to 20 GHz means the servers can adjust their processing speed. A maximum link latency of 1.5 milliseconds ensures that the system responds quickly, which is critical in industrial settings. The operation is divided into 1000 time slots, which are small intervals of time used to schedule tasks and allocate resources. The replay memory buffer is where data is temporarily stored while being processed. A size of 3000 units means the system can hold a significant amount of data at once, which helps manage tasks efficiently and conserve energy. ReLU is a mathematical function used in neural networks within the system to make decisions, such as task scheduling or resource allocation. A discount factor of 0.95 means the system values long-term efficiency slightly more than short-term gains, encouraging energy-efficient decisions over time. The learning rate is 0.001, ensuring the system learns gradually and avoids making large, energy-inefficient changes based on new data. The batch size of 32 is a balance between computational efficiency and energy use, allowing the system to learn effectively without overloading the servers or consuming excessive power.

Table 3. Simulation parameters.

Parameters	Value
Number of MEC servers	4
Number of IIoT devices	[50, 100, 150]
Task size	[5, 30] MB
Upper-bound bandwidth	20 MHz
CPU frequency of MEC server	[5, 20] GHz
Maximum link latency	1.5 ms
Number of time slots	1000
Replay memory buffer size	3000
Activation function	ReLU
Discount factor on reward	0.95
Learning rate	0.001
Batch size	32

5.2. Performance Evaluation

With the above setup simulation infrastructure, we deploy our management scheme and reference approaches in the controller to evaluate the performance. DQG-PD leverages the virtualization and the proposed DQN agent to guide the allocation process. Priority on industrial services and demand are observed from the infrastructure plane. The target networks discover the action batches on offloading decisions resource (bandwidth and computing resources) virtualization properties by exploration. We compared this with (1) meta-heuristic balancing the resources and demand and (2) single-agent DRL with a reward emphasized for service with higher priority (low PDB and PELR).

To evaluate the performance of our contribution domains, we set 3 slices of different priority levels with demanding conditions. The traffic rates are configured to be high congestion following four different settings to capture the bottleneck and constrained resource evaluation. For the performance metrics, we focus on (1) the reward (resources, costs, energy, and reliability) of the agent policies and (2) overall ratios based on demand and priority levels in requesting services.

We captured the reward scoring metrics of the exploited DQG-PD algorithm under different learning rates (α) and discount factors (γ). The results highlight the trade-off between the speed of learning and the quality of the solution (optimality). The combination of $\alpha = 0.001$ and $\gamma = 0.95$ is chosen as the optimal setting due to its demonstrated performance in balancing the trade-off. This means that the learning rate (α), which determines how quickly the algorithm updates its knowledge, is set to a slower pace (0.01) to ensure stability. Meanwhile, the discount factor (γ), which controls how much future rewards are considered, is relatively high (0.95), emphasizing the importance of long-term rewards. These settings guide the algorithm toward optimal decision-making over time despite the initial exploration delays. Over 1000 episodes, we illustrate the comparison between the proposed DQG-PD algorithm and two other schemes (MT-HRT-PD and SADRL-PD) in terms of total reward scores, respectively. During the early phases (exploration), the DQG-PD experiences higher latencies, which is expected when the algorithm is still exploring various possibilities. However, by the 500th episode, it converges with a near-optimal solution (>80 setting metric), reducing the delay significantly and outperforming the other reference schemes, which overcomes MT-HRT-PD by 18.12 scores and SADRL-PD by 35.56 scores.

In terms of total reward, which is cumulated based on positive and negative scores as the algorithm converges during each episode, DQG-PD demonstrates superior performance. The reward is based on the algorithm's efficiency in multi-batch processing. By the final episode, DQG-PD reached a reward of 98.76, while MT-HRT-PD and SADRL-PD achieved lower scores of 68.13 and 43.62, respectively. This highlights the algorithm's overall better performance across different metrics, such as cost efficiency, resource usage, and latency reduction.

Each sub-reward is captured to compare different objectives to different approach performances. Figure 3 depicts the sub-reward on resources, which states the overloading queues and requests to the server. Whether the resources are overloaded or underutilized, negative rewards are accumulated. Regarding resource optimization, we illustrate how DQG-PD allocates virtual service nodes and forwards graphs to physical nodes and links efficiently. Dynamic resource allocation improves overall resource utilization, reducing bottlenecks and ensuring that the demand for created industrial service slices is met. DQG-PD achieved 26.61 positive scores in this category, outperforming MT-HRT-PD and SADRL-PD by 13.85 and 8.88 scores, respectively. The algorithm significantly enhances resource efficiency by adapting to real-time conditions and prioritizing mission-critical service demands.

In terms of costs in Figure 4, we balance between the number of servers deployed, service types, and consequences of packet loss in each slice. DQG-PD achieved positive scores of 19.27, which accounts for 19.51% of the overall reward, compared to other schemes, where MT-HRT-PD and SADRL-PD obtained 7.59 and 14.33 scores, respectively. The cost sub-reward is crucial as it focuses on reducing the financial burden of industrial service deployments, including minimizing resource provisioning costs, network bandwidth usage, and overall infrastructure expenditures. DQG-PD ensures efficient management of each prioritized slice, thereby lowering operational costs while maintaining the required QoS and QoE.

The energy sub-reward focuses on reducing the consumption amount that the service operates and network traffic experiences while traversing a complete performance. The DQG-PD algorithm optimizes the resource placement and routing decisions to reduce end-to-end latency with prioritized demand considered on energy metrics, improving overall service responsiveness and green computation. DQG-PD achieved 28.98 scores, which accounted for 29.34% of the total reward metrics, emphasizing the energy-focused and bettering performances compared to MT-HRT-PD (9.13 higher) and SADRL-PD (19.85 higher) in Figure 5.

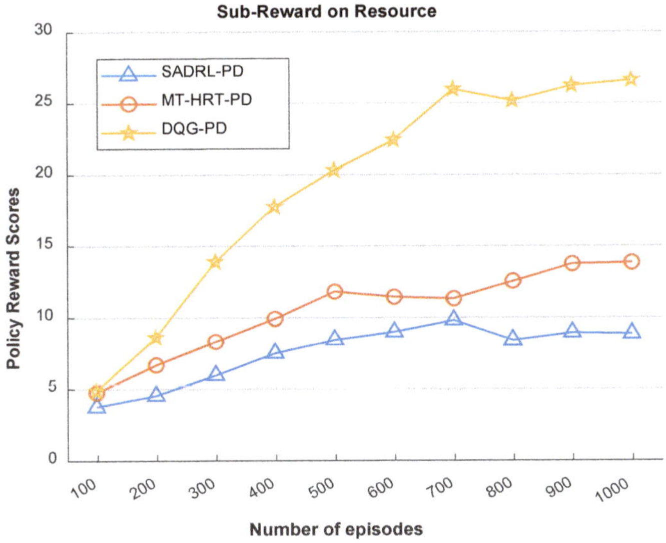

Figure 3. Sub-reward on resource.

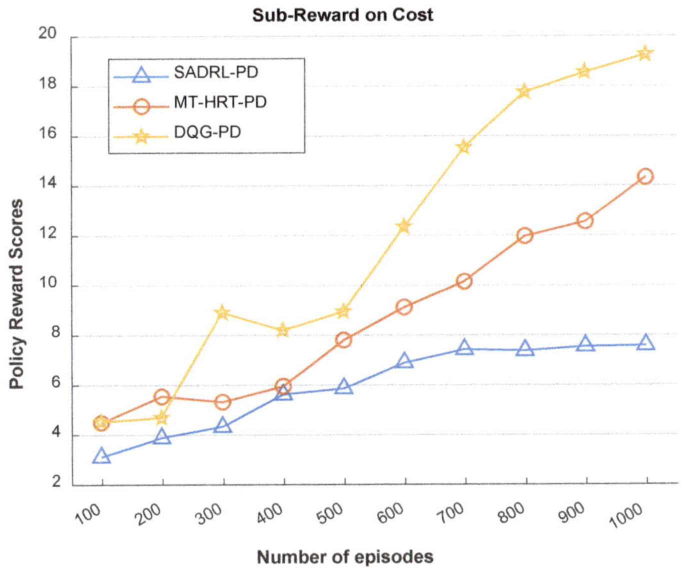

Figure 4. Sub-reward on cost.

Finally, the sub-reward on reliability metrics, as shown in Figure 6, emphasizes the algorithm's robustness and fault tolerance. DQG-PD achieved a high-reliability score (22.68), significantly outperforming MT-HRT-PD (by 2.58) and SADRL-PD (by 4.66). This reliability is achieved by incorporating dynamic service configurations, which adapt to network failures, traffic surges, demanding congestion, or disruptions. DQG-PD ensures high availability by considering factors such as backup paths, failover mechanisms, and redundancy, thereby enhancing service reliability and minimizing downtime due to network issues.

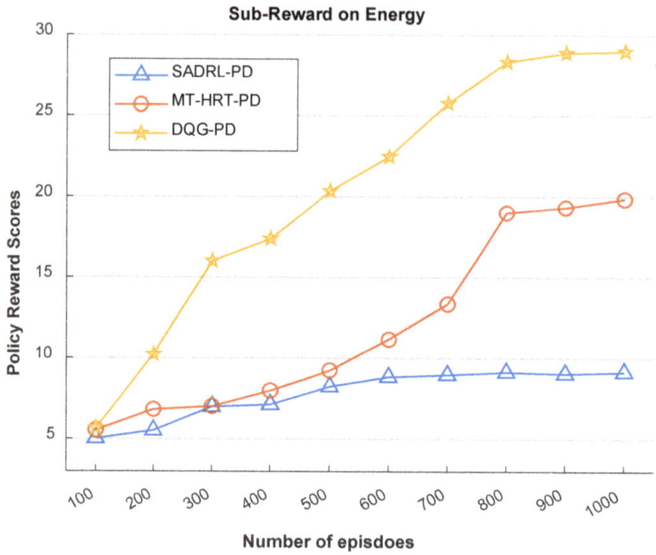

Figure 5. Sub-reward on energy.

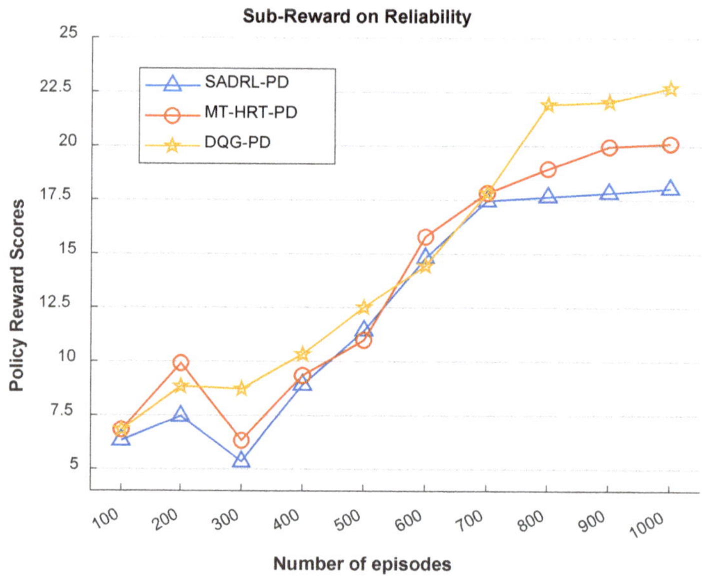

Figure 6. Sub-reward on reliability.

To deeply evaluate the efficiencies in controlling different demanding factors and priority levels, we captured the completion ratios of combined service requests in different topology sizes and the number of virtual chains (3, 6, 9, and 12). First, our simulation estimated the acceptance ratios when the service is requested. After accepting the requests, we executed the services through the chain to detect any possible failure. If failure is detected, we configure the restoration properties to revive the service execution. However, the properties of slice criticality vary. If restoration is successful, we capture our results as completion ratios. Over 1000 time slots, we configured the traffic rates to stimulate the control policies. DQG-PD achieved 99.87%, which is 10.78% and 19.54% higher than MT-HRT-PD and SADRL-PD, respectively, as shown in Figure 7. The elaboration of each phase can be discussed as follows:

- The high acceptance ratio demonstrates the controller's scalability and ability to effectively accommodate a larger number of service requests.
- The restoration ratio measures how well the system recovers from service failures, ensuring uninterrupted service and high availability, particularly for high incoming task requests.
- In total, we concluded the performance into completion ratios, which demonstrates its effectiveness in completing tasks even under heavy loads.

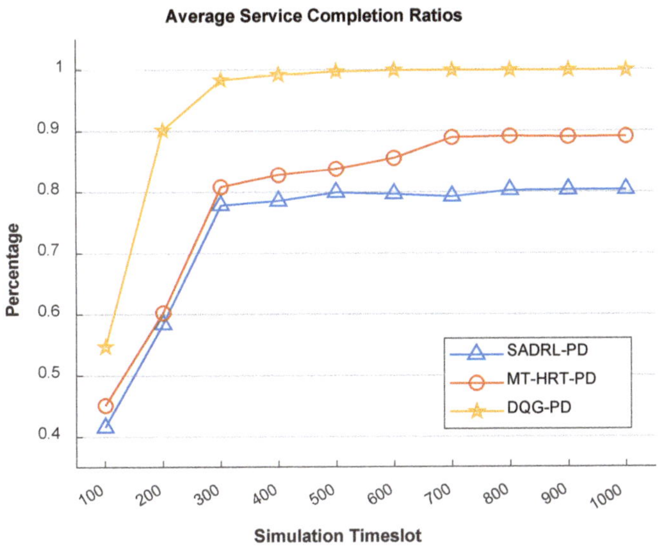

Figure 7. Average service complete ratios show how efficient the service percentage is over the different time slots.

6. Conclusions

This paper presented priority/demanding-based resource management to address optimal resource consumption and energy, which enabled O-RAN to aim at a cognitive IIoT network. The proposed scheme leveraged MEC-incorporated O-RAN to handle resource efficiency and resource management. DQG-PD leverages virtualization and DQN agents to direct resource allocation to industrial services according to priority demand. DQG-PD framework ensures sufficient system components for automated policy orchestration, including resource management, energy, cost, and service priorities. In this framework, DQG-PD enhances resource management and energy efficiency for resource-constrained IIoT devices and scales into a system for heterogeneity critical in real-time. Our results demonstrate that the DQG-PD algorithm significantly enhances computation, resource utilization, and energy efficiency when compared to existing reference schemes.

For future work, we leverage advancing federated learning to minimize the trade-off between communication overhead and joint the cross-protocol with asynchronous. Furthermore, addressing security and privacy concerns will be key focus areas in expanding the effectiveness of our approach.

Author Contributions: Conceptualization, S.R. and S.K. (Seungwoo Kang); methodology, S.R. and P.T.; software, P.T. and S.R.; validation, S.R., I.S. and G.C. formal analysis, I.S. and S.R.; investigation, S.K. (Seokhoon Kim); resources, S.K. (Seokhoon Kim); data curation, P.T. and S.R.; writing—original draft preparation, S.R.; writing—review and editing, S.R., I.S., S.K. (Seungwoo Kang) and P.T.; visualization, S.R. and G.C.; supervision, S.K. (Seokhoon Kim); project administration, S.K. (Seokhoon Kim); funding acquisition, S.K. (Seokhoon Kim). All authors have read and agreed to the published version of the manuscript.

Funding: This work was supported by Institute of Information & Communications Technology Planning and Evaluation (IITP) grant funded by the Korean government (MSIT) (No. RS-2022-00167197, Development of Intelligent 5G/6G Infrastructure Technology for The Smart City); in part by the National Research Foundation of Korea (NRF), Ministry of Education, through the Basic Science Research Program under Grant NRF-2020R1I1A3066543; in part by BK21 FOUR (Fostering Outstanding Universities for Research) under Grant 5199990914048; and in part by the Soonchunhyang University Research Fund.

Data Availability Statement: Derived data supporting the findings of this study are available from the corresponding author on request.

Conflicts of Interest: The authors declare no conflicts of interest.

References

1. Westergren, U.H.; Mähler, V.; Jadaan, T. Enabling Digital Transformation: Organizational Implementation of the Internet of Things. *Inf. Manag.* **2024**, *61*, 103996. [CrossRef]
2. Niboucha, R.; Saad, S.B.; Ksentini, A.; Challal, Y. Zero-Touch Security Management for MMTC Network Slices: DDoS Attack Detection and Mitigation. *IEEE Internet Things J.* **2022**, *10*, 7800–7812. [CrossRef]
3. Eloranta, V.; Turunen, T. Platforms in Service-Driven Manufacturing: Leveraging Complexity by Connecting, Sharing, and Integrating. *Ind. Mark. Manag.* **2016**, *55*, 178–186. [CrossRef]
4. Leng, J.; Sha, W.; Wang, B.; Zheng, P.; Zhuang, C.; Liu, Q.; Wuest, T.; Mourtzis, D.; Wang, L. Industry 5.0: Prospect and Retrospect. *J. Manuf. Syst.* **2022**, *65*, 279–295. [CrossRef]
5. Xu, X.; Lu, Y.; Vogel-Heuser, B.; Wang, L. Industry 4.0 and Industry 5.0—Inception, Conception and Perception. *J. Manuf. Syst.* **2021**, *61*, 530–535. [CrossRef]
6. Chi, H.R.; Wu, C.K.; Huang, N.-F.; Tsang, K.-F.; Radwan, A. A Survey of Network Automation for Industrial Internet-of-Things toward Industry 5.0. *IEEE Trans. Ind. Inform.* **2023**, *19*, 2065–2077. [CrossRef]
7. Mao, W.; Zhao, Z.; Chang, Z.; Min, G.; Gao, W. Energy-Efficient Industrial Internet of Things: Overview and Open Issues. *IEEE Trans. Ind. Inform.* **2021**, *17*, 7225–7237. [CrossRef]
8. Mao, Y.; You, C.; Zhang, J.; Huang, K.; Letaief, K.B. A Survey on Mobile Edge Computing: The Communication Perspective. *IEEE Commun. Surv. Tutor.* **2017**, *19*, 2322–2358. [CrossRef]
9. Polese, M.; Bonati, L.; D'Oro, S.; Basagni, S.; Melodia, T. Understanding O-RAN: Architecture, Interfaces, Algorithms, Security, and Research Challenges. *IEEE Commun. Surv. Tutor.* **2023**, *25*, 1376–1411. [CrossRef]
10. Liang, X.; Wang, Q.; Al-Tahmeesschi, A.; Chetty, S.B.; Grace, D.; Ahmadi, H. Energy Consumption of Machine Learning Enhanced Open RAN: A Comprehensive Review. *IEEE Access* **2024**, *12*, 81889–81910. [CrossRef]
11. Chih-Lin, I.; Kuklinski, S.; Chen, T.; Ladid, L. A Perspective of O-RAN Integration with MEC, SON, and Network Slicing in the 5G Era. *IEEE Netw.* **2020**, *34*, 3–4. [CrossRef]
12. Lu, J.; Feng, W.; Pu, D. Resource Allocation and Offloading Decisions of D2D Collaborative UAV-Assisted MEC Systems. *KSII Trans. Internet Inf. Syst.* **2024**, *18*, 211–232.
13. Ojaghi, B.; Adelantado, F.; Verikoukis, C. SO-RAN: Dynamic RAN Slicing via Joint Functional Splitting and MEC Placement. *IEEE Trans. Veh. Technol.* **2023**, *72*, 1925–1939. [CrossRef]
14. Ateya, A.A.; Algarni, A.D.; Hamdi, M.; Koucheryavy, A.; Soliman, N.F. Enabling Heterogeneous IoT Networks over 5G Networks with Ultra-Dense Deployment—Using MEC/SDN. *Electronics* **2021**, *10*, 910. [CrossRef]
15. Ros, S.; Tam, P.; Song, I.; Kang, S.; Kim, S. Handling Efficient VNF Placement with Graph-Based Reinforcement Learning for SFC Fault Tolerance. *Electronics* **2024**, *13*, 2552. [CrossRef]
16. Shi, X.; Zhang, Z.; Cui, Z.; Cai, X. Many-Objective Joint Optimization for Dependency-Aware Task Offloading and Service Caching in Mobile Edge Computing. *KSII Trans. Internet Inf. Syst.* **2024**, *18*, 1238–1259.
17. Yi, D.; Zhou, X.; Wen, Y.; Tan, R. Toward Efficient Compute-Intensive Job Allocation for Green Data Centers: A Deep Reinforcement Learning Approach. In Proceedings of the 2019 IEEE 39th International Conference on Distributed Computing Systems (ICDCS), Dallas, TX, USA, 7–10 July 2019. [CrossRef]
18. Popli, S.; Jha, R.K.; Jain, S. A Survey on Energy Efficient Narrowband Internet of Things (NBIoT): Architecture, Application and Challenges. *IEEE Access* **2019**, *7*, 16739–16776. [CrossRef]
19. Kim, D.-Y.; Park, J.; Kim, S. Data Transmission in Backscatter IoT Networks for Smart City Applications. *J. Sens.* **2022**, *2022*, e4973782. [CrossRef]
20. Ren, Y.; Guo, A.; Song, C. Multi-Slice Joint Task Offloading and Resource Allocation Scheme for Massive MIMO Enabled Network. *KSII Trans. Internet Inf. Syst.* **2023**, *17*, 794–815.

21. Ros, S.; Tam, P.; Song, I.; Kang, S.; Kim, S. A Survey on State-of-The-Art Experimental Simulations for Privacy-Preserving Federated Learning in Intelligent Networking. *Electron. Res. Arch.* **2024**, *32*, 1333–1364. [CrossRef]
22. Nagappan, K.; Rajendran, S.; Alotaibi, Y. Trust Aware Multi-Objective Metaheuristic Optimization Based Secure Route Planning Technique for Cluster Based IIoT Environment. *IEEE Access* **2022**, *10*, 112686–112694. [CrossRef]
23. Kang, S.; Ros, S.; Song, I.; Tam, P.; Math, S.; Kim, S. Real-Time Prediction Algorithm for Intelligent Edge Networks with Federated Learning-Based Modeling. *Comput. Mater. Contin.* **2023**, *77*, 1967–1983. [CrossRef]
24. Mao, M.; Lee, A.; Hong, M. Deep Learning Innovations in Video Classification: A Survey on Techniques and Dataset Evaluations. *Electronics* **2024**, *13*, 2732. [CrossRef]
25. Patel, D.; Won, M. Experimental Study on Low Power Wide Area Networks (LPWAN) for Mobile Internet of Things. In Proceedings of the 2017 IEEE 85th Vehicular Technology Conference (VTC Spring), Sydney, NSW, Australia, 4–7 June 2017. [CrossRef]
26. Qin, Z.; Li, F.Y.; Li, G.Y.; McCann, J.A.; Ni, Q. Low-Power Wide-Area Networks for Sustainable IoT. *IEEE Wirel. Commun.* **2019**, *26*, 140–145. [CrossRef]
27. Zhou, F.; Feng, L.; Kadoch, M.; Yu, P.; Li, W.; Wang, Z. Multiagent RL Aided Task Offloading and Resource Management in Wi-Fi 6 and 5G Coexisting Industrial Wireless Environment. *IEEE Trans. Ind. Inform.* **2022**, *18*, 2923–2933. [CrossRef]
28. Tam, P.; Ros, S.; Song, I.; Kim, S. QoS-Driven Slicing Management for Vehicular Communications. *Electronics* **2024**, *13*, 314. [CrossRef]
29. Hazra, A.; Adhikari, M.; Amgoth, T.; Srirama, S.N. Intelligent Service Deployment Policy for Next-Generation Industrial Edge Networks. *IEEE Trans. Netw. Sci. Eng.* **2021**, *9*, 3057–3066. [CrossRef]
30. Zhang, J.; Wu, J.; Chen, Z.; Chen, Z.; Gan, J.; He, J.; Wang, B. Spectrum- and Energy- Efficiency Analysis under Sensing Delay Constraint for Cognitive Unmanned Aerial Vehicle Networks. *KSII Trans. Internet Inf. Syst.* **2022**, *16*, 1392–1413.
31. Ernest, H.; Madhukumar, A.S. Computation Offloading in MEC-Enabled IoV Networks: Average Energy Efficiency Analysis and Learning-Based Maximization. *IEEE Trans. Mob. Comput.* **2023**, *23*, 6074–6087. [CrossRef]
32. Lim, H.; Hwang, T. Energy-Efficient Beamforming and Resource Allocation for Multi-Antenna MEC Systems. *IEEE Access* **2022**, *10*, 18008–18022. [CrossRef]
33. Sun, M.; Xu, X.; Huang, Y.; Wu, Q.; Tao, X.; Zhang, P. Resource Management for Computation Offloading in D2D-Aided Wireless Powered Mobile-Edge Computing Networks. *IEEE Internet Things J.* **2021**, *8*, 8005–8020. [CrossRef]
34. Tong, Z.; Cai, J.; Mei, J.; Li, K.; Li, K. Dynamic Energy-Saving Offloading Strategy Guided by Lyapunov Optimization for IoT Devices. *IEEE Internet Things J.* **2022**, *9*, 19903–19915. [CrossRef]
35. Guo, F.; Zhang, H.; Li, X.; Ji, H. Victor Joint Optimization of Caching and Association in Energy-Harvesting-Powered Small-Cell Networks. *IEEE Trans. Veh. Technol.* **2018**, *67*, 6469–6480. [CrossRef]
36. Qazzaz, M.M.; Kułacz, Ł.; Kliks, A.; Zaidi, S.A.; Dryjanski, M.; McLernon, D. Machine Learning-Based XApp for Dynamic Resource Allocation in O-RAN Networks. *arXiv* **2024**, arXiv:2401.07643. [CrossRef]
37. Mungari, F. An RL Approach for Radio Resource Management in the O-RAN Architecture. In Proceedings of the 2021 18th Annual IEEE International Conference on Sensing, Communication, and Networking (SECON), Rome, Italy, 6–9 July 2021. [CrossRef]
38. Yu, P.; Yang, M.; Xiong, A.; Ding, Y.; Li, W.; Qiu, X.; Meng, L.; Kadoch, M.; Cheriet, M. Intelligent-Driven Green Resource Allocation for Industrial Internet of Things in 5G Heterogeneous Networks. *IEEE Trans. Ind. Inform.* **2022**, *18*, 520–530. [CrossRef]
39. *3GPP TS 23.203 V17.2.0*; Technical Specification Group Services and System Aspects. Policy and Charging Control Architecture; ETSI: Valbonne, France, 2021.
40. Tam, P.; Math, S.; Kim, S. Optimized Multi-Service Tasks Offloading for Federated Learning in Edge Virtualization. *IEEE Trans. Netw. Sci. Eng.* **2022**, *9*, 4363–4378. [CrossRef]

processes

MDPI

Review

Review of Intelligent Modeling for Sintering Process Under Variable Operating Conditions

Jie Hu [1,2,3,*], Hongxiang Li [1,2,3], Junyong Liu [1,2,3] and Sheng Du [1,2,3]

1 School of Automation, China University of Geosciences, Wuhan 430074, China; hongxingli@cug.edu.cn (H.L.); liujunyong@cug.edu.cn (J.L.); dusheng@cug.edu.cn (S.D.)
2 Hubei Key Laboratory of Advanced Control and Intelligent Automation for Complex Systems, Wuhan 430074, China
3 Engineering Research Center of Intelligent Technology for Geo-Exploration, Ministry of Education, Wuhan 430074, China
* Correspondence: hujie@cug.edu.cn

Abstract: The steel industry serves as a cornerstone of a nation's industrial system, with sintering playing a pivotal role in the steelmaking process. In an effort to enhance the intelligence of the sintering process and improve production efficiency, numerous scholars have carried out extensive research on data analysis and intelligent modeling techniques. These studies have made significant contributions to expanding production capacity, optimizing cost efficiency, and enhancing the quality of products, and supporting the sustainable development of the steel industry. This paper begins with an analysis of the sintering production process, explores the distinctive characteristics of the sintering process, and discusses the methods for identifying the operating conditions of sintering. It also provides an overview of the current state of research on both mechanism modeling and data-driven modeling approaches for the sintering process. Finally, the paper summarizes the existing challenges in sintering process modeling and offers perspectives on the future direction of research in this field.

Keywords: sintering process; mechanism modeling; data-driven modeling; operating conditions; prediction

MSC: 68T20

1. Introduction

The steel industry is a cornerstone sector in a nation's industrial system, with profound implications for a country's economy, society, and national defense. Its robust development is directly linked to the overall development level and competitiveness of the country. The steel production process encompasses several key stages, including the coking process, sintering process, ironmaking process, steelmaking process, and rolling process. Among these, sintering plays a crucial role as a heat-induced agglomeration process that mixes iron ore powder, recycled ironmaking products, fluxes, slagging agents, and solid fuels, contributing significantly to the steelmaking process [1].

However, due to the complexity of the sintering system and technical limitations in industrial settings, many key parameters are difficult to measure directly or exhibit significant measurement delays [2]. For instance, accurate prediction of the sintering endpoint is critical for the quality of sintered [3]; the composition and drum strength of the sintered ore directly affect the quality of downstream smelting [4]; and dynamic monitoring of sintering flue gas composition is central to achieving green production and emission control [5]. To address the challenges posed by these hard-to-measure parameters, sintering modeling has become a key research tool for solving such problems [6]. By constructing mathematical models or leveraging data-driven techniques, sintering modeling can accurately predict difficult-to-measure or delayed parameters, providing real-time guidance for the production process.

Currently, mainstream modeling approaches are primarily classified into two distinct categories: mechanism modeling and data-driven modeling [7]. Mechanism modeling, based on a deep understanding of thermodynamics, fluid mechanics, and chemical reaction kinetics, can accurately describe the physical and chemical phenomena during the sintering process and provide logically sound explanations of the effects of parameter variations [8]. In contrast, data-driven modeling relies on historical data and statistical patterns, using methods like machine learning to capture nonlinear relationships between inputs and outputs. This approach has shown considerable effectiveness in parameter prediction and model training speed.

This paper seeks to offer a comprehensive review of the advancements in research within the field of sintering modeling, with a focus on analyzing the current status, application scenarios, advantages, and limitations of mechanistic and data-driven modeling. By summarizing the characteristics and recent developments of different modeling methods, this paper seeks to offer insights and references for both theoretical research and industrial practices in sintering modeling. Section 2 provides an overview of the sintering process and its defining characteristics; Section 3 discusses the methods for identifying operating conditions in the sintering process; Section 4 reviews both mechanistic and data-driven modeling approaches for the sintering process; and Section 5 presents a comprehensive summary and outlook on existing modeling methods.

2. Analysis of Sintering Process

Sintering is a metal smelting process widely applied in the steel manufacturing sector [9,10]. The primary objective of this process is to agglomerate fine ore particles into larger masses, forming sintered ore suitable for blast furnace smelting [11]. The process involves complex production procedures and inherent characteristics, making it essential to conduct a thorough analysis of the process and its features before undertaking sintering process modeling studies.

2.1. Description of Iron Ore Sintering Process

Modern sintering methods typically place greater emphasis on improving energy efficiency, reducing environmental impact, and manufacturing precision components. In response to the high demand for steel, the most commonly used method in steel production is the belt-type, forced-air sintering process. Beneath the sintering machine, there are two rows of wind boxes. The exhaust fans below these wind boxes continuously extract air from within, allowing air to flow into the material layer above and ultimately exit through the wind boxes. This process provides sufficient oxygen for the combustion of fuel within the mixed charge, ensuring stable operation of the sintering process. Currently, most sintering plants use sintering machines with a functional area of 360 m^2. The process flow diagram of the sintering process is shown in Figure 1.

The sintering process mainly consists of several steps, including mixing and granulation, ignition, sintering, and cooling. Generally, the sintering process takes approximately 120 min. The batching process involves mixing iron ore powder with fuel (coke powder), return fines, and fluxes (limestone and dolomite) to form the raw material mixture. Water is added to the raw mix, which is then subjected to both primary and secondary mixing and granulation to form uniform particles with appropriate moisture content and particle size distribution. The optimal particle size of sinter charged into the blast furnace is typically between 5 and 20 mm. The proper distribution of these particles is crucial for improving the permeability of the material layer. During secondary mixing, the raw mix undergoes steam preheating, which helps raise the initial temperature of the mix. The water-mixed granules are then transported by a conveyor belt to the charging bin. The primary goal of the first mixing stage is to achieve uniformity and moisture adjustment. During this stage, various raw materials are evenly mixed to ensure that different components, such as iron ore powder, fluxing agents, fuel, and return fines, are fully blended into a homogeneous mixture. The second mixing stage, in addition to further homogenization and fine-tuning of moisture content, primarily aims to create a sinter mix with a specific particle size distribution and appropriate moisture levels. This mixture has sufficient permeability, facilitating efficient airflow during the sintering process. To prevent smaller granules from being carried away by the wind boxes, the granules are distributed on the sintering grate in a manner where the particle size gradually increases from top to bottom. This is achieved through a nine-roll spreading machine. Additionally, to protect the sintering grate from high temperatures, a layer of coarse sinter is typically spread on the grate as a bed material before charging. In normal sintering production, the material layer has a thickness of approximately 700 mm, and the solid fuel on the surface of the raw mix is initiated beneath the ignition chamber. During the sintering ignition process, key factors include ignition temperature, oxygen supply, particle size, and the chemical composition of the material. In ironmaking plants, operators primarily determine whether sintering has achieved complete combustion by measuring the temperature in the flue gas duct of the sintering process using thermocouples, in combination with their professional judgment and operational experience. The sintering grate is equipped with 24 wind boxes that initiate ventilation for sintering. As the sintering machine advances, the raw mixture undergoes melting and combustion in a top-to-bottom progression. This process culminates in the formation of sintered ore with specific strength characteristics at the burn-through point (BTP).

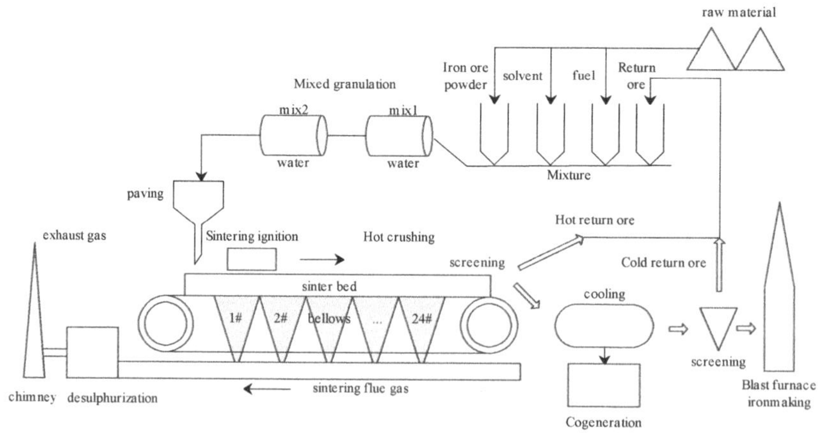

Figure 1. 360 square metre sintering machine.

The fully mixed raw mix burns within the material layer, generating high temperatures of approximately 1300 °C. Proper temperature promotes the reaction of minerals in the sinter, resulting in the formation of sinter with good mechanical strength. However, excessively high temperatures may cause over-melting of the minerals, which negatively affects the strength and structure of the sinter, increasing its brittleness and, consequently, its fragility. This high-temperature environment induces physical changes and chemical reactions in the sintering mixture, leading to the formation of distinct layers within the material bed. The material layer can be divided into several zones from bottom to top, including the raw material layer, the over-wet layer, the preheating and drying layer, the combustion zone, and the sintered ore layer. The combustion zone represents the area with the highest temperature and the most intense reaction activity. As the combustion zone moves downward, high-temperature molten material agglomerates into blocks, forming the molten layer. The introduction of cooling air causes the sintered ore to cool and form the sintered layer.The preheating and drying layer is in direct proximity to the combustion zone and is exposed to the high-temperature exhaust gases produced therein. The free moisture in the material layer quickly evaporates, and the evaporated water vapor comes into contact with the colder material layer below, forming the over-wet layer. From the layering phenomenon of the sintering bed described above, it can be concluded that the combustion of carbon is integral to the quality and yield of the sintered ore.

Sintering production generally occurs in a high-alkalinity environment. The addition of fluxes such as limestone and quicklime to the sintering mixture ensures the alkaline conditions necessary for the sintering process. Because of the substantial presence of calcium oxide (CaO), a series of chemical reactions take place with the raw mixture, leading to the formation of diverse minerals. The main minerals formed include magnetite (Fe_3O_4), dicalcium silicate ($2CaO{\cdot}SiO_2$), calcium ferrite ($CaO{\cdot}Fe_3O_2$), calcium ferrite ($2CaO{\cdot}Fe_2O_3$), and tricalcium silicate ($3CaO{\cdot}SiO_2$). The properties of these minerals are summarized in Table 1.

Table 1. Main mineral properties in sintered ore.

Components	Melting Point/(°C)	Compressive Strength/(Mpa)	Reductability/(%)
Fe_3O_4	1590	3.69	26.7
$CaO{\cdot}Fe_3O_2$	1216	3.76	40.1
$2CaO{\cdot}Fe_3O_2$	1436	1.42	28.5
$2CaO{\cdot}SiO_2$	2130	-	-
$3CaO{\cdot}SiO_2$	1410	0.67	-

The primary mineral in sintered ore is calcium ferrite ($CaO{\cdot}Fe_2O_3$). As shown in Table 1, ($CaO{\cdot}Fe_2O_3$) exhibits the highest compressive strength, the best reducibility, and the point of minimum melting temperature. The higher the content of $CaO{\cdot}Fe_2O_3$, the more favorable it is for producing high-quality molten iron during the ironmaking process. These stable minerals enhance the mechanical strength of the sinter. Additionally, limestone improves the flowability of the ore, reduces ore fines during the sintering process, and minimizes the formation of small particles. Therefore, it is essential to produce a sufficient amount of $CaO{\cdot}Fe_2O_3$ during the sintering process to ensure better steelmaking performance.

2.2. Analysis of Sintering Characteristics

The sintering process is characterized by strong nonlinearity and significant time delays, with many factors within the process exhibiting strong interdependencies. An examination of the sintering process identifies the following key characteristics:

1. Multiple types of parameters. Raw material parameters: coke powder ratio, return fines, and the contents of CaO, SiO_2, MgO, total iron (TFe). Operational parameters: Grate speed and material layer thickness. State parameters: Wind box negative pressure, BTP, BTP temperature, average vertical combustion rate, sintering rise point position, and sintering rise point temperature.
2. Nonlinearity. The sintering process involves numerous physical and chemical reactions, encompassing the evaporation and decomposition of water, redox reactions, and solid-phase reactions of sintering materials. Various factors affect the comprehensive coke ratio, such as the chemical composition of the raw mix, its permeability, and the sintering endpoint position. These parameters display time-dependent and uncertain behaviors, with many of them being unmeasurable in real-time, resulting in significant nonlinearity among the sintering variables. Consequently, developing accurate mathematical models for the sintering process proves to be a difficult task.
3. Time delay. There is a time delay between the detection of raw material composition and the subsequent production of sintered ore. The production rate of sintered ore is a key factor influencing the comprehensive coke ratio. Delays in detecting sintered ore production affect the coke ratio, which complicates the selection of suitable data for use as inputs in time-series predictions. Nevertheless, this delay is primarily attributed to sensor detection, with measurement intervals generally remaining fixed. This challenge can be mitigated by shifting the input and output data either forward or backward to account for the delay prior to making model predictions.
4. Strong coupling between parameters. The sintering process is governed by numerous parameters, primarily encompassing raw material, state, and operational factors. Raw material and operational parameters exert an indirect influence on the target parameters by altering the state parameters. These parameters are highly interdependent, such that a variation in one parameter induces simultaneous changes across multiple others.
5. Multiple operating modes. In actual sintering production, various types of charge recipes are used to guide production, with each recipe representing a distinct operating mode. When predicting indicators such as carbon efficiency, a single integrated predictive model is inadequate for comprehensively forecasting carbon efficiency under different operating modes.

3. Identification Methods for Sintering Process Conditions

The sintering process is a continuous and extended-duration manufacturing procedure characterized by complex material and energy conversion and transfer mechanisms, with variable operating conditions that exhibit intricate operational features [12]. Production data under different operating conditions exhibit distinct characteristics, and relying on a single model to describe the sintering process may lead to inaccurate results, thereby affecting the prediction of key parameters [13]. Therefore, in sintering process modeling research, it is essential to first conduct effective identification of the variable operating conditions, and then propose appropriate modeling methods based on this identification.

Cluster analysis is frequently employed in industrial processes to categorize operational conditions based on industrial data [14,15]. As a multivariate statistical technique, cluster analysis classifies data into distinct operational states, ensuring that data within the same state share similar attributes. For instance, reference [16] utilized the fuzzy c-means clustering algorithm to classify distinct operational conditions in the nylon polymerization process, while reference [17] proposed a hierarchical clustering method based on the Ward algorithm for automatic classification of various operational conditions in photovoltaic power plants. In the identification of sintering process conditions, reference [18] employed the K-means clustering algorithm to distinguish various operational states within the sintering process. Changes in operational conditions lead to variations in the comprehensive coke ratio, which was used to validate the effectiveness of the K-means algorithm. However, this method is hindered by the high number of computational steps and long processing times, making it unsuitable for real-time application in sintering operations. The fuzzy clustering algorithm can address some of these limitations. reference [19] applied the fuzzy C-Means clustering algorithm, and reference [20] proposed a weighted kernel fuzzy C-Means clustering algorithm to identify multiple operational states in the sintering process. However, these methods require the pre-definition of the quantity of clusters (i.e., the count of operational conditions). This poses a challenge, as the number of operational conditions in the sintering process cannot be predetermined. To overcome this, reference [21] introduced a quantification error modeling approach [22], and subsequently proposed a fuzzy C-Means clustering algorithm based on the quantification error model for the automatic identification of multiple operational conditions in sintering. In reference [23], a multi-dimensional characterization method for sintering conditions was proposed, based on polycrystalline indicators. This method integrates polycrystalline indicators with radar charts to define and calculate performance and balance indicators for sintering conditions, providing a comprehensive and accurate assessment of operational states. In reference [24], the affinity propagation clustering algorithm was applied to effectively classify different operational conditions, and support vector machine were used to recognize these conditions. Fuzzy C-Means clustering methods enable the accurate classification of production data under stable and smooth sintering production modes, particularly in cases where the number of process parameters is relatively small. However, most of these methods do not consider the real-time status information of actual sintering operations, which may limit their applicability in practical engineering settings.

Most methods for identifying operational conditions in the sintering process are limited to considering either production status information or the classification of operational conditions based on different production data characteristics. These approaches lack a comprehensive method that simultaneously accounts for both production status information and the varied characteristics of different production data for intelligent recognition of operational conditions. Images of the rear section of the sintering machine can reflect the actual production status and contain rich information about the sintering operational conditions. These images also carry production-related data such as yield, quality, and energy consumption. Timely and effective acquisition of the tail section images of the sintering machine is therefore a crucial prerequisite for the accurate and efficient recognition of sintering operational conditions. Thus, in practical sintering production, a deep analysis of the tail section images and the distinct characteristics of production data is essential. Research into intelligent recognition methods that consider both production status and the different characteristics of production data is beneficial for accurately describing the dynamic changes in the sintering production state. This approach will also provide a foundation for developing high-precision dynamic prediction models for carbon consumption in the sintering process.

4. Modeling Methods for the Sintering Process

During real-world sintering processes, it is essential to measure and monitor several critical parameters in real-time to maintain production safety, operational stability, and energy efficiency. However, due to limitations such as the high cost of sensors and the challenging industrial conditions, accurate measurement of most parameters is difficult and time-consuming. Therefore, modeling the sintering process and predicting certain key parameters is of significant importance for the monitoring, optimization and regulation of the sintering production process.

4.1. Mechanism Modeling

The sintering process involves several complex steps, including raw material mixing, segregation, ignition, and sintering, accompanied by intricate physicochemical changes. This process is characterized by numerous process parameters, such as temperature, pressure, flow rate, and velocity, along with extensive material and energy exchanges and transfers. Mechanism-based models are primarily derived from the physicochemical characteristics of the material strata involved in the process [25], as well as the laws of energy conservation and mass balance [26]. These models can clearly and accurately describe the interrelationships between various parameters of the sintering process.

In mechanism modeling research, many scholars have proposed corresponding analytical models, summarizing valuable theoretical findings with practical applications. For instance, focusing on individual mixed particles, reference [25] proposed a mechanism-based model to characterize the rate of combustion of solid coke during the sintering process, both under single-addition or distributed addition conditions. Additionally, a fuel particle model was established, with the combustion process and heat transfer identified as key factors influencing sintering productivity. Reference [27] proposed a transient heat and mass transfer model, which explains temperature changes within the material layer after ignition during sintering. A further model [28] describes the combustion behavior of solid fuel layers during sintering. Based on the local non-equilibrium thermodynamic relationships in the sintering process, another model [29] was developed to describe heat transfer, subject to five specific assumptions to ensure accurate heat transfer effects. Reference [30] established an unsteady-state, two-dimensional mechanism model, based on the analysis of key chemical reactions and physical processes involved in iron ore sintering, using reasonable assumptions. Additionally, a thermal reaction mechanism model, grounded in the principles of energy conservation, was developed to forecast the ignition temperature during the sintering process [26]. Reference [31] introduced a mechanism model to provide a detailed description of coke particle combustion in the sintering process. Reference [32] reflected the impact of liquid-phase formation during coke combustion on the sintering temperature field. Reference [9] outlined the direct influence of coke particle combustion behavior and gas flow velocity on the temperature, width, and velocity of the flame front within the sintering bed. It incorporated a granulation model into the thermal treatment framework to characterize coke combustion, and integrated two endothermic reactions, thereby enhancing the accuracy of temperature change predictions within the sintering bed.

While these mechanism-based models are theoretically rigorous and effectively reveal the inherent relationships between parameters in the sintering process, they require precise measurement of process parameters for the various materials involved, and their development relies on numerous assumptions. However, the sintering process represents a complex industrial system governed by numerous parameters, time delays, varying operating conditions, and nonlinearity. Some critical process parameters cannot be measured directly, limiting the application of mechanism-based modeling methods in characterizing the dynamics of these complex industrial systems. As a result, these models face significant challenges when applied to real-world industrial settings.

4.2. Data-Driven Modeling

With the advancement of database technology and artificial intelligence, scholars both domestically and internationally have begun to explore data-driven predictive models to address the challenges that mechanism models face in predicting complex industrial process parameters. Data-driven modeling methods involve studying the implicit mathematical relationships between production data, thus avoiding the complexities of mechanism analysis [33]. These models utilize actual production data to compute the relationships between various process parameters. These models are especially effective for complex and dynamic industrial processes, facilitating the creation of data-driven models that are customized to meet the specific requirements of industrial operations. Commonly used data-driven models include support vector machines (SVM) [34,35], feedforward neural networks [36,37], deep belief networks (DBN) [38], autoencoders [39], recurrent neural networks (RNN) [40], and convolutional neural networks (CNN) [41]. In the context of sintering, data-driven models primarily focus on predicting certain key parameters, which can then serve as the basis for process control or optimization. Sintering parameter prediction mainly targets parameters that cannot be directly measured or those for which measurement involves time delays, such as sintering endpoints, sintered product composition indicators, sintering flue gas composition, and sintering ore drum index, among others.

4.2.1. FeO Prediction Method

In the sintering process, the ferrous oxide (FeO) content refers to the mass fraction of FeO in the sintered ore. It is one of the key indicators used to evaluate the quality of the sintered ore, directly reflecting the extent of reduction reactions and fuel consumption efficiency during the sintering process. The optimal range for FeO content is typically closely related to the actual requirements of blast furnace smelting. During sintering, the FeO content cannot be quantified in real-time through online sensors and is usually determined through laboratory chemical analysis. Given the inherent time lag in measurement, it is challenging to adjust process parameters promptly to optimize production. Therefore, the ability to accurately predict and control the FeO content in sintered ore is crucial for optimizing the sintering process and improving smelting efficiency.

Reference [42] introduces a data-driven approach for forecasting the FeO content in sintered ore, utilizing multi-source data and $LSTM$. This approach incorporates multi-source features, including image data, vibration, and temperature parameters, to effectively reflect the FeO content in the sintered ore. Reference [43] introduced an innovative framework for dynamic time feature expansion and extraction, utilizing recursive neural network regression to forecast critical quality variables, such as FeO, for sintered ore quality prediction. Reference [24] proposed a multi-model ensemble framework for predicting FeO content in the iron ore sintering process, utilizing affinity propagation clustering to effectively classify different operating conditions and employing support vector machine (SVM) algorithms to identify these conditions. Reference [44] developed a method for predicting FeO content, integrating heat transfer mechanisms with a data-driven model, in which sintered ore is classified into three categories based on the peak temperature. Three variants of Long Short-Term Memory ($LSTM$) models, known for their robust adaptability to dynamic and nonlinear data across varying conditions, are utilized to forecast the FeO content during the sintering process. Reference [45] introduced the use of a restricted Boltzmann machine (RBM) to design a supervised RBM ($SRBM$), integrating quality variables into the visible layer to direct the model's learning of quality-relevant features. A stack of multiple $SRBMs$ is used to form a supervised DBN, which facilitates the prediction of FeO content by progressively learning quality-related features across layers. Reference [46] introduced an online measurement approach for *FeO* content, utilizing infrared images of the sinter machine's rear section in conjunction with CNN. Reference [47] through the compression of observed images, image features are combined with numerical data corresponding to sampling time. A multi-source information fusion model, *MIF-Autoformer*, which integrates deep convolutional neural networks with $Autoformer$, is proposed for soft sensing-based modeling of sintering quality. Finally, reference [48] proposed an online composition monitoring model utilizing deep neural networks (DNN) alongside an advanced component prediction model based on $LSTM$, designed to support field operators in real-time management of variations in sintered ore composition. Reference [49] proposed a novel semi-supervised dynamic feature extraction framework based on sequence pre-training and fine-tuning to predict the FeO content in sintered ore. Reference [50] presented an implicit subspace identification regression neural network based on orthogonal basis decomposition and reconstruction, this approach employs a recursive Fourier transform-like encoding block to extract features that capture long-term memory through orthogonal basis decomposition. Subsequently, a stochastic gradient-based identification algorithm is used to approximate the true system and model the *FeO* content.

4.2.2. BTP Prediction Method

BTP denotes the specific location or time point in the sintered ore bed where the temperature required for combustion reactions and melting is reached. During the sintering process, the BTP marks the completion of the combustion of fuel and thermal energy transfer within the material bed, making it one of the critical process parameters in sintering. Predicting the BTP can help optimize the bed height, sintering speed, and fuel ratio, thereby improving production efficiency and product quality. However, the BTP is difficult to measure directly, typically relying on manual experience, thermocouple monitoring, or offline experimental methods. These approaches face challenges such as measurement delays, insufficient accuracy, or operational complexity, preventing real-time adjustment of process parameters. By predicting the location and time point of the sintering end, dynamic control of the sintering process can be achieved, providing a basis for enhancing process stability. Reference [51] proposed a probabilistic spatiotemporal perception network named *BTPNet*. Within the encoder network, a multi-channel temporal convolution network ($MTCN$) is employed to extract temporal features. Additionally, a novel architecture unit, called the variable interaction awareness module ($VIAM$), is introduced to capture spatial features, thereby enabling accurate multi-step prediction of the BTP. In reference [52], an integration of process expertise and multiple feature selection techniques is used to identify key feature variables associated with BTP. A forecasting model for BTP and burn through temperature (BTT) is established using a gradient boosting decision tree ($GBDT$) algorithm. Grid search and cross-validation methods are employed to fine-tune the parameters of the ensemble algorithm, and a system model based on training data is developed. Moreover, a decision model is incorporated into the result generated by the predictive model, enhancing the system's prediction accuracy. Reference [53] developed a multi-step prediction model known as the denoising spatiotemporal Encoder-Decoder, which forecasts BTP in advance. Mechanistic analysis is conducted to identify the key BTP variables, and formulate BTP prediction as a sequence-to-sequence modeling task. Reference [54] utilized mechanistic and mutual information analyses to identify key process variables that are directly associated with BTP. The weighted kernel just-in-time learning ($WKJITL$) method is subsequently employed to extract historical production data analogous to the BTP query data, facilitating local learning-based modeling. Additionally, a fuzzy broad learning system ($FBLS$) is introduced as an effective approach for BTP soft sensor prediction. Finally, reference [55] proposed a decomposition-driven encoder-decoder model that leverages a self-attention mechanism to capture long-range dependencies between variables, and is applied for multi-step BTP prediction in the sintering process. Reference [56] proposed a 3-D convolution-based multi-step BTP prediction model that captures spatiotemporal features, resolved the spatial interdependencies among process variables, while addressing the limitations inherent in existing loss functions, which primarily rely on Euclidean distance and fail to capture the dynamic information in multi-step prediction sequences.

4.2.3. Carbon Efficiency Prediction Method

Carbon efficiency denotes the efficiency with which carbon energy is utilized during the sintering process. Commonly used indicators for carbon efficiency include the comprehensive coke ratio (CCR) and the ratio of carbon monoxide (CO) to carbon dioxide (CO_2), denoted as CO/CO_2. CCR serves as an indicator of carbon utilization efficiency, representing the amount of carbon consumed to produce one ton of sintered ore. A lower CCR indicates a reduced carbon consumption per ton of sintered ore, implying higher carbon energy utilization efficiency. The CO/CO_2 ratio reflects the completeness of carbon combustion; a higher $\mathrm{CO}/\mathrm{CO}_2$ ratio indicates lower combustion efficiency, with a higher proportion of CO in the exhaust gases. Conversely, a lower ratio signifies a reduced CO content in the exhaust, indicating higher combustion efficiency. Both of these indicators are challenging to measure simply and stably using sensors, and they can only be assessed after the entire sintering process is completed. Therefore, if these indicators are to be used for optimizing or scheduling sintering production, it is necessary to predict them before the completion of the sintering process.

Reference [57] asserted that the primary energy consumption process in sintering is the combustion of carbon, and it predicts carbon efficiency. This study uses the CCR as the indicator for measuring carbon efficiency and proposes a prediction model based on a particle swarm optimization (PSO) algorithm combined with a backpropagation (BP) neural network to analyze carbon efficiency. Reference [58] investigated the multi-time-scale characteristics of carbon efficiency by developing a model that integrates intelligent multi-time-scale techniques and neural networks. This model is capable of optimizing process variables across both short-term and long-term time frames. It uses CCR and the ratio of CO to CO_2 in the exhaust gases as indicators of carbon efficiency, establishing prediction models for state variables using both a single neural network and a linear combination of neural networks. The study indicates that the carbon efficiency prediction method has practical significance. Reference [18] selected CCR as the indicator for carbon efficiency and designs a method for modeling and optimization that is grounded in operational modes. This study utilizes the K-means clustering technique to identify distinct operational modes within the sintering process. For each identified mode, a CCR prediction model is developed, incorporating two BP neural networks. The model predicts the optimal operating mode based on CCR, reducing the CCR problem to a two-step optimization problem, which is solved using PSO. The method is validated using real industrial data, demonstrating the predictive performance of the model.Reference [59] proposed a carbon efficiency prediction model combining Elman and recurrent neural networks.

Over the past few years, a flexible and efficient modeling approach, known as the width learning model [60], has gained attention in the industrial sector. Reference [20], Grounded in the principles of the sintering process, this approach identifies the key sintering parameters that affect carbon efficiency and proposes a weighted fuzzy C-means clustering algorithm to recognize various operating conditions. Subsequently, a width learning model is developed for each operating condition. Finally, the nearest-neighbor criterion is employed to determine the optimal width learning model for predicting the carbon efficiency time series. Reference [61] introduced a specialized kernel-based fuzzy C-means clustering algorithm to classify real operational data under multiple conditions, which is then utilized to model the iron ore sintering process. Additionally, the width learning model's broad network structure is used to model carbon efficiency predictions under different operating conditions. Reference [54] developed a soft sensor model for sintering endpoint prediction based on a weighted kernel instant learning and fuzzy width learning system. The method involves using the weighted kernel instant learning approach to gather historical production data comparable to the sintering endpoint query data for local learning-based modeling and adopts the fuzzy width learning system as an effective approach for predicting sintering endpoint soft measurements. Reference [62] designed a dynamic carbon consumption prediction model for sintering, where broad learning models are developed for different operating conditions. Reference [63] proposed a novel adaptive weighted broad echo state learning system ($AWBESLS$) for dynamic carbon consumption prediction in the sintering process, which adaptively assigned weights to production data to mitigate the influence of outliers and used an echo state network (ESN) to capture the dynamic states of the process.

Moreover, many scholars have applied SVM [64] to model the sintering process. Reference [21] develops a multi-level carbon efficiency prediction model based on mechanism analysis, identifying the sintering process parameters that influence the comprehensive coke ratio. For different operating conditions, the least squares support vector machine (LS-SVM) is used, and a differential evolution algorithm is proposed to optimize the parameters and weights of the LS-SVM sub-models to improve their generalization ability. The results indicate that the prediction accuracy is within acceptable limits and meets the demands of real-world sintering production" or "the practical requirements of sintering production. Reference [65] constructed an optimization model to minimize blending costs, constrained by the best granulation and mineralization performance of the mixture, and uses an LS-SVM-based prediction model along with the basic properties of raw materials to predict the fuel consumption, drum strength, and productivity of sintered ore. This model comprehensively considers sintering performance, optimizes raw material costs, and achieves low-carbon, low-cost sintering. Reference [66] proposed a CO/CO_2 soft measurement model based on a hybrid kernel relevance vector machine for incomplete output data through data augmentation. Reference [67] proposed a sintering energy consumption prediction model using extreme learning machine and support vector regression.

4.2.4. Other Parameters and Summary

Reference [68] developed a quality prediction model for the tumbler strength in the sintering process using a BP neural network algorithm with momentum and variable learning rates. Reference [59] taking into account the unique characteristics of the process, a real-time dynamic forecasting model for the CCR, which reflects carbon efficiency, was developed. This model leverages predictive error information and is grounded in the principles of generalized learning to enhance accuracy and adaptability. Reference [69] applied linear regression and artificial neural network (ANN) algorithms to predict the productivity of the sintering machine and the composition of input materials. In reference [70], a hybrid ensemble model was proposed to predict key operational parameters, including solid fuel consumption, gas fuel consumption, BTP, and tumbler index (TI). This model integrates the extreme learning machine with an enhanced AdaBoost. RT algorithm, leveraging their complementary strengths to achieve higher predictive accuracy and robustness. Reference [71] developed a novel fusion network by integrating the local feature extraction capabilities of *CNN*, the sequential data processing strengths of *LSTM*, and the adaptive focus provided by the attention mechanism. This attention-augmented *CNN-LSTM* fusion network demonstrated substantial improvements in the accuracy of ignition temperature predictions, highlighting its effectiveness in capturing both spatial and temporal dependencies within the data. Reference [72] combined a local thermal non-equilibrium model and proposed a data-driven Tumble strength prediction approach. Reference [73] proposed a knowledge-data dual-driven graph neural network (*KDGNN*) to address the limitation of data-driven models that neglect domain knowledge, and was applied for end-to-end prediction of tumbler strength. Reference [74] constructed a data-driven prediction model with multiple time scales to predict the iron grade of sintered ore.

In summary, data-driven modeling methods do not require precise mechanistic knowledge or comprehensive expert knowledge. Instead, these methods build models using large amounts of data and continuously refine model parameters to improve their ability to fit real-world processes, ultimately establishing a data-driven model. The advantages of data-driven modeling methods include strong adaptability in handling highly coupled, nonlinear, and time-varying complex sintering reaction processes. However, their limitations include model accuracy being constrained by sample data and algorithms, with a heavy reliance on empirical data. Compared to mechanistic models, data-driven approaches are better at describing nonlinear, complex industrial processes, are more efficient, and have greater versatility. As a result, they have gradually become the preferred modeling method for sintering process modeling.

4.2.5. Summary of Data-Driven Models

The limitations of data-driven methods in the context of the sintering process can be discussed as follows: First, the composition of raw materials, operating conditions, and environmental factors in industrial processes are often highly complex and variable. Changes in the composition, moisture, and particle size of the raw materials during the sintering process can lead to diverse and complicated sintering production data. Relying solely on data-driven methods to build models may struggle to capture all the complex relationships between features and variables. Second, many industrial processes, such as sintering, may not have sufficient historical data, or the quality of available data may be poor. Missing data, noise, or incorrect labeling can affect the accuracy and robustness of models. In the iron ore sintering process, if there is insufficient data from various raw material compositions or operating conditions, the model may fail to fully learn the complex characteristics of the sintering process, resulting in inaccurate predictions or overfitting. Third, many data-driven methods, particularly deep learning, are "black box" models, meaning that they are difficult to interpret and understand. In industrial applications, engineers typically want to understand and control the decision-making process of the model. In the sintering process, data-driven models may not provide sufficient transparency, making it challenging to integrate them with traditional engineering expertise, thus reducing the model's operability. Fourth, in practical production, raw materials and operating conditions frequently change. A model trained under specific raw materials and operating conditions may perform poorly when faced with new raw materials or conditions. In the sintering process, variations in ore composition, particle size, or moisture content can cause data-driven models to make biased predictions regarding sintering outcomes, especially if the training data does not cover all possible raw material combinations.

To address these challenges and enable data-driven methods to adapt to a broader range of industrial applications, the following strategies can be employed: First, perform in-depth feature engineering by selecting features closely related to the sintering process, such as ore composition, particle size distribution, heating rate, and moisture content, and use them as input features for the model. Through data analysis and domain knowledge, key features that significantly influence the sintering process can be identified and extracted. Second, increase the number of sensors and monitoring devices to improve the frequency and accuracy of data collection. Installing real-time monitoring equipment, such as temperature, humidity, and gas composition sensors, during the sintering process will provide more high-quality training data. Data cleaning and preprocessing can be used to remove or correct anomalous data, while data augmentation techniques can simulate and generate new data to compensate for the lack of data in actual production, thereby enhancing the model's ability to adapt to different operating conditions and raw materials. Third, choose machine learning algorithms with a certain level of interpretability, such as decision trees or random forests, which can provide a visual representation of the decision-making process. This helps engineers understand the model's behavior and facilitates the integration of the model with practical operational experience. By combining data-driven models with expert knowledge, operational rules or optimization strategies that meet industrial needs can be derived, allowing for a better integration of data-driven predictions with traditional engineering expertise to support decision-making. Fourth, through incremental learning, the model can continuously receive new data and update itself during production, adapting to changes in raw material proportions and operating conditions. Transfer learning can be applied by leveraging pre-trained models developed for specific scenarios, enabling rapid adjustments to new environments or conditions. This approach allows for fine-tuning existing models based on different raw materials and operating conditions, reducing the need for extensive training data.

5. Summary and Prospect

In recent years, substantial advancements have been achieved in the modeling and prediction of the sintering process, particularly in improving process efficiency, optimizing energy utilization, and achieving green production. The research has primarily focused on the prediction of sinter ore composition, BTP forecasting, carbon efficiency optimization, and the modeling of key performance parameters, resulting in a variety of innovative methods and technological applications. Overall, the research on sintering process modeling and prediction has evolved from data fusion and feature extraction to dynamic optimization, gradually achieving an integrated approach that combines machine learning with industrial mechanisms. Through dynamic operational condition classification, model optimization, and soft-sensing modeling, these studies have provided crucial technical support for the intelligent control of the sintering process, energy utilization optimization, and low-carbon production.

5.1. Problems

Currently, the modeling of the sintering process primarily adopts a data-driven approach. The typical workflow involves obtaining actual production data from the sintering plant, followed by data preprocessing such as anomaly detection and correction. Afterward, feature engineering is performed, and models are selected either based on operational condition identification or through direct modeling. Below are some potential issues that may arise:

1. Data limitations affecting prediction accuracy. One of the key challenges of data-driven modeling methods lies in the necessity of having sufficient training data to train the model. In turn, machine learning techniques based on data-driven approaches are used to construct and design the prediction model's structure and parameters. While

data-driven models perform well in predicting the sintering process, when labeled data is difficult to obtain, traditional supervised data-driven models fail to achieve the desired prediction accuracy.

2. Insufficient consideration of real-world sintering conditions. Existing models for the sintering process often fail to adequately account for the multi-parameter, nonlinear, time delay, strong coupling, and multi-condition characteristics of the sintering process. These complexities make it difficult to develop accurate models. Additionally, a single modeling approach may not yield high-precision prediction models for all indicators, highlighting the limitations of conventional methods in capturing the full complexity of the process.
3. Time asymmetry between process influencing factors impacting model accuracy. The sintering process is a continuous, long-duration industrial production process, where iron ore powder undergoes steps such as mixing, granulation, distribution, and sintering, taking approximately one hour to complete. The parameters that need to be predicted during the sintering process are closely related to prior process parameters. For example, in the prediction of carbon efficiency, factors like carbon ratio and moisture content influence the carbon combustion trajectory in subsequent sintering materials, which in turn affects the composition of the exhaust gases. As process parameters are detected simultaneously in the sintering process, but there is a time difference-referred to as time asymmetry-between the various parameters influencing the sintering process at any given moment, this creates modeling challenges and negatively impacts the accuracy of the computational models.

5.2. Prospects

The modeling technology for the sintering process has advanced to a new level, achieving some successes in practical applications. However, several challenges remain, such as the complexity of process coupling and the difficulty of calculation, the inability to fully incorporate all characteristic evaluation factors into the model, incomplete data for model updates, and the inability of simulation calculations for specific problems to meet actual research needs. With the continuous upgrade of computer networks and industrial information technologies, big data and intelligent sintering production have become crucial components of future innovations in intelligent manufacturing. The following points may serve as directions for improvement in steel sintering process modeling:

1. Incorporating more methods into data-driven models. In recent years, large models have been rapidly developed. By leveraging the powerful data pattern discovery capabilities of these models, it may be possible to predict certain parameters that are difficult to measure or forecast.
2. Fully considering the actual conditions of sintering production. Most studies on energy consumption modeling in the sintering process have treated various process parameters at different time scales as inputs to energy consumption models. However, these studies have not adequately accounted for the diverse operating conditions and time delays characteristic of the sintering process. A single modeling approach cannot achieve high-precision prediction models for all indicators. Therefore, research on hybrid modeling methods, combining multiple models, multi-level structures, and intelligent modeling techniques across different time scales, is needed. This represents a new approach to achieving high-precision prediction of sintering energy consumption.
3. Considering multiple objectives in operational parameter settings. In actual sintering production, operational parameters must not only meet the demands of a single objective but also ensure smooth production and guarantee the quality and yield

of sintered ore. With the flourishing development of multi-objective optimization algorithms, the next step will be to consider both the constraints of smooth production and the uncertainty of state parameters under multi-level and multi-objective conditions. Research will focus on intelligent optimization techniques for the global carbon efficiency optimization of the sintering process, as well as the optimization of raw material parameters and operational settings, based on advanced multi-objective optimization algorithms.

4. Integrating the model into the real-time control system can significantly enhance operational efficiency. By combining the hybrid model with the real-time control system, it is possible to predict key parameters such as carbon consumption and gas emissions at various stages of the sintering process. These predictions can then be used to adjust operational parameters of the sintering equipment in real time, such as temperature, airflow rate, and raw material proportions. This predictive feedback control approach effectively prevents energy waste and improves the overall efficiency of the sintering process. Moreover, integrating the data-driven hybrid model with an expert system enables adaptive adjustments to complex operating conditions, enhancing the intelligence of the system while building upon traditional control systems.

Author Contributions: Conceptualization, J.H.; methodology, J.H.; software, J.H.; validation, J.H. and H.L.; formal analysis, J.H. and H.L.; investigation, J.H.; resources, J.H. and H.L.; data curation, J.H. and J.L.; writing—original draft preparation, J.H. and H.L.; writing—review and editing, J.H. and H.L.; visualization, J.H. and J.L.; supervision, J.H.; project administration, J.H. and S.D.; funding acquisition, J.H. and S.D. All authors have read and agreed to the published version of the manuscript.

Funding: This work was supported in part by the National Natural Science Foundation of China under Grant 62303431, in part by the Natural Science Foundation of Wuhan under Grant 2024040801020281, in part by the Hubei Provincial Natural Science Foundation of China under Grant 2024AFB589, in part by the "CUG Scholar" Scientific Research Funds at China University of Geosciences (Wuhan) under Grant No.2021009, and in part by the Fundamental Research Funds for the Central Universities, China University of Geosciences (Wuhan) under Grant CUG2106210.

Data Availability Statement: The results/data/figures in this manuscript have not been published elsewhere, nor are they under consideration by another publisher. The original contributions presented in the study are included in the article, further inquiries can be directed to the corresponding author.

Conflicts of Interest: The authors declare no conflicts of interest.

References

1. Wang, F.; Shi, X.; Ping, X.; Gao, J.; Zhang, J.; Zhang, H. Influence of sinter parameters on CO emission in iron ore sintering process. *Metals* **2022**, *12*, 1202. [CrossRef]
2. Gao, Q.; Wang, H.; Pan, X.; Jiang, X.; Zheng, H.; Shen, F. A forecast model of the sinter tumble strength in iron ore fines sintering process. *Powder Technol.* **2021**, *390*, 256–267. [CrossRef]
3. Wu, M.; Xu, C.; She, J.; Cao, W. Neural-network-based integrated model for predicting burn-through point in lead–zinc sintering process. *J. Process Control* **2012**, *22*, 925–934. [CrossRef]
4. Umadevi, T.; Naik, D.K.; Sah, R.; Brahmacharyulu, A.; Marutiram, K.; Mahapatra, P.C. Studies on parameters affecting sinter strength and prediction through artificial neural network model. *Miner. Process. Extr. Metall.* **2016**, *125*, 32–38. [CrossRef]
5. Legemza, J.; Findorak, R.; Frohlichova, M. Utilization of charcoal in the iron-ore sintering process. *Sci. Iran.* **2016**, *23*, 990–997. [CrossRef]
6. Reiterer, M.W.; Ewsuk, K.G. An analysis of four different approaches to predict and control sintering. *J. Am. Ceram. Soc.* **2009**, *92*, 1419–1427. [CrossRef]
7. Yang, C.; Yang, C. Deep fusion of time series and visual data through temporal features: A soft-sensor model for FeO content in sintering process. *Expert Syst. Appl.* **2024**, 126243. [CrossRef]

8. Wang, G.; Wen, Z.; Lou, G.; Dou, R.; Li, X.; Liu, X.; Su, F. Mathematical modeling of and parametric studies on flue gas recirculation iron ore sintering. *Appl. Therm. Eng.* **2016**, *102*, 648–660. [CrossRef]
9. Zhao, J.P.; Loo, C.E.; Dukino, R.D. Modelling fuel combustion in iron ore sintering. *Combust. Flame* **2015**, *162*, 1019–1034. [CrossRef]
10. Wu, S.; Liu, Y.; Du, J.; Mi, K.; Lin, H. New concept of iron ores sintering basic characteristics. *Chin. J. Eng.* **2002**, *24*, 254–257.
11. Lu, L. Important iron ore characteristics and their impacts on sinter quality—A review. *Mining, Metall. Explor.* **2015**, *32*, 88–96. [CrossRef]
12. Cheng, Z.; Yang, J.; Zhou, L.; Liu, Y.; Wang, Q. Sinter strength evaluation using process parameters under different conditions in iron ore sintering process. *Appl. Therm. Eng.* **2016**, *105*, 894–904. [CrossRef]
13. Wang, J.; Li, X.; Li, Y.; Wang, K. BTP prediction of sintering process by using multiple models. In Proceedings of the 26th Chinese Control and Decision Conference (2014 CCDC), Shenyang, China, 31 May–2 June 2014; pp. 4008–4012.
14. Diaz-Rozo, J.; Bielza, C.; Larranaga, P. Clustering of data streams with dynamic Gaussian mixture models: An IoT application in industrial processes. *IEEE Internet Things J.* **2018**, *5*, 3533–3547. [CrossRef]
15. Zhao, J.; Liu, Y.; Wang, L.; Wang, D. A generalized heterogeneous type-2 fuzzy classifier and its industrial application. *IEEE Trans. Fuzzy Syst.* **2019**, *28*, 2287–2301. [CrossRef]
16. Aumi, S.; Corbett, B.; Mhaskar, P.; Clarke-Pringle, T. Data-based modeling and control of nylon-6, 6 batch polymerization. *IEEE Trans. Control Syst. Technol.* **2012**, *21*, 94–106. [CrossRef]
17. Jasiński, M.; Sikorski, T.; Leonowicz, Z.; Kaczorowska, D.; Suresh, V.; Szymanda, J.; Jasinska, E. Different working conditions identification of a PV power plant using hierarchical clustering. In Proceedings of the 2020 12th International Conference on Electronics, Computers and Artificial Intelligence (ECAI), Bucharest, Romania, 25–27 June 2020; pp. 1–8.
18. Chen, X.; Chen, X.; Wu, M.; She, J. Modeling and optimization method featuring multiple operating modes for improving carbon efficiency of iron ore sintering process. *Control Eng. Pract.* **2016**, *54*, 117–128. [CrossRef]
19. Hu, J.; Wu, M.; Chen, X.; She, J.; Cao, W.; Chen, L.; Ding, H. Hybrid prediction model of carbon efficiency for sintering process. *IFAC-PapersOnLine* **2017**, *50*, 10238–10243. [CrossRef]
20. Hu, J.; Wu, M.; Chen, L.; Zhou, K.; Zhang, P.; Pedrycz, W. Weighted kernel fuzzy C-means-based broad learning model for time-series prediction of carbon efficiency in iron ore sintering process. *IEEE Trans. Cybern.* **2020**, *52*, 4751–4763. [CrossRef] [PubMed]
21. Hu, J.; Wu, M.; Chen, X.; Du, S.; Zhang, P.; Cao, W.; She, J. A multilevel prediction model of carbon efficiency based on the differential evolution algorithm for the iron ore sintering process. *IEEE Trans. Ind. Electron.* **2018**, *65*, 8778–8787. [CrossRef]
22. Kolesnikov, A.; Trichina, E.; Kauranne, T. Estimating the number of clusters in a numerical data set via quantization error modeling. *Pattern Recognit.* **2015**, *48*, 941–952. [CrossRef]
23. Fang, Y.; Gui, W.; Jiang, Z.; Pan, D.; Yu, H. Comprehensive working condition evaluation of the sintering process based on polymorphic indicators. *Adv. Eng. Inform.* **2023**, *58*, 102220. [CrossRef]
24. Li, X.; Wang, B.; Yu, Z.; Xing, X.-D. Identification of working conditions and prediction of FeO content in sintering process of iron ore fines. *J. Iron Steel Res. Int.* **2024**, *31*, 2090–2100. [CrossRef]
25. Hou, P.; Choi, S.; Choi, E.; Kang, H. Improved distribution of fuel particles in iron ore sintering process. *Ironmak. Steelmak.* **2011**, *38*, 379–385. [CrossRef]
26. Du, S.; Wu, M.; Chen, X.; Cao, W. An intelligent control strategy for iron ore sintering ignition process based on the prediction of ignition temperature. *IEEE Trans. Ind. Electron.* **2019**, *67*, 1233–1241. [CrossRef]
27. Giri, B.K.; Roy, G.G. Mathematical modelling of iron ore sintering process using genetic algorithm. *Ironmak. Steelmak.* **2012**, *39*, 59–66. [CrossRef]
28. Yang, W.; Choi, S.; Choi, E.S.; Ri, D.W.; Kim, S. Combustion characteristics in an iron ore sintering bed-evaluation of fuel substitution. *Combust. Flame* **2006**, *145*, 447–463. [CrossRef]
29. Zhou, G.; Zhong, W.; Zhao, H.; Jin, B.-S.; Wang, T.-C.; Liu, F. Heat transfer of spent ion exchange resin in iron ore sintering process. *Appl. Therm. Eng.* **2015**, *88*, 258–264. [CrossRef]
30. Zhang, B.; Zhou, J.; Li, M. Prediction of sinter yield and strength in iron ore sintering process by numerical simulation. *Appl. Therm. Eng.* **2018**, *131*, 70–79. [CrossRef]
31. Hou, P.; Choi, S.; Yang, W.; Choi, E.; Kang, H. Application of intra-particle combustion model for iron ore sintering bed. *Mater. Sci. Appl.* **2011**, *2*, 370. [CrossRef]
32. Ohno, K.; Noda, K.; Nishioka, K.; Maeda, T.; Shimizu, M. Effect of coke combustion rate equation on numerical simulation of temperature distribution in iron ore sintering process. *ISIJ Int.* **2013**, *53*, 1642–1647. [CrossRef]
33. Sun, Y.; Haghighat, F.; Fung, B.C.M. A review of the state-of-the-art in data-driven approaches for building energy prediction. *Energy Build.* **2020**, *221*, 110022. [CrossRef]

34. Chen, K.; Yu, J. Short-term wind speed prediction using an unscented Kalman filter based state-space support vector regression approach. *Appl. Energy* **2014**, *113*, 690–705. [CrossRef]
35. Jain, R.K.; Smith, K.M.; Culligan, P.J.; Taylor, J.E. Forecasting energy consumption of multi-family residential buildings using support vector regression: Investigating the impact of temporal and spatial monitoring granularity on performance accuracy. *Appl. Energy* **2014**, *123*, 168–178. [CrossRef]
36. Geng, Z.; Dong, J.; Chen, J.; Han, Y. A new self-organizing extreme learning machine soft sensor model and its applications in complicated chemical processes. *Eng. Appl. Artif. Intell.* **2017**, *62*, 38–50. [CrossRef]
37. Lee, S.; Choeh, J.Y. Predicting the helpfulness of online reviews using multilayer perceptron neural networks. *Expert Syst. Appl.* **2014**, *41*, 3041–3046. [CrossRef]
38. Hu, C.H.; Pei, H.; Si, X.S.; Du, D.-B.; Pang, Z.-N.; Wang, X. A prognostic model based on DBN and diffusion process for degrading bearing. *IEEE Trans. Ind. Electron.* **2019**, *67*, 8767–8777. [CrossRef]
39. Ren, L.; Sun, Y.; Cui, J.; Zhang, L. Bearing remaining useful life prediction based on deep autoencoder and deep neural networks. *J. Manuf. Syst.* **2018**, *48*, 71–77. [CrossRef]
40. Li, X.; Han, C.; Lu, G.; Yan, Y. Online dynamic prediction of potassium concentration in biomass fuels through flame spectroscopic analysis and recurrent neural network modelling. *Fuel* **2021**, *304*, 121376. [CrossRef]
41. Cruz, Y.J.; Rivas, M.; Quiza, R.; Villalonga, A.; Haber, R.E.; Beruvides, G. Ensemble of convolutional neural networks based on an evolutionary algorithm applied to an industrial welding process. *Comput. Ind.* **2021**, *133*, 103530. [CrossRef]
42. Bai, X.; Chen, C.; Liu, W.; Zhang, H. Data-driven prediction of sinter composition based on multi-source information and LSTM network. In Proceedings of the 2021 40th Chinese Control Conference (CCC), Shenyang, China, 26–28 July 2021; pp. 1–6.
43. Li, Y.; Yang, C.; Sun, Y. Dynamic time features expanding and extracting method for prediction model of sintering process quality index. *IEEE Trans. Ind. Informat.* **2021**, *18*, 1737–1745.
44. Jiang, Z.; Huang, L.; Jiang, K.; Xie, Y. Prediction of FeO content in sintering process based on heat transfer mechanism and data-driven model. In Proceedings of the 2020 Chinese Automation Congress (CAC), Beijing, China, 6–8 November 2020; pp. 4846–4851.
45. Yuan, X.; Gu, Y.; Wang, Y.; Chen, Z.; Sun, B.; Yang, C. FeO content prediction for an industrial sintering process based on supervised deep belief network. *IFAC-PapersOnLine* **2020**, *53*, 11883–11888. [CrossRef]
46. Zhang, N.; Chen, X.; Huang, X.; Fan, X.; Gan, M.; Ji, Z.; Sun, Z.; Peng, Z. Online measurement method of FeO content in sinter based on infrared machine vision and convolutional neural network. *Measurement* **2022**, *202*, 111849. [CrossRef]
47. Yang, C.; Yang, C.; Zhang, X.; Zhang, J. Multisource Information Fusion for Autoformer: Soft Sensor Modeling of FeO Content in Iron Ore Sintering Process. *IEEE Trans. Ind. Inform.* **2023**, *19*, 11584–11595. [CrossRef]
48. Liu, S.; Liu, X.; Lyu, Q.; Li, F. Comprehensive system based on a DNN and LSTM for predicting sinter composition. *Appl. Soft Comput.* **2020**, *95*, 106574. [CrossRef]
49. Li, Y.; Yang, C.; Sun, Y. Sintering quality prediction model based on semi-supervised dynamic time feature extraction framework. *Sensors* **2022**, *22*, 5861. [CrossRef] [PubMed]
50. Wang, S.; Yang, C.; Lou, S. Facilitating Ferrous Oxide Prediction: Enabling Sintering Forecasting With Orthogonal Basis-Based Implicit Subspace Identification. *IEEE Trans. Ind. Informat.* **2024**, *Early Access*. [CrossRef]
51. Yan, F.; Yang, C.; Zhang, X.; Yang, C.; Ruan, Z. BTPNet: A probabilistic spatial-temporal aware network for burn-through point multistep prediction in sintering process. *IEEE Trans. Neural Netw. Learn. Syst.* **2024**, *35*, 1–9. [CrossRef]
52. Liu, S.; Lyu, Q.; Liu, X.; Sun, Y.; Zhang, X. A prediction system of burn-through point based on gradient boosting decision tree and decision rules. *ISIJ Int.* **2019**, *59*, 2156–2164. [CrossRef]
53. Yan, F.; Yang, C.; Zhang, X. DSTED: A denoising spatial-temporal encoder-decoder framework for multistep prediction of burn-through point in sintering process. *IEEE Trans. Ind. Electron.* **2022**, *69*, 10735–10744. [CrossRef]
54. Hu, J.; Wu, M.; Cao, W.; Pedrycz, W. Soft-Sensing of Burn-Through Point Based on Weighted Kernel Just-in-Time Learning and Fuzzy Broad-Learning System in Sintering Process. *IEEE Trans. Ind. Informat.* **2024**, *20*, 7316–7324. [CrossRef]
55. Xie, Y.; He, B.; Zhang, X.; Song, Z.; Kano, M. EnvFormer: A Decomposition-based Transformer for Multi-step Burn-through Point Prediction in Sintering Process. *IEEE Trans. Instrum. Meas.* **2024**, *73*, 2531310. [CrossRef]
56. Yan, F.; Yang, C.; Zhang, X.; Gao, L. A 3-D convolution-based burn-through point multistep prediction model for sintering process. *IEEE Trans. Ind. Electron.* **2023**, *71*, 4219–4229. [CrossRef]
57. Chen, X.; Chen, X.; She, J.; Wu, M. Hybrid multistep modeling for calculation of carbon efficiency of iron ore sintering process based on yield prediction. *Neural Comput. Appl.* **2017**, *28*, 1193–1207. [CrossRef]
58. Zhou, K.; Chen, X.; Wu, M.; Cao, W.; Hu, J. A new hybrid modeling and optimization algorithm for improving carbon efficiency based on different time scales in sintering process. *Control Eng. Pract.* **2019**, *91*, 104–105. [CrossRef]

59. Chen, X.; Chen, X.; She, J.; Wu, M. A hybrid time series prediction model based on recurrent neural network and double joint linear–nonlinear extreme learning network for prediction of carbon efficiency in iron ore sintering process. *Neurocomputing* **2017**, *249*, 128–139. [CrossRef]
60. Chen, C.L.P.; Liu, Z.; Feng, S. Universal approximation capability of broad learning system and its structural variations. *IEEE Trans. Neural Netw. Learn. Syst.* **2018**, *30*, 1191–1204. [CrossRef] [PubMed]
61. Hu, J.; Wu, M.; Chen, L.; Pedrycz, W. A novel modeling framework based on customized kernel-based fuzzy C-means clustering in iron ore sintering process. *IEEE/ASME Trans. Mechatron.* **2021**, *27*, 950–961. [CrossRef]
62. Hu, J.; Wu, M.; Cao, W.; Pedrycz, W. Dynamic modeling framework based on automatic identification of operating conditions for sintering carbon consumption prediction. *IEEE Trans. Ind. Electron.* **2023**, *71*, 3133–3141. [CrossRef]
63. Hu, J.; Wu, M.; Pedrycz, W. Adaptive Weighted Broad Echo State Learning System-Based Dynamic Modeling of Carbon Consumption in Sintering Process. *IEEE Trans. Neural Netw. Learn. Syst.* **2024**, *Early Access*. [CrossRef]
64. Torres-Barrán, A.; Alaíz, C.M.; Dorronsoro, J.R. Faster SVM training via conjugate SMO. *Pattern Recognit.* **2021**, *46*, 101–112. [CrossRef]
65. Huang, X.; Fan, X.; Chen, X.; Gan, M.; Ji, Z.; Zheng, R. A novel blending principle and optimization model for low-carbon and low-cost sintering in ironmaking process. *Powder Technol.* **2019**, *355*, 629–636. [CrossRef]
66. Hu, J.; Li, H.; Li, H.; Wu, M.; Cao, W.; Pedrycz, W. Relevance vector machine with hybrid kernel-based soft sensor via data augmentation for incomplete output data in sintering process. *Control Eng. Pract.* **2024**, *145*, 105850. [CrossRef]
67. Wang, J.; Qiao, F.; Zhao, F.; Sutherland, J.W. A data-driven model for energy consumption in the sintering process. *J. Manuf. Sci. Eng.* **2016**, *138*, 101001. [CrossRef]
68. Shao, H.; Yi, Z.; Chen, Z.; Zhou, Z.; Deng, Z. Application of artificial neural networks for prediction of sinter quality based on process parameters control. *Trans. Inst. Meas. Control* **2020**, *42*, 422–429. [CrossRef]
69. Mallick, A.; Dhara, S.; Rath, S. Application of machine learning algorithms for prediction of sinter machine productivity. *Mach. Learn. Appl.* **2021**, *6*, 100186. [CrossRef]
70. Wang, S.H.; Li, H.F.; Zhang, Y.J.; Zou, Z.-S. A hybrid ensemble model based on ELM and improved AdaBoost. RT algorithm for predicting the iron ore sintering characters. *Comput. Intell. Neurosci.* **2019**, *2019*, 4164296. [CrossRef]
71. Xiong, D.L.; Ning, H.Y.; Xie, M.; Pan, C.-Y.; Chen, L.-J.; Yu, Z.-W.; Long, H.-M. A Predictive Model for Sintering Ignition Temperature Based on a CNN-LSTM Neural Network with an Attention Mechanism. *Processes* **2024**, *12*, 2185. [CrossRef]
72. Ye, J.; Ding, X.; Chen, C.; Guan, X.; Cao, X. Tumble strength prediction for sintering: Data-driven modeling and scheme design. In Proceedings of the 2020 Chinese Automation Congress (CAC), Shanghai, China, 6–8 November 2020; pp. 5500–5505.
73. Yan, F.; Yang, C.; He, W.; Mu, J.; Guo, H. Knowledge and Data Dual-Driven Graph Network for Tumbler Strength Prediction in Sintering Process. *IEEE Trans. Instrum. Meas.* **2024**, *73*, 2525414. [CrossRef]
74. Chen, X.; Shi, X.; Tong, C. Multi-time-scale TFe prediction for iron ore sintering process with complex time delay. *Control Eng. Pract.* **2019**, *89*, 84–93. [CrossRef]

processes

Review

Biosensors for Detecting Food Contaminants—An Overview

António Inês and Fernanda Cosme *

CQ-VR, Chemistry Research Centre-Vila Real, Department of Biology and Environment, University of Trás-os-Montes e Alto Douro, ECVA, Quinta de Prados, 5000-801 Vila Real, Portugal; aines@utad.pt
* Correspondence: fcosme@utad.pt

Abstract: Food safety is a pressing global concern due to the risks posed by contaminants such as pesticide residues, heavy metals, allergens, mycotoxins, and pathogenic microorganisms. While accurate, traditional detection methods like ELISA, HPLC, and mass spectrometry are often time-consuming and resource-intensive, highlighting the need for innovative alternatives. Biosensors based on biological recognition elements such as enzymes, antibodies, and aptamers, offer fast, sensitive, and cost-effective solutions. Using transduction mechanisms like electrochemical, optical, piezoelectric, and thermal systems, biosensors provide versatile tools for detecting contaminants. Advances in DNAzyme- and aptamer-based technologies enable the precise detection of heavy metals, while enzyme- and protein-based biosensors monitor metal-induced changes in biological activity. Innovations like microbial biosensors and DNA-modified electrodes enhance detection accuracy. Biosensors are also highly effective in identifying pesticide residues, allergens, mycotoxins, and pathogens through immunological, enzymatic, and nucleic acid-based techniques. The integration of nanomaterials and bioelectronics has significantly improved the sensitivity and performance of biosensors. By facilitating real-time, on-site monitoring, these devices address the limitations of conventional methods to ensure food quality and regulatory compliance. This review highlights the transformative role of biosensors and how biosensors are improved by emerging technologies in food contamination detection, emphasizing their potential to mitigate public health risks and enhance food safety throughout the supply chain.

Keywords: biosensors; safety; food; beverage; heavy metals: pesticides; mycotoxins; allergens; foodborne pathogens

1. Introduction

Food safety is an increasingly pressing global concern due to risks such as pesticide residues, food allergies, heavy metals, and pathogenic microorganisms. The extensive use of pesticides, combined with environmental pollution, has resulted in the persistent presence of contaminants in food sources [1]. In addition, food allergies have become a significant public health problem, particularly in developed countries [2]. Spoilage caused by pathogenic microorganisms is another critical food safety challenge [3].

Furthermore, heavy metals in the environment adversely affect plant growth, reduce crop quality, and accumulate in plants. This contamination affects human health through the food chain, as heavy metals have been associated with genetic damage and an increased risk of cancer. Ensuring food quality is critical because contaminants such as allergens, pathogenic microorganisms, heavy metals, and herbicides pose significant health risks [4]. With consumers becoming increasingly vigilant, the detection of contaminants in food has

 https://doi.org/10.3390/pr13020380

become essential [5]. Consequently, the identification of food contaminants has become a priority [6].

Traditional detection methods, including enzyme-linked immunosorbent assay (ELISA), high-performance liquid chromatography (HPLC), mass spectroscopy (MS), and gas chromatography (GC), are highly accurate but are often expensive, time-consuming, and require skilled operators. Therefore, there is an urgent need for faster, simpler, and highly sensitive detection techniques [7]. The development of sensitive and reliable on-site technologies for the detection and monitoring of food contaminants is essential to ensuring food safety and protection from harmful substances.

Biosensors are particularly well suited for monitoring contaminants. These devices, a subset of chemical sensors, use biological recognition elements for analyte detection. Recent research has focused on developing a variety of chemical and biological sensing devices, based on different operating principles, to identify hazardous substances in food. These advances represent a significant area of interest in the field of food safety [8].

Biosensors, which integrate biological components such as enzymes, DNA, RNA, antigens, living cells, or antibodies with electronic sensing elements like conductance, intensity, electromagnetic radiation phase, electric current, mass, viscosity, electric potential, temperature, and impedance [9–11], provide rapid and accurate results through their combined functionality. The incorporation of biosensors into food quality monitoring systems presents an innovative strategy to enhancing food safety and quality assurance [12].

Next-generation biosensor arrays incorporate artificial intelligence algorithms, enabling their specialization, selectivity, responsiveness, and consistency. With artificial intelligence support, these biosensors more accurately identify biological analytes, enhancing performance and reliability. Artificial intelligence is transforming food systems by providing tailored solutions through machine learning, natural language processing, computer vision, and reinforcement learning [13]. These advancements improve food safety through real-time detection and prevention of contamination. The application of machine learning has significantly increased the efficiency of various sensors used in food safety evaluation. By integrating machine learning with noninvasive biosensors, it is now possible to monitor food safety more efficiently, with a particular focus on the stability of bio-recognition molecules [14].

Machine learning enhances biosensors, transforming them into intelligent systems capable of predicting analytes using stable training models [15]. It improves biosensor specificity during data analysis and helps detect subtle patterns within sensor data. Additionally, machine learning enhances the sensor's ability to monitor multiple analytes in complex food matrices, increasing the functionality and versatility of biosensors.

These devices allow rapid, accurate, and on-site detection of contaminants, revolutionizing the management of food safety risks throughout the supply chain [16–18]. The performance of a biosensor is evaluated based on several parameters, including sensitivity, selectivity, specificity, reproducibility, size, diagnostic speed, scalability for large-scale production, and cost effectiveness [19]. Various biosensors enhanced by machine learning have been employed to analyze a range of food contaminants [20].

This review aims to provide an overview of how biosensors are used in the food industry to monitor and control contaminants, as well as how biosensors have been enhanced by emerging technologies, ultimately enhancing the safety and quality of food products.

2. Biosensors for Detecting Food Contaminates

Biosensors can be categorized based on various criteria, with two common approaches being the type of transduction system employed, such as electrochemical, optical, piezoelectric, and thermal biosensors, and the type of biorecognition element (biocomponent or bioreceptor). The biocomponent or bioreceptor, such as isolated enzymes, whole cells, tissues, or aptamers, is essential in biosensors, enabling selective analyte detection and interaction. The energy released from this interaction is converted into a measurable electrical signal [21]. Common biological elements include enzymes and antibodies. Biosensors are further classified into biocatalytic sensors and affinity sensors based on their interaction with the analyte.

Biocatalytic sensors, or metabolism sensors, catalyze analyte conversion and measure the resulting changes, such as product formation or reaction inhibition [22]. In contrast, affinity sensors detect specific, irreversible binding between the analyte and the biological component, resulting in a measurable physicochemical change. Figure 1 illustrates the main biosensors used in the food industry to detect food contaminants.

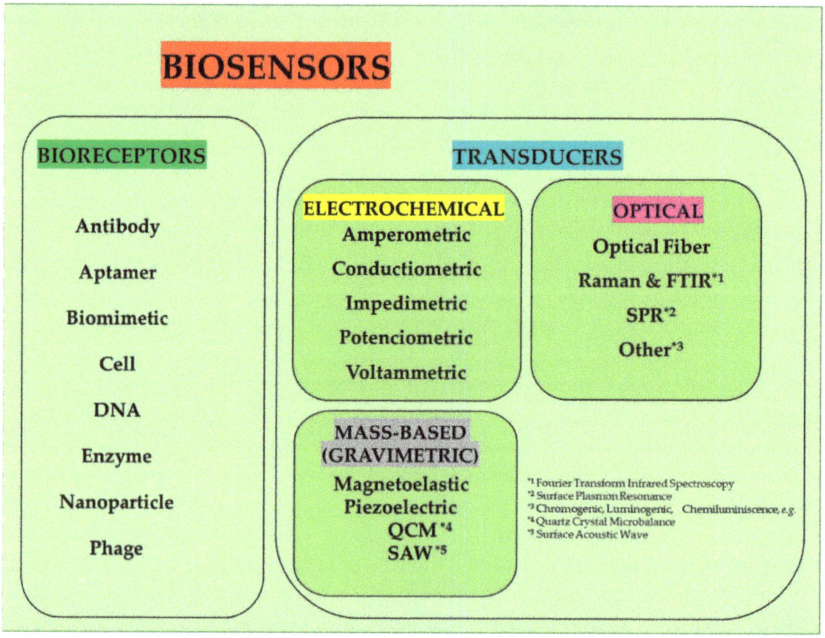

Figure 1. Classification of biosensors.

Electrochemical biosensors, the first type of biosensor to be commercialized, are being studied extensively. These devices detect changes in electrical properties, such as current or potential, caused by chemical reactions between the bioreceptor and the analyte. These changes are converted into signals that correspond to the analyte concentration. Advantages of electrochemical biosensors include minimal sample preparation, high sensitivity with small volumes, and automation capabilities. However, challenges such as poor reproducibility and stability remain [23]. Electrochemical biosensors are further classified based on signal type as– potentiometric biosensors (measure potential differences using ion-selective electrodes, providing analyte concentration data), amperometric biosensors (measure current changes in the medium, offering high sensitivity and fast responses, though they are susceptive to interference from unwanted electroactive species), conductometric biosensors (detect conductivity changes due to biochemical reactions, operate at low voltages, and do not require reference electrodes), and ion-selective field-effect transistor biosensors (detect ion activity through potential changes at the gate electrode, offering direct ion detection) [24–29].

Optical biosensors detect changes in light properties resulting from interactions between the bioreceptor and the analyte, correlating these changes to analyte concentration. These devices measure changes in light intensity and offer advantages such as resistance to electromagnetic interference, compactness, simplicity, noninvasiveness, and suitability for in vivo use. Based on their optical configuration, optical biosensors are classified as intrinsic or extrinsic. Intrinsic configurations involve direct light passage through the sample, while extrinsic configurations use external pathways. Absorption-based biosensors measure analyte concentrations by detecting light absorption at specific wavelengths, utilizing single or multiple optical fibers [30]. Surface plasmon resonance biosensors detect refractive index changes, caused by analyte binding, at a metal–dielectric interface using surface plasmon propagation [31,32]. Fluorescence-based biosensors detect frequency changes in emitted radiation, often using fluorescent labels and fluorescence resonance energy transfer. Luminescence-based biosensors rely on light emitted from exothermic reactions, with bioluminescence occurring naturally in biological systems.

Piezoelectric biosensors integrate a biorecognition element with a piezoelectric material, typically quartz crystals, which are preferred for their availability, heat resistance, and stability in aqueous solutions [33]. These biosensors detect changes in mass, density, or viscosity on the surface of a piezoelectric crystal, based on affinity interaction recordings. Known for their simplicity and low cost, they are highly practical for real-world applications [33]. Piezoelectric biosensors work by generating an electrical potential when subjected to mechanical stress and by deforming elastically in response to an electric field.

Thermal biosensors, also known as calorimetric or thermometric biosensors, measure temperature changes resulting from bioreceptor–analyte interactions, which correlate with analyte concentration. These sensors use thermistors or thermopiles as transducers [24,34] and offer several advantages, including label-free detection, minimal recalibration, and resistance to sample interference [24]. Thermal biosensors are widely used for their ability to measure thermal changes proportional to molar enthalpy and product formation, and they are particularly valuable in biochemical reactions. Enzyme-based designs are often emphasized in research due to their exothermic reactions.

Immunosensors are designed to detect analyte–antibody interactions and are categorized into three main types: luminescent or colorimetric sensors, surface plasmon resonance sensors, and electrochemical sensors. Antibodies, or immunoglobulins, are Y-shaped proteins produced by B lymphocytes in response to foreign substances. Their specificity makes them ideal for biosensors, where they bind tightly to antigens, forming complexes. Antibodies used in biosensors can be monoclonal, polyclonal, or recombinant. Monoclonal antibodies target a single epitope, while polyclonal antibodies bind to multiple epitopes, offering stronger binding but higher cross-reactivity. Recombinant antibodies are genetically engineered. Key features for use in biosensors include high sensitivity and minimal cross-reactivity [35]. Effective antibody immobilization is essential for biosensor performance, with methods such as covalent binding, non-covalent techniques, and affinity coupling being employed. Factors like temperature, pH, and ionic strength also affect antibody activity and sensor accuracy [36].

Aptamers, typically single-stranded RNA or DNA molecules consisting of 2–60 nucleotides, bind specifically to targets such as organic molecules and cells [37]. Aptasensors, biosensors that use aptamers as biorecognition elements, were first introduced in 1996 [38] and have since found a variety of applications. Aptamers offer several advantages, including high stability and affinity, simplicity, cost effectiveness, and excellent reproducibility across different production batches.

Enzymes, biocatalysts that accelerate chemical reactions, are highly specific to certain substrates, making them ideal for use in sensors. Various enzymes, including cholinesterase, urease, glucose oxidase, and others, are widely used in enzymatic inhibition analysis, a well-established method [39].

Proteins such as phytochelatins and metallothioneins can also act as biorecognition elements when immobilized on a transducer surface [40].

Whole cell-based biosensors use living cells, such as microorganisms or plant cells, which can be natural or recombinant [41]. These biosensors are inexpensive, easy to cultivate, and resilient to changes in pH, temperature, or ionic strength. They can perform multistep reactions and regenerate by allowing the cells to regrow, often without the need for sample preparation. However, they have slower response time and are more susceptible to interference from contaminants.

Biosensors are indispensable tools for detecting and measuring contaminants in food, offering rapid, sensitive, and selective analysis. They are critical for identifying and monitoring a wide range of food contaminants, including heavy metals [10,42], pesticides [43], herbicides [44], allergens [45,46], mycotoxins [46–49], histamine [50], and other indicators of food quality, as shown in Table 1. Consequently, biosensors are essential for ensuring the safety and quality of food throughout the supply chain. More recently, in Food Safety 4.0, intelligent biosensors played a key role in transforming traditional methods into data-driven solutions. Intelligent biosensors are advanced devices that combine biosensing technologies with digital systems. These sensors detect hazards like pathogens, contaminants, allergens, and quality issues early, enabling proactive risk management. Integrated throughout the food supply chain, these biosensors provide real-time data that empower stakeholders to make informed decisions, ensuring food safety and quality from production to consumption [51].

Regarding the limits of detection (LOD) of some heavy metals, the optical aptasensors for Pb^{2+} presented values from 0.07 to 100 nM, and the electrochemical aptasensor presented an LOD from 0.00000051 to 2.9 nM; for Hg^{+}, the values varied from 0.026 to 10.5 nM, and from 0.0001 to 25 nM for optical and for electrochemical aptasensors, respectively [53]. The optimized surface plasmon resonance biosensor developed by [68] enabled the biosensor to achieve an LOD as low as 0.2 μg/mL for egg allergen detection in red wines. Zhou et al. [115], in a revision of the most recent progresses in photoelectrochemical biosensors and their applications for monitoring mycotoxins in food, presented the LOD values for AFB1 (from 0.00032 to 5.0 pg/mL), OTA (from 0.02 to 2.0 pg/mL), and for fumonisin B1 (4.7 pg/mL). The meta-nano-channel biosensor, developed by Ron et al. [75], employed for the specific and label-free sensing of Botulinum neurotoxin BoNT, presented a limit of detection in the fg/mL range 10–100 ng/mL, with good linearity and a tuneable sensitivity. Zaraee et al. [85] presented a rapid, label-free, and cost-effective optical biosensor for the detection of *E. coli* with an LOD of 2.2 CFU/mL. A limit of detection of 6 CFU/mL was shown by Bagheryan et al. [99] in a Diazonium-based impedimetric aptasensor for the rapid, label-free detection of *Salmonella* Typhimurium in food samples. A polydopamine-enhanced vertically ordered mesoporous silica film anti-fouling electrochemical aptasensor, developed by Jin et al. [109], for the purpose of indicator-free *Vibrio parahaemolyticus* discrimination, using a stable inherent Au signal, presented 10^3 CFU/mL as a limit of detection. An LOD of 10 copies/μL of genomic DNA for *Listeria monocytogenes* and the possibility of distinguishing (high sensibility) *Listeria monocytogenes* from *Salmonella*, *Escherichia coli* O157:H7, and *Staphylococcus aureus* was achieved in a paper-based bipolar electrode electrochemiluminescence (pBPE-ECL) analysis system used for the sensitive detection of pathogenic bacteria. This system was developed by Liu et al. [90].

Table 1. Biosensors for detecting food contaminant.

Target Food Contaminate	Biosensor	Reference
	Heavy metals	
Heavy metal—Hg^{2+}; Ag^{+}; Pb^{2+}	Aptamers	[52–55];
Cadmium	Immunochromatography sensor	[56]
	Enzyme-linked immunosensor	[57–59]
Heavy metals	Conductometric biosensor	[60]
Pb^{2+}; Cu^{2+}	Enzyme biosensor	[61,62]
Heavy metals	DNAzymes	[59,63]
Heavy metals	Nucleic acid	[59,63]
	Pesticides	
Carbamate	Acetylcholinesterase biosensor	[64,65]
Organophosphorus	Non-enzymatic electrochemical sensors	[66]
Pesticide	Enzyme-based biosensor—acetylcholinesterase	[44]
Pesticide	Molecularly imprinted polymer-based biosensor	[43]
	Allergens	
Allergen	Antibody-based biosensor	[45]
Allergen	Nucleic acid-based biosensor	[46]
Egg ovalbumin	Electrochemical immunosensor	[67]
Egg ovalbumin	Surface plasmon resonance biosensor	[68]
	Fungal toxins—Mycotoxins	
Patulin	Immunochemical sensor	[69]
Aflatoxin B	Bio-electrochemical assay	[70]
Fusarium	Molecularly imprinted polymer-based biosensor	[48]
Ochratoxin A	Electrochemical immunosensor	[71–73]
	Immunosensor with fluorescence	[74]
Mycotoxins	Enzyme-based biosensor	[47]
	Bacterial toxins	
Botulinum neurotoxin (*Clostridium botulinum* toxin)	Meta-Nano-Channel (MNC) Field-Effect Transistor (FET) biosensor	[75]
Botulinum neurotoxin (*Clostridium botulinum* toxin)	Surface Acoustic Wave Immunosensor	[76]
	Foodborne Pathogens–Bacteria	
Campylobacter jejuni	Mechanical Biosensor QCM	[77,78]
Cronobacter sakazakii	Optical Biosensor Colorimetric	[79,80]
Cronobacter sakazakii	Optical Biosensor SPR Antibody	[81]
Cronobacter sakazakii	Electrochemical Biosensor Antibody	[82]
Escherichia coli O157:H7	Optical Biosensor Antibody	[83]
Escherichia coli	Optical Biosensor Antibody	[84]
Escherichia coli	Optical Biosensor Interferometric	[85]
Escherichia coli O157:H7	Electrochemical Biosensor Antibody	[78]
Escherichia coli	Electrochemical Chemiluminescence (ELC) Biosensors Aptamer-Based ECL Sensors	[86]
Listeria monocytogenes	Optical Biosensor Chemiluminescence	[87]
Listeria monocytogenes	Electrochemical Biosensor	[88]
Listeria monocytogenes	Electrochemical Biosensor Antibody	[89]
Listeria monocytogenes	Electrochemical Chemiluminescence (ELC) Biosensors Paper-Based Bipolar electrode ECL	[90]
Mycobacterium tuberculosis	Mechanical Biosensor Multi-Channel Series Piezoelectric Guartz Crystal (MSPQC)	[91,92]
Pseudomonas	Optical Biosensor Surface Plasmon Resonance (SPR)	[93,94]
Salmonella	Optical Biosensor Antibody	[95]

Table 1. *Cont.*

Target Food Contaminate	Biosensor	Reference
Salmonella enterica subsp. *enterica* Enteritidis	Optical Biosensor Antibody	[84]
Salmonella enterica subsp. *enterica* Typhimurium	Optical Biosensor Aptamer	[96]
Salmonella enterica subsp. *enterica* Typhimurium	Optical Biosensor localized Surface Plasmon Resonance (LSPR)	[97]
Salmonella enterica subsp. *enterica* Typhimurium	Electrochemical Impedimetric	[98,99]
Salmonella	Mechanical Biosensor Quartz Crystal Microbalance (QCM)	[100,101]
Staphylococcus aureus	Electrochemical Potentiometric	[102,103]
Staphylococcus aureus	Mechanical Biosensor QCM	[104,105]
Streptococcus agalactiae	Electrochemical Amperometric	[106,107]
Vibrio parahaemolyticus	Electrochemical Biosensor	[108]
Vibrio parahaemolyticus	Electrochemical Biosensor	[109]
Vibrio parahaemolyticus	Electrochemical Biosensor	[110]
Vibrio parahaemolyticus	Electrochemical Chemiluminescence (ELC) Biosensors ECL Immunosensor	[111]
Vibrio parahaemolyticus	SERS Biosensor	[112]
Vibrio vulnificus	Colorimetric Biosensor	[113]
	Foodborne Pathogens–Virus	
Norovirus	Electrochemical biosensor	[114]
	Histamine	
Histamine	Molecularly imprinted polymer-based biosensor	[48]

2.1. Heavy Metals

The most common metallic contaminants include chromium (Cr), cadmium (Cd), lead (Pb), arsenic (As), mercury (Hg), copper (Cu), and zinc (Zn) [116]. Heavy metals such as cadmium, lead, and mercury pose significant health risks when present in food products [117]. To protect public health, it is crucial to regulate heavy metals like lead, cadmium, and chromium in food sources [118]. Biosensors offer highly sensitive and rapid methods for detecting heavy metal contamination in food samples. For instance, a sensor has been developed to simultaneously detect lead and cadmium in fruits and vegetables [119]. These devices enable real-time monitoring with exceptional precision and selectivity, making them invaluable for ensuring food safety.

Various methods have been employed for the in situ detection of heavy metal ions, including amperometric sensors [120], electrochemical sensors [63], acoustic sensors [121], and inhibition-based biosensors [122]. Together, these techniques significantly enhance the capabilities of biosensors, allowing for the efficient and reliable monitoring of heavy metal contamination.

Biomaterials with biological activity and a specific affinity for heavy metals are widely used to modify electrodes used for detection [53]. Electrochemical sensors integrate sensitive biomaterials, such as nucleic acids, enzymes, antigens/antibodies, or whole cells, with an electrochemical transducer in order to convert biochemical signals into electronic signals [123–125]. Among these biomaterials, nucleic acids and enzymes are the most extensively studied for electrode modification in heavy metal detection [63,126].

Microbial biosensors provide a cost-effective and highly sensitive solution for the detection of heavy metal ions. For instance, microbial fluorescence-based biosensors [127,128] use reporter genes, activated in response to specific biochemical interactions between cellular reporters and inducer molecules. The integration of a chemostat-like microfluidic platform with microbial biosensors allows for molecular analytical detection on a chip [129]. Furthermore, optical DNA biosensors combined with evanescent wave analysis offer rapid, in situ detection of heavy metal ions [130]. Certain heavy metals bind to nucleic acid bases, forming metal ion-guided pairings like thymine (T)-Hg^{2+}-T and cytosine (C)-Ag^{+}-C [130]. This feature has garnered an interest in functional nucleic acids, including DNAzymes and aptamers, used for electrode modification in heavy metal detection [125]. DNAzymes are highly stable and specific molecules with a strong binding affinity, making them effective tools for heavy metal detection [59,131]. For instance, Tang et al. [132] developed a DNAzyme-based electrochemical sensor using rolling circle amplification to detect Pb^{2+} in water.

Aptamers, single-stranded DNA, RNA, or peptide sequences, exhibit a high affinity and specificity for target molecules. As cost-effective and easily produced alternatives to antibodies, aptamers are highly sensitive and specific. They have been effectively used to detect heavy metals such as lead (Pb), mercury (Hg), and cadmium (Cd) in food [125]. For example, Miao et al. [133] developed a DNA-modified Fe_3O_4@Au nanoparticle-based electrochemical sensor to detect Hg^{2+} and Ag^{+} in water, juice, and wine. Similarly, an aptamer-based electrochemical sensor was designed for the detection of arsenic (As^{3+}) in water using the $(GT)_{21}$-ssDNA sequence for specific recognition [134]. Aptamer-based sensors for Pb^{2+} [135] and Cd^{2+} [136] further highlight their applicability in heavy metal detection.

Enzymes, as biocatalysts, accelerate chemical reactions and exhibit high specificity for substrates, making them ideal for use in sensors. Certain heavy metals interact strongly with enzymes, altering their activity. These changes in enzyme activity can be monitored indirectly by measuring the corresponding electrical signals [125]. Enzymatic biosensors have been developed for detecting specific heavy metals in food, such as a urease inhibition-based sensor for identifying Pb^{2+} and Hg^{2+} ions in water [137]. Enzyme-based biosensors detect heavy metals through the activation or inhibition of enzyme activity, often caused by interactions between metal ions and thiol groups in enzymes. Common enzymes used include glucose oxidase, urease, and alkaline phosphatase, although selectivity challenges exist, as some enzymes can interact with multiple metals.

Protein-based biosensors detect metal–protein complex formation without labeling, measuring changes in electrical capacitance or impedance. Capacitive protein-based biosensors are particularly sensitive to low heavy metal concentrations and outperform cell-based devices in detection capabilities.

This suite of biosensor technologies collectively represents a powerful toolkit for detecting and monitoring heavy metal contaminants in food, ensuring safety, and maintaining quality throughout the food supply chain.

2.2. Pesticides

Pesticides, also known as plant protection products, are used to enhance crop yields and protect crops from diseases and infestations [138]. They include herbicides, insecticides, fungicides, plant growth regulators, and repellents. Pesticides can be chemically classified into groups such as organochlorines, organophosphates, carbamates, pyrethrin, and pyrethroids [139]. Among these, organophosphates, organochlorines, and carbamates are the most problematic classes. Pesticides can accumulate in vegetables, fruits, and meat throughout the food chain [140], and their residues in food products pose significant health risks to consumers.

Biosensors have proven to be highly effective tools for detecting pesticide residues, thereby ensuring food safety and compliance with regulatory standards [141]. These devices facilitate the rapid, sensitive, and selective detection of pesticide and herbicide residues [142]. By enabling real-time monitoring, biosensors have demonstrated great success in detecting trace levels of pesticides, enhancing food safety protocols and protecting consumers from the health hazards associated with such contaminants [143–145].

Biomaterials, such as enzymes [146], antibodies [147], and aptamers [148], are employed to identify and measure pesticides at ultra-low concentrations. These biomaterials exhibit highly sensitive and consistent interactions with pesticide molecules. Biosensors that are specifically designed and optimized to detect particular pesticides, employ a variety of techniques, including the optical [149–151], electrochemical [152–154], calorimetric [155], and piezoelectric [156] methods, based on enzyme inhibition.

For example, electrochemical biosensors for pesticide detection utilize enzymes, whole cells, or antibody–antigen interactions (immunosensors) [157,158]. Immunosensors have proven to be highly effective at rapid monitoring in agricultural applications [159]. Sensing systems for herbicide detection include molecular imprinting fluorescent chemosensors [160] and chemiluminescence immunoassays [161].

Electrochemical sensors are categorized into enzymatic and non-enzymatic types [162]. Enzymatic sensors rely on enzyme-catalyzed reactions at the electrode surface, while non-enzymatic sensors depend on the direct electrochemical activity of the analyte on noble metal electrodes [163]. Enzymatic sensors generally offer higher selectivity than their non-enzymatic counterparts [164]. Numerous enzymatic sensors using acetylcholinesterase have been developed, leveraging the strong binding affinity between organophosphorus pesticides and the enzyme's active sites [165,166].

Biosensors based on acetylcholinesterase inhibition are particularly effective for detecting organophosphate pesticides [167–169]. Organophosphates inhibit acetylcholinesterase by phosphorylating the serine residue at the enzyme's active site, preventing the hydrolysis of acetylcholine [158].

Enzymatic biosensors are widely studied for their stability, sensitivity, and accuracy, making them particularly effective for detecting pesticides [170]. These biosensors utilize acetylcholinesterase to detect enzymatic inhibition caused by organophosphates and carbamates. The inhibition occurs when these compounds bind to the enzyme's active site, blocking the hydrolysis of acetylcholine into choline and acetate [171]. Enzyme-based biosensors offer advantages such as high specificity, sensitivity, selectivity, availability, and versatility. They are classified into two types: direct and indirect. Direct biosensors measure analyte concentration or product formation during enzymatic reactions, while indirect biosensors detect enzyme inhibition caused by the interaction with the target analyte [167,172,173].

Although enzyme-based biosensors are highly specific, this specificity limits their ability to detect multiple analytes. Efforts are ongoing to address this limitation [174]. For instance, Borah et al. developed an amperometric biosensor based on the inhibition of the enzyme glutathione S-transferase [175]. Another approach involves integrating multiple enzymes, each sensitive to different pesticide types, into a single biosensing platform [157].

Electrochemical biosensors using whole cells have been proposed as an alternative to enzyme-based systems for pesticide detection [158]. Microbial cells are a cost-effective and stable option, eliminating the need for labor-intensive isolation and purification processes. Large quantities of cells can be easily cultivated [176,177]. Physiological changes in these cells, induced by exposure to toxicants (e.g., alterations in respiratory chain activity), are used to evaluate acute biotoxicity.

Using biosensors for pesticide detection provides efficient analysis, enhanced precision, low-concentration detection, continuous monitoring, and cost advantages over conventional methods, making them highly valuable for a variety of pesticide detection applications [178–180].

2.3. Allergens

Food allergies have become a significant food safety concern, with prevalence rates estimated at 1% to 3% in adults and 4% to 6% in children, primarily due to hidden allergens in processed foods [181]. These allergies result from the type I hypersensitivity reaction of the immune system to ingested allergens, posing life-threatening risk [182]. Clinical studies have documented 160 food allergens, with approximately 90% of allergic reactions attributed to eight major allergens: eggs, milk, shellfish, fish, peanuts, tree nuts, soybeans, and wheat [182].

Approximately 100 countries worldwide have legislation regarding the declaration of allergenic ingredients [183]. Since 1985, the Codex Alimentarius has included food allergens, with the General Standard for the Labelling of Prepackaged Foods mandating the declaration of eight 'priority' allergens: cereals containing gluten (such as wheat, rye, barley, oats, spelt, or their hybridized strains); crustaceans and their products; eggs and egg products; fish and fish products; peanuts, soybeans and their products; milk and milk products (including lactose); tree nuts and their products; and sulphites at concentrations of 10 mg/kg or more) [184].

In Europe, most countries adhere to European Union (EU) legislation, which mandates the declaration of 14 allergens. This list includes the Codex-8, with peanut and soya named separately, and adds celery, mustard, sesame, lupine, and mollusks [185]. In the United States, allergen declaration is mandated by the Food Allergen Labeling and Consumer Protection Act [186], which includes the Codex allergens but names only wheat among cereals, excluding other gluten-containing grains. The Food Allergy Safety, Treatment, Education, and Research Act amended the Food Allergen Labelling and Consumer Protection Act to add sesame to the ninth major food allergen, effective 1 January 2023 [187]. For tree nuts, the specific nut must be declared; for crustacea and fish, the species must be identified [188].

In Canada, priority allergens include the Codex-8, along with mollusks and mustard [189]. In Australia and New Zealand, a "contains statement" is mandatory for priority allergens. These include wheat, fish, crustaceans, mollusks, eggs, milk, lupine, peanuts, soy, sesame, almonds, Brazil nuts, cashews, hazelnut, macadamia nuts, pecans, pistachios, pine nuts, and walnuts, as well as barley, oats, and rye when they contain gluten. Sulphites must be declared if added at levels of 10 mg/kg or more [190].

Japan, the first country to regulate both intentional and unintentional allergen presence, categorizes allergens into those for mandatory disclosure (wheat, buckwheat, eggs, milk, peanut, shrimp, crab, and walnuts) and those for recommended disclosure (almonds, abalone, squid, salmon roe, oranges, cashews, kiwifruit, beef, sesame, salmon, mackerel, soybean, chicken, banana, pork, macadamia nuts, peach, yam, apple, and gelatin) [191]. South Korea has a similar approach, with a distinct list of mandatory allergens including eggs (confined to those from poultry), milk, buckwheat, peanuts, soybeans, wheat, mackerel, crab, shrimp, pork, peach, tomato, sulfurous acid (when present at 10 mg/kg or more), walnuts, chicken, beef, squid, clams (including oyster, abalone, and mussels), and pine nuts [192].

Food allergens originate from both animal and plant sources, with around 40% derived from organisms that produce five or more allergens. These allergens are often concentrated within a limited number of biochemically active protein families [193,194]. To safeguard individuals with food allergies, effective analytical methods capable of detecting trace amounts of allergenic ingredients in processed foods are essential. Biosensors provide the rapid and accurate detection of allergens, thereby enhancing food safety for sensitive populations [195].

Biosensors utilize recognition elements, such as antibodies or aptamers, to specifically target allergenic proteins from common sources like nuts and shellfish [196,197]. For instance, immunosensors leverage antibodies designed to detect specific allergenic proteins, enabling the quick and sensitive analysis of food samples [195]. DNA-based biosensors identify genetic sequences linked to allergenic components, offering a reliable approach for allergen detection [197].

Electrochemical biosensors have significantly advanced allergen detection due to their high sensitivity, selectivity, and user-friendliness [198]. Innovations in nanoscience and bioelectronics have further enhanced their performance by integrating biological receptors with nanomaterials such as metal nanoparticles, graphene, and quantum dots, which increase electrode surface activity and electron transfer efficiency [199,200].

Electrochemical immunosensors, which combine antibodies with electrochemical sensors, are widely employed for detecting food allergen [201,202]. These sensors detect allergenic proteins through antigen–antibody binding, generating electrical signals proportional to analyte concentration [203,204]. Their high selectivity arises from precise immunological interactions [195].

Nucleic acid-based electrochemical biosensors are prized for their compatibility with miniaturization and microfabrication, as well as their simplicity in detecting food allergens [205,206]. Despite the limited electrochemical activity of DNA probes and aptamers, innovative approaches to probe immobilization, signal amplification, and performance improvement are driving their development [167].

Sundhoro et al. [207] pioneered the use of molecularly imprinted polymers to detect the soybean allergen marker genistein in complex foods. The sensor demonstrated performance comparable to, or better than, portable allergen detection tools like lateral flow devices and ELISA, offering high selectivity, rapid detection, and cost effectiveness. However, its sensitivity still falls short of advanced methods like mass spectrometry and PCR.

Freitas et al. [208] designed an electrochemical dual immunosensor to simultaneously detect peanut allergens, Ara h 1, and Ara h 6, with detection limits as low as 0.05%. The sensor's performance was validated through recovery studies and comparisons with ELISA, confirming its reliability and effectiveness in complex food matrices.

2.4. *Mycotoxins*

Mycotoxins are low-molecular-weight, heat-stable secondary metabolites produced by toxic molds belonging to the genera *Aspergillus*, *Penicillium*, *Alternaria*, and *Fusarium*. These toxins, found in the mycelium and spores of molds, include aflatoxins, ochratoxins, fumonisins, citrinin, patulin, zearalenone, trichothecenes, tremorgenic toxins, and ergot alkaloids. Mycotoxins pose significant risks to public health [209]. Their toxicity depends on factors such as species, mechanisms of action, metabolism, and the defense responses of organisms consuming contaminated food [210]. Due to these risks, most countries have established regulatory limits for mycotoxin levels in food, with thresholds varying by product type [211].

Ochratoxin A (OTA) has been identified in various crops, including cereals, grapes, coffee, and cocoa, as well as in derived food products, such as beer, wine, and vinegar. Biosensors for OTA offer rapid response times, cost-effective production, and reliable accuracy for on-site analysis [212]. OTA detection methods are broadly categorized into two approaches: (i) rapid screening tests providing qualitative results, and (ii) confirmatory tests offering precise quantitative measurements [213,214].

Portable biosensors, such as optical immunosensors, optical aptasensors, surface plasmon resonance biosensors, and photoelectrochemical biosensors, have been developed for detecting OTA in foods and beverages [215].

Optical methods for mycotoxin detection such as colorimetric, fluorescent, chemiluminescent, and surface plasmon resonance are valued for their simplicity, speed, reliability, and high sensitivity [216]. These biosensors combine a biological sensing element with an optical transducer to detect analytes binding to a bio-recognition element immobilized on a substrate [217]. This interaction generates an electronic signal proportional to the analyte concentration [218]. Commonly used biorecognition elements include enzymes, substrates, antibodies, and nucleic acids, with enzymatic systems often employed to convert analytes into measurable products [216].

Optical biosensors operate in two modes: label-free detection, where an analyte–transducer interaction generates a direct signal, and label-based detection, where labels produce colorimetric, fluorescent, or luminescent signals [219]. Optical biosensors for OTA detection represent a leading nanotechnological alternative to traditional methods, offering rapid, sensitive, and specific analysis with minimal noise, low detection limits, and multiplexing capabilities. Label-free biosensors require minimal sample volumes and are suitable for real-time, on-site monitoring [220]. These devices use transducers to convert biorecognition interactions into measurable optical signals, such as absorption, transmission, or polarization [221].

Photoelectrochemical biosensors detect OTA by converting the chemical energy of a semiconductor into electricity under light illumination, generating a photocurrent or photovoltage. These biosensors are cost-effective and high sensitivity, but their reliance on electrochemical processes and a light source limits portability [115].

Electrochemical immunosensors have been effectively employed to detect aflatoxin B1 in pistachios [222]. Immunological biosensors, which use antibodies specific to mycotoxins, and DNA-based biosensors, which target genetic sequences associated with mycotoxin-producing molds, show significant promise for detecting these contaminants [223].

Colorimetric and luminescent sensors convert visible or UV light into analytical signals [224]. For example, a colorimetric sensor for aflatoxin B1 detection employed a direct competitive ELISA principle, with a color change measured spectrometrically at 620 nm. This method achieved sensitivity as low as 0.2 ng/mL, outperforming microtiter plate ELISA [225].

Enzyme-based biosensors frequently utilize acetylcholinesterase due to its high susceptibility to mycotoxins, particularly aflatoxin B1, which inhibits its activity [226,227]. This inhibition is reversible, as the toxins bind non-covalently to the enzyme [228]. Among enzymatic inhibition methods, aflatoxins are among the most sensitive toxins [229]. Cholinesterase has been demonstrated to be effective for detecting aflatoxin B1 [230].

A portable biosensor for Aflatoxin detection using surface plasmon resonance technology has been developed. This sensor utilizes surface plasmon resonances in ~50 nm metallic films and surface functionalization for selectivity. Moon et al. [231] employed this device for in situ monitoring of aflatoxin B1 in grains. However, its high cost and lack of reusability necessitate further research to improve practicality.

Zearalenone, a nonsteroidal estrogenic mycotoxin produced by *Fusarium* fungi, poses significant risks in food [232]. Researchers have developed a label-free amperometric immunosensor using mesoporous carbon and trimetallic nanorattles for its detection. Panini et al. [233] created a microfluidic immunoassay with anti-Zearalenon antibodies.

Fumonisins, another class of mycotoxins from the *Fusarium* species, have been detected using competitive lateral-flow immunoassays. Mirasoli et al. [234] designed such an assay for total fumonisins in maize, integrating enzyme-catalyzed chemiluminescence detection and a portable charge-coupled device camera.

Lu and Gunasekaran [235] introduced an electrochemical immunosensor capable of simultaneously detecting two mycotoxins, fumonisin B1 and deoxynivalenol, in a single assay.

Deoxynivalenol, another *Fusarium*-derived mycotoxin [236], was detected using a biosensor by Romanazzo et al. [237]. This system employed an enzyme-linked immunomagnetic assay with immunomagnetic beads and magnetized screen-printed electrodes as transducers.

Patulin, a mycotoxin from the *Penicillium expansum*, *Aspergillus*, *Penicillium*, and *Paecilomyces* species, presents a significant health concern [238]. Detection methods include a competitive SPR-based immunoassay that utilizes laser-induced interactions to generate a detectable resonance shift. Funari et al. [239] developed a piezoelectric biosensor, immobilizing oriented antibodies on a quartz crystal's gold surface using photonics-based techniques.

Many biosensors utilizing machine learning have been designed to detect mycotoxins, valued for their exceptional accuracy and precision [14].

2.5. Foodborne Pathogens

Foodborne pathogens are a major cause of food contamination during production, processing, and distribution. Consequently, the rapid and sensitive detection of pathogenic microorganisms is crucial to prevent food spoilage and foodborne illnesses. Numerous biosensor platforms have been developed to detect these pathogens [240–242]. The bacteria species most commonly responsible for outbreaks include *Salmonella*, *Escherichia coli*, *Campylobacter* spp., *Vibrio cholerae*, *Listeria monocytogenes*, and *Shigella* [243,244].

The primary function of a biosensor is to convert biochemical reactions into measurable electrical signals. Biosensors employing immunological, enzymatic, and molecular recognition elements are widely utilized to specifically identify genetic sequences, surface antigens, or metabolic by-products of pathogens [245]. DNA-based biosensors, which use nucleic acid probes, have demonstrated efficacy in detecting the genetic sequences of pathogens such as *Escherichia coli* and *Salmonella* [246–249].

Immunosensors, employing antibodies as recognition elements, are capable of identifying the surface antigens of bacteria like *Salmonella*, *Campylobacter* spp., *Listeria monocytogenes*, and *Escherichia coli* [250,251]. Biosensors have been specifically designed for the detection of pathogens including *Salmonella* [97,252–256], *L. monocytogenes* [251,257,258], *E. coli* [259–262], *Campylobacter* [263], *C. perfringens* [264], *Staphylococcus aureus* [257,265], and *Toxoplasma gondi* [266].

An electrochemical DNA biosensor for the selective identification of *Salmonella enterica* subsp. *enterica* serovar Typhi (*S.* Typhi) in real samples was proposed and fabricated by Bacchu et al. [267]. According to the authors, this biosensor showed excellent discrimination capability to some mismatched bases and to different bacterial cultures belonging to the same and distant genera. This DNA biosensor also presented a lower limit of detection and the capacity to be reused more than six to seven times.

Angelopoulou et al. [268] were able to simultaneous detect two bacteria, namely *Salmonella enterica* subsp. *enterica* serovar Typhimurium and *Escherichia coli* O157:H7, using, for the first time, a label-free optical immunosensor based on the arrays of Mach–Zehnder Interferometers monolithically integrated onto silicon chips.

Da Silva et al. [269], in a review, emphasize the importance of the application of electrochemical point-of-care devices for the monitoring of potentially harmful and/or toxic species that can be found in water resources, as well as waterborne pathogens (protozoa, bacteria, and viruses), allowing for faster on-site analysis. For the detection of the protozoa *Giardia lamblia* and *Entamoeba histolytica*, a metronidazole-probe sensor, based on an imprinted biocompatible nanofilm for the rapid and sensitive detection of anaerobic protozoan was used. The method was developed by Roy et al. [270]. For the detection of *Cryptosporidium*, a novel three-dimensional microTAS chip for the ultra-selective single-base mismatched *Cryptosporidium* DNA biosensor was used. The method was developed by Ilkhani et al. [271].

Concerning viruses that can be transmitted by food and water consumption, norovirus and hepatitis A virus are found to be the main cause of foodborne infections. Baek et al. [114] developed an electrochemical biosensor applied to detect human norovirus, prepared by standard procedure from an oyster. According to the authors, this biosensor can be used as a very sensitive and selective point-of-care bioanalytical platform for the detection of human norovirus in various food samples. The DNA sensor, developed by Manzano et al. [272], can be adapted to a portable format to be adopted as an easy-to-use and low-cost method for screening hepatitis A virus (HAV) in contaminated food and water.

The majority of these methods are based on immunosensors (antibody-based) or DNA-based sensors. Peptides have also been investigated as recognition biomolecules in the development of biosensors, offering high sensitivity, low-cost, and rapid response times. Some of these biosensors hold potential portable devices for on-site analyses, enhancing the detection of bacterial pathogens in food [51].

Nowadays, as a tool for helping improve risk management and ensure the highest standards of food safety, we can deal with intelligent biosensors, which offer attractive, smarter solutions, including real-time monitoring, predictive analytics, enhanced traceability, and consumer empowerment [273]. IoT-based intelligent biosensors for detecting *Vibrio parahaemolyticus* and smartphone-based intelligent biosensors for detecting ochratoxin A (OTA) in wine, instant coffee, and *Salmonella enterica* subsp. *enterica* Typhimurium are already available.

3. Strengths and Limitations of Biosensors for Detecting Food Contaminants

Biosensors offer significant advantages in detecting food contaminants, making them a valuable tool in ensuring food safety. A key strength of biosensors is their rapid detection capability, enabling real-time or near-real-time monitoring, which is crucial for timely intervention. Their high sensitivity and specificity, achieved through the use of biological recognition elements such as enzymes, antibodies, and aptamers, allow for the detection of contaminants at very low concentrations with remarkable accuracy [18].

Another notable advantage of biosensors is their cost effectiveness. Compared to traditional detection methods, biosensors are more affordable, lowering the overall cost of food safety monitoring. Additionally, many biosensors are designed to be portable and suitable for on-site application, eliminating the need for complex laboratory equipment or specialized expertise. Their versatility is highlighted by their ability to detect a wide array of contaminants, including heavy metals, pesticides, allergens, mycotoxins, and pathogens, making them suitable for diverse applications in food safety. The integration of nanomaterials further enhances the performance of biosensors, improving their sensitivity, stability, and overall efficacy [18,274].

Despite these strengths, biosensors also have limitations that need to be addressed to fully realize their potential. Sensitivity constraints can be an issue, especially when detecting low concentrations of certain contaminants. The complexity of food matrices can also interfere with the accuracy and reliability of biosensor readings, posing challenges for some applications. Furthermore, the stability and shelf life of biological components, such as enzymes and antibodies, can be limited, affecting their long-term usability [18,275].

Calibration and standardization are essential to ensure consistent and reliable results, due to the variability in biosensor performance. The process of obtaining regulatory approval and validation can be time-consuming, delaying the adoption of biosensors in the food industry. There are also integration challenges, as the integration of biosensors into existing food safety management systems requires overcoming compatibility issues and training personnel [18].

The application of machine learning to biosensors has grown significantly, but key challenges must be addressed to maximize its potential. A major hurdle is data availability, as large and diverse datasets from biosensors are expensive and difficult to obtain. Strategies for managing missing data are critical. The complexity of biological molecules also complicates data acquisition and analysis, requiring precise identification and segmentation. The quality and quantity of data used for training are critical, as they affect the algorithm's performance. Ensuring the datasets can accurately identify target compounds is essential [14].

In summary, while biosensors have transformative potential for food safety monitoring due to their rapid, sensitive, and cost-effective nature, addressing their limitations is essential. Overcoming these challenges will be critical for the broader implementation and reliability of biosensors in detecting food contaminants, ultimately improving food safety and protecting public health.

4. Conclusions

Food safety is significantly threatened by contaminants such as heavy metals, pesticides, allergens, mycotoxins, and pathogenic microorganisms, all of which pose serious health risks. Heavy metals, including lead, mercury, and cadmium, are among the most hazardous contaminants. Detection methods include DNA-modified electrodes, enzymatic inhibition sensors, and aptamer-based systems. Biosensors for pesticide detection use various biomaterials, including enzymes, antibodies, and aptamers, to detect trace residues, with electrochemical biosensors, particularly enzymatic ones, being commonly used for detecting organophosphates and carbamates. Innovations include integrating multiple enzymes and using whole-cell biosensors. For allergen detection, biosensors utilizing antibodies, aptamers, and nucleic acids identify allergenic proteins. Nanotechnology-enhanced electrochemical sensors have improved both sensitivity and portability, although some systems still face sensitivity challenges. Mycotoxins, toxic compounds produced by molds, are detected using optical and electrochemical biosensors, such as immunosensors and aptasensors. For on-site analysis, advanced approaches, such as label-free biosensors, provide high sensitivity. The detection of foodborne pathogens has been revolutionized by immunosensors and DNA-based biosensors, allowing for the specific, efficient, and rapid identification of pathogens, thereby reducing the risks associated with foodborne illnesses. While traditional biosensors are valued for their simplicity, portability, and cost effectiveness, improvements in reproducibility and stability are necessary to meet the food industry's demands. Enhancing the ability to trace and extract features from complex food matrices is essential for identifying contaminants and optimizing processes. Integrating machine learning enhances biosensor reliability and performance, addressing challenges like single-molecule detection and signal noise. Advanced machine learning techniques and improved computing hardware can increase sensitivity and pattern recognition. Developing portable biosensors, utilizing artificial intelligence, internet of things, and nanomaterials, will improve food safety monitoring. A multidisciplinary platform with high efficiency and portability is crucial for addressing global food safety and health issues. In summary, biosensors have the potential to transform multiple food safety applications, ensuring regulatory compliance and protecting public health with greater efficiency. Intelligent biosensors will be a powerful tool in improving risk management and ensuring the highest standards of food safety and quality both now and in the future.

Funding: This research was funded by the Chemistry Research Centre-Vila Real (CQ-VR) (UIDB/00616/2020 and UIDP/00616/2020). (https://doi.org/10.54499/UIDP/00616/2020 and https://doi.org/10.54499/UIDB/00616/2020).

Conflicts of Interest: The authors declare no conflicts of interest.

References

1. Mali, H.; Shah, C.; Patel, D.H.; Trivedi, U.; Subramanian, R.B. Bio-Catalytic System of Metallohydrolases for Remediation of Neurotoxin Organophosphates and Applications with a Future Vision. *J. Inorg. Biochem.* **2022**, *231*, 111771. [CrossRef]
2. Renz, H.; Allen, K.J.; Sicherer, S.H.; Sampson, H.A.; Lack, G.; Beyer, K.; Oettgen, H.C. Food Allergy. *Nat. Rev. Dis. Primers.* **2018**, *4*, 17098. [CrossRef] [PubMed]
3. Kulawik, P.; Rathod, N.B.; Ozogul, Y.; Ozogul, F.; Zhang, W. Recent Developments in the Use of Cold Plasma, High Hydrostatic Pressure, and Pulsed Electric Fields on Microorganisms and Viruses in Seafood. *Crit. Rev. Food Sci. Nutr.* **2022**, *63*, 9716–9730. [CrossRef] [PubMed]
4. Pan, M.; Yin, Z.; Liu, K.; Du, X.; Liu, H.; Wang, S. Carbon-Based Nanomaterials in Sensors for Food Safety. *Nanomaterials* **2019**, *9*, 1330. [CrossRef] [PubMed]
5. Viswanathan, S.; Radecka, H.; Radecki, J. Electrochemical Biosensors for Food Analysis. *Monatsh. Chem.* **2009**, *140*, 891–899. [CrossRef]
6. Song, M.; Khan, I.M.; Wang, Z. Research Progress of Optical Aptasensors Based on Aunps in Food Safety. *Food Anal. Methods* **2021**, *14*, 2136–2151. [CrossRef]
7. Liang, S.; Sutham, P.; Wu, K.; Mallikarjunan, K.; Wang, J.P. Giant Magnetoresistance Biosensors for Food Safety Applications. *Sensors* **2022**, *22*, 5663. [CrossRef]
8. Scognamiglio, V.; Arduini, F.; Palleschi, G.; Rea, G. Biosensing technology for sustainable food safety. *TrAC Trend. Anal. Chem.* **2014**, *62*, 1–10. [CrossRef]
9. Thakur, M.S.; Ragavan, K.V. Biosensors in food processing. *J. Food Sci. Technol.* **2013**, *50*, 625–641. [CrossRef] [PubMed]
10. Odobašić, A.; Šestan, I.; Begić, S. *Biosensors for Determination of Heavy Metals in Waters*; IntechOpen: London, UK, 2019. [CrossRef]
11. Kazemi-Darsanaki, R.; Azizzadeh, A.; Nourbakhsh, M.; Raeisi, G.; Azizollahi Aliabadi, M. Biosensors: Functions and applications. *J. Biol. Today's World* **2013**, *2*, 53–61. [CrossRef]
12. Najeeb, J.; Ali, J.; Ali, M.A.; Aslam, M.F.; Raza, A. Biosensors: Their fundamentals, designs, types and most recent impactful application: A review. *J. Biosens. Bioelectron.* **2017**, *8*, 235.
13. Dhal, S.B.; Kar, D. Leveraging artificial intelligence and advanced food processing techniques for enhanced food safety, quality, and security: A comprehensive review. *Discov. Appl. Sci.* **2025**, *7*, 75. [CrossRef]
14. Hassan, M.M.; Xu, Y.; Sayada, J.; Zareef, M.; Shoaib, M.; Chen, X.; Li, H.; Chen, Q. Progress of machine learning-based biosensors for the monitoring of food safety: A review. *Biosens. Bioelectron.* **2025**, *267*, 116782. [CrossRef]
15. Mostajabodavati, S.; Mousavizadegan, M.; Hosseini, M.; Mohammadimasoudi, M.; Mohammadi, J. Machine learning-assisted liquid crystal-based aptasensor for the specific detection of whole-cell *Escherichia coli* in water and food. *Food Chem.* **2024**, *448*, 139113. [CrossRef] [PubMed]
16. Lv, M.; Liu, Y.; Geng, J.; Kou, X.; Xin, Z.; Yang, D. Engineering nanomaterials-based biosensors for food safety detection. *Biosens. Bioelectron.* **2018**, *106*, 122–128. [CrossRef] [PubMed]
17. Mishra, G.K.; Barfidokht, A.; Tehrani, F.; Mishra, R.K. Food safety analysis using electrochemical biosensors. *Foods* **2018**, *7*, 141. [CrossRef]
18. Nath, S. Advancements in food quality monitoring: Integrating biosensors for precision detection. *Sustain. Food Technol.* **2024**, *2*, 976–992. [CrossRef]
19. Arugula, M.A.; Simonian, A. Novel trends in affinity biosensors: Current challenges and perspectives. *Meas. Sci. Technol.* **2014**, *25*, 032001–032022. [CrossRef]
20. Hassan, M.M.; Xu, Y.; Zareef, M.; Li, H.; Rong, Y.; Chen, Q. Recent advances of nanomaterial-based optical sensor for the detection of benzimidazole fungicides in food: A review. *Crit. Rev. Food Sci. Nutr.* **2023**, *63*, 2851–2872. [CrossRef]
21. Rotariu, L.; Lagarde, F.; Jaffrezic-Renault, N.; Bala, C. Electrochemical biosensors for fast detection of food contaminants–trends and perspective. *TrAC Trends Anal. Chem.* **2016**, *79*, 80–87. [CrossRef]
22. Marazuela, M.D.; Moreno-Bondi, M.C. Fiber-optic biosensors—An overview. *Anal. Bioanal. Chem.* **2002**, *372*, 664–682. [CrossRef]
23. Gautam, P.; Suniti, S.; Prachi, K.; Amrita, D.; Madathil, B.; Nair, A.N. A review on recent advances in biosensors for detection of water contamination. *Int. J. Environ. Sci.* **2012**, *2*, 1565–1574.
24. Wang, Y.; Xu, H.; Zhang, J.; Li, G. Electrochemical sensors for clinical analysis. *Sensors* **2008**, *8*, 2043–2081. [CrossRef] [PubMed]
25. Martinkova, P.; Kostelnik, A.; Valek, T.; Pohanka, M. Main streams in the construction of biosensors and their applications. *Int. J. Electrochem. Sci.* **2017**, *12*, 7386–7403. [CrossRef]
26. Monosik, R.; Stredansky, M.; Sturdik, E. Biosensors—Classification, characterization and new trends. *Acta Chim. Slovaca* **2012**, *5*, 109–120. [CrossRef]
27. Dzyadevych, S.V.; Jaffrezic-Renault, N. Conductometric biosensors. In *Biological Identification*; Schaudies, R.P., Ed.; Woodhead Publishing: Cambridge, UK, 2014; Volume 6, pp. 153–193.

28. Jaffrezic-Renault, N.; Dzyadevych, S.V. Conductometric microbiosensors for environmental monitoring. *Sensors* **2008**, *8*, 2569–2588. [CrossRef]
29. Eivazzadeh-Keihan, R.; Pashazadeh-Panahi, P.; Baradaran, B.; de la Guardia, M.; Hejazi, M.; Sohrabi, H.; Mokhtarzadeh, A.; Maleki, A. Recent progress in optical and electrochemical biosensors for sensing of Clostridium botulinum neurotoxin. *TrAC Trends Anal. Chem.* **2018**, *103*, 184–197. [CrossRef]
30. Bosch, M.E.; Sanchez, A.J.R.; Rojas, F.S.; Ojeda, C.B. Recent development in optical fibre biosensors. *Sensors* **2007**, *7*, 797–859. [CrossRef]
31. Wijaya, E.; Lenaerts, C.; Maricot, S.; Hastanin, J.; Habraken, S.; Vilcot, J.-P. Surface plasmon resonance-based biosensors: From the development of different SPR structures to novel surface functionalization strategies. *Curr. Opin. Solid State Mater. Sci.* **2011**, *15*, 208–224. [CrossRef]
32. Asal, M.; Ozen, O.; Sahinler, M.; Polatoglu, I. Recent developments in enzyme, DNA and immuno-based biosensors. *Sensors* **2018**, *18*, 1924. [CrossRef]
33. Pohanka, M. The piezoelectric biosensors: Principles and applications, a review. *Int. J. Electrochem. Sci.* **2017**, *12*, 496–506. [CrossRef]
34. Perumal, V.; Hashim, U. Advances in biosensors: Principle, architecture and applications. *J. Appl. Biomed.* **2014**, *12*, 1–5. [CrossRef]
35. Ferrigno, P.K. Non-antibody protein-based biosensors. *Essays Biochem.* **2016**, *60*, 19–25.
36. Sharma, S.; Byrne, H.; O'Kennedy, R.J. Antibodies and antibody-derived analytical biosensors. *Essays Biochem.* **2016**, *60*, 9–18.
37. Peltoma, R.; Benito-Peña, E.; Moreno-Bondi, M.C. Bioinspired recognition elements for mycotoxin sensors. *Anal. Bioanal. Chem.* **2018**, *410*, 747–771. [CrossRef] [PubMed]
38. Pfeiffer, F.; Mayer, G. Selection and biosensor application of aptamers for small molecules. *Front. Chem.* **2016**, *4*, 25. [CrossRef] [PubMed]
39. Chauhan, R.; Singh, J.; Sachdev, T.; Basu, T.; Malhotra, B.D. Recent advances in mycotoxins detection. *Biosens. Bioelectron.* **2016**, *81*, 532–545. [CrossRef] [PubMed]
40. Cornelis, R.; Crews, H.; Caruso, J.; Heumann, K. *Handbook of Elemental Speciation: Techniques and Methodology*; John Wiley & Sons, Ltd.: Chichester, UK, 2003; ISBN 0-471-49214-0.
41. Gu, M.B.; Mitchell, R.J.; Kim, B.C. Whole-cell-based biosensors for environmental biomonitoring and application. *Adv. Biochem. Engin./Biotechnol.* **2004**, *87*, 269–305.
42. Shakya, A.K.; Singh, S. State of the art in fiber optics sensors for heavy metals detection. *Opt. Laser Technol.* **2022**, *153*, 108246. [CrossRef]
43. Umapathi, R.; Park, B.; Sonwal, S.; Rani, G.M.; Cho, Y.; Huh, Y.S. Advances in optical-sensing strategies for the on-site detection of pesticides in agricultural foods. *Trends Food Sci. Technol.* **2022**, *119*, 69–89. [CrossRef]
44. Loguercio, L.F.; Thesing, A.; Demingos, P.; de Albuquerque, C.D.; Rodrigues, R.S.; Brolo, A.G.; Santos, J.F. Efficient acetylcholinesterase immobilization for improved electrochemical performance in polypyrrole nanocomposite-based biosensors for carbaryl pesticide. *Sens. Actuators B Chem.* **2021**, *339*, 129875. [CrossRef]
45. Fu, L.; Cherayil, B.J.; Shi, H.; Wang, Y.; Zhu, Y.; Fu, L.; Cherayil, B.J.; Shi, H.; Wang, Y.; Zhu, Y. Detection and quantification methods for food allergens. In *Food Allergy: From Molecular Mechanisms to Control Strategies*; Springer: Singapore, 2019; pp. 69–91.
46. Pilolli, R.; Monaci, L.; Visconti, A. Advances in biosensor development based on integrating nanotechnology and applied to food-allergen management. *TrAC Trends Anal. Chem.* **2013**, *47*, 12–26. [CrossRef]
47. Majdinasab, M.; Ben Aissa, S.; Marty, J.L. Advances in colorimetric strategies for mycotoxins detection: Toward rapid industrial monitoring. *Toxins* **2020**, *13*, 13. [CrossRef]
48. Sergeyeva, T.; Yarynka, D.; Dubey, L.; Dubey, I.; Piletska, E.; Linnik, R.M.; Antonyuk, T.; Ternovska, O.; Brovko, S. Piletsky Sensor based on molecularly imprinted polymer membranes and smartphone for detection of *Fusarium contamination* in cereals. *Sensors* **2020**, *20*, 4304. [CrossRef]
49. Jia, Y.; Zhao, S.; Li, D.; Yang, J.; Yang, L. Portable chemiluminescence optical fiber aptamer-based biosensors for analysis of multiple mycotoxins. *Food Control* **2023**, *144*, 109361. [CrossRef]
50. Mahnashi, M.H.; Mahmoud, A.M.; Alhazzani, K.; Alanazi, A.; Algahtani, M.M.; Alaseem, A.M.; Alqahtani, Y.S.; El-Wekil, M.M. Enhanced molecular imprinted electrochemical sensing of histamine based on signal reporting nanohybrid. *Microchem. J.* **2021**, *168*, 106439. [CrossRef]
51. Chen, Y.; Wang, Y.; Zhang, Y.; Wang, X.; Zhang, C.; Cheng, N. Intelligent Biosensors Promise Smarter Solutions in Food Safety 4.0. *Foods* **2024**, *13*, 235. [CrossRef] [PubMed]
52. Wang, S.; Si, S. Aptamer biosensing platform based on carbon nanotube long-range energy transfer for sensitive, selective and multicolor fluorescent heavy metal ion analysis. *Anal. Methods* **2013**, *5*, 2947–2953. [CrossRef]
53. Guo, W.; Zhang, C.; Ma, T.; Liu, X.; Chen, Z.; Li, S.; Deng, Y. Advances in aptamer screening and aptasensors' detection of heavy metal ions. *J. Nanobiotechnology* **2021**, *19*, 1–19. [CrossRef] [PubMed]

54. Wang, L.; Peng, X.; Fu, H.; Huang, C.; Li, Y.; Liu, Z. Recent advances in the development of electrochemical aptasensors for detection of heavy metals in food. *Biosens. Bioelectron.* **2020**, *147*, 111777. [CrossRef]
55. Sawan, S.; Errachid, A.; Maalouf, R.; Jaffrezic-Renault, N. Aptamers functionalized metal and metal oxide nanoparticles: Recent advances in heavy metal monitoring. *TrAC Trends Anal. Chem.* **2022**, *157*, 116748. [CrossRef]
56. Sasaki, K.; Yongvongsoontorn, N.; Tawarada, K.; Ohnishi, Y.; Arakane, T.; Kayama, F.; Abe, K.; Oguma, S.; Ohmura, N. Cadmium purification and quantification using immunochromatography. *J. Agric. Food Chem.* **2009**, *57*, 4514–4519. [CrossRef]
57. Liu, G.L.; Wang, J.F.; Li, Z.Y.; Liang, S.Z.; Wang, X.N. Immunoassay for cadmium detection and quantification. *Biomed. Environ. Sci.* **2009**, *22*, 188–193. [CrossRef]
58. Liu, G.; Wang, J.; Li, Z.; Liang, S.; Liu, S.; Wang, X. Development of direct competitive enzyme-linked immunosorbent assay for the determination cadmium residue in farm produce. *Appl. Biochem. Biotechnol.* **2009**, *159*, 708–717. [CrossRef]
59. Cui, L.; Wu, J.; Ju, H. Electrochemical sensing of heavy metal ions with inorganic, organic and bio-materials. *Biosens. Bioelectron.* **2015**, *63*, 276–286. [CrossRef] [PubMed]
60. Soldatkin, O.O.; Kucherenko, I.S.; Pyeshkova, V.M.; Kukla, A.L.; Jaffrezic-Renault, N.; El'Skaya, A.V.; Dzyadevych, S.V.; Soldatkin, A.P. Novel conductometric biosensor based on three-enzyme system for selective determination of heavy metal ions. *Bioelectrochemistry* **2012**, *83*, 25–30. [CrossRef]
61. Moyo, M.; Okonkwoa, J.O.; Agyei, N.M. An amperometric biosensor based on horseradish peroxidase immobilized onto maize tassel-multiwalled carbon nanotubes modified glassy carbon electrode for determination of heavy metal ions in aqueous solution. *Enzyme Microb. Technol.* **2014**, *56*, 28–34. [CrossRef] [PubMed]
62. Syshchyk, O.; Skryshevsky, V.A.; Soldatkin, O.O.; Soldatkin, A.P. Enzyme biosensor systems based on porous silicon photoluminescence for detection of glucose, urea, and heavy metals. *Biosens. Bioelectron.* **2015**, *66*, 89–94. [CrossRef]
63. Wu, Q.; Bi, H.-M.; Han, X.-J. Research progress of electrochemical detection of heavy metal ions. *Chin. J. Anal. Chem.* **2021**, *49*, 330–340. [CrossRef]
64. Cesarino, I.; Moraes, F.C.; Lanza, M.R.; Machado, S.A. Electrochemical detection of carbamate pesticides in fruit and vegetables with a biosensor based on acetylcholinesterase immobilised on a composite of polyaniline-carbon nanotubes. *Food Chem.* **2012**, *135*, 873–879. [CrossRef]
65. Zhang, Y.; Arugula, M.A.; Wales, M.; Wild, J.; Simonian, A.L. A novel layerby-layer assembled multienzyme/CNT biosensor for discriminative detection between organophosphorus and nonorganophosphrus pesticides. *Biosens. Bioelectron.* **2014**, *67*, 287–295. [CrossRef] [PubMed]
66. Hossain, M.I.; Hasnat, M.A. Recent advancements in non-enzymatic electrochemical sensor development for the detection of organophosphorus pesticides in food and environment. *Heliyon* **2023**, *9*, e19299. [CrossRef]
67. Čadková, M.; Metelka, R.; Holubová, L.; Horák, D.; Dvořáková, V.; Bílková, Z.; Korecká, L. Magnetic beads-based electrochemical immunosensor for monitoring allergenic food proteins. *Anal. Biochem.* **2015**, *484*, 4–8. [CrossRef]
68. Pilolli, R.; Monaci, L. Challenging the limit of detection for egg allergen detection in red wines by surface plasmon resonance biosensor. *Food Anal. Methods* **2016**, *9*, 2754–2761. [CrossRef]
69. de Champdorè, M.; Bazzicalupo, P.; De Napoli, L.; Montesarchio, D.; Di Fabio, G.; Cocozza, I.; Parracino, A.; Rossi, M.; D'Auria, S. A new competitive fluorescence assay for the detection of patulin toxin. *Anal. Chem.* **2007**, *79*, 751–757. [CrossRef] [PubMed]
70. Rejeb, I.B.; Arduini, F.; Amine, A.; Gargouri, M.; Palleschi, G. Development of a bio-electrochemical assay for AFB1 detection in olive oil. *Biosens. Bioelectron.* **2009**, *24*, 1962–1968. [CrossRef]
71. Vidal, J.C.; Bonel, L.; Ezquerra, A.; Duato, P.; Castillo, J.R. An electrochemical immunosensor for ochratoxin A determination in wines based on a monoclonal antibody and paramagnetic microbeads. *Anal. Bioanal. Chem.* **2012**, *403*, 1585–1593. [CrossRef] [PubMed]
72. Alarcon, S.H.; Micheli, L.; Palleschi, G.; Compagnone, D. Development of an electrochemical immunosensor for ochratoxin A. *Anal. Lett.* **2004**, *37*, 1545–1558. [CrossRef]
73. Liu, X.; Yang, Z.; Zhang, Y.; Yu, R. A novel electrochemical immunosensor for ochratoxin A with hapten immobilization on thionine/gold nanoparticle modified glassy carbon electrode. *Anal. Methods* **2013**, *5*, 1481–1486. [CrossRef]
74. Varriale, A.; Staiano, M.; Iozzino, L.; Severino, L.; Anastasio, A.; Cortesi, M.L.; D'Auria, S. FCS-based sensing for the detection of ochratoxin and neomycin in food. *Prot. Pept. Lett.* **2009**, *16*, 1425–1428. [CrossRef]
75. Ron, I.; Bhattacharyya, I.M.; Samanta, S.; Tiwari, V.S.; Greental, D.; Shima-Edelstein, R.; Pikhay, E.; Roizin, Y.; Akabayov, B.; Shalev, G. Label-free and specific detection of active Botulinum neurotoxin in 0.5 μL drops with the meta-nano-channel field-effect biosensor. *Sens. Actuators B Chem.* **2023**, *393*, 134171. [CrossRef]
76. Grabka, M.; Jasek, K.; Witkiewicz, Z. Surface Acoustic Wave Immunosensor for Detection of Botulinum Neurotoxin. *Sensors* **2023**, *23*, 7688. [CrossRef] [PubMed]
77. Wang, H.; Wang, L.; Hu, Q.; Wang, R.; Li, Y.; Kidd, M. Rapid and sensitive detection of *Campylobacter jejuni* in poultry products using a nanoparticle-based piezoelectric immunosensor integrated with magnetic immunoseparation. *J. Food Prot.* **2018**, *81*, 1321–1330. [CrossRef] [PubMed]

78. Masdor, N.A.; Altintas, Z.; Shukor, M.Y.; Tothill, I.E. Subtractive inhibition assay for the detection of *Campylobacter jejuni* in chicken samples using surface plasmon resonance. *Sci. Rep.* **2019**, *9*, 13642. [CrossRef] [PubMed]
79. Kim, H.S.; Kim, Y.J.; Chon, J.W.; Kim, D.H.; Yim, J.H.; Kim, H.; Seo, K.H. Two-stage label-free aptasensing platform for rapid detection of *Cronobacter sakazakii* in powdered infant formula. *Sens. Actuators B Chem.* **2017**, *239*, 94–99. [CrossRef]
80. Shukla, S.; Lee, G.; Song, X.; Park, J.H.; Cho, H.; Lee, E.J.; Kim, M. Detection of Cronobacter sakazakii in powdered infant formula using an immunoliposome-based immunomagnetic concentration and separation assay. *Sci. Rep.* **2016**, *6*, 34721. [CrossRef]
81. Rodriguez-Emmenegger, C.; Avramenko, O.A.; Brynda, E.; Skvor, J.; Alles, A.B. Poly(HEMA) brushes emerging as a new platform for direct detection of food pathogen in milk samples. *Biosens. Bioelectron.* **2011**, *26*, 4545–4551. [CrossRef]
82. Dou, W.; Tang, W.; Zhao, G. A disposable electrochemical immunosensor arrays using 4-channel screen-printed carbon electrode for simultaneous detection of *Escherichia coli* O157:H7 and *Enterobacter sakazakii*. *Electrochim. Acta* **2013**, *97*, 79–85. [CrossRef]
83. Liu, L.; Chao, Y.; Cao, W.; Wang, Y.; Luo, C.; Pang, X.; Fan, D.; Wei, Q. A label-free amperometric immunosensor for detection of zearalenone based on trimetallic Au-core/AgPt-shell nanorattles and mesoporous carbon. *Anal. Chim. Acta* **2014**, *847*, 29–36. [CrossRef] [PubMed]
84. Wang, B.; Wang, Q.; Cai, Z.; Ma, M. Simultaneous, rapid and sensitive detection of three food-borne pathogenic bacteria using multicolor quantum dot probes based on multiplex fluoroimmunoassay in food samples. *LWT Food Sci. Technol.* **2015**, *61*, 368–376. [CrossRef]
85. Zaraee, N.; Bhuiya, A.M.; Gong, E.S.; Geib, M.T.; Ünlü, N.L.; Ozkumur, A.Y.; Ünlü, M.S. Highly Sensitive and Label-free Digital Detection of Whole Cell E. coli with Interferometric Reflectance Imaging. *arXiv* **2019**, arXiv:1911.06950.
86. Hao, N.; Zhang, X.; Zhou, Z.; Hua, R.; Zhang, Y.; Liu, Q.; Qian, J.; Henan, L.; Wang, K. AgBr nanoparticles/3D nitrogen-doped graphene hydrogel for fabricating all-solid-state luminol-electrochemiluminescence *Escherichia coli* aptasensors. *Biosens. Bioelectron.* **2017**, *97*, 377–383. [CrossRef]
87. Shang, Q.; Su, Y.; Liang, Y.; Lai, W.; Jiang, J.; Wu, H.; Zhang, C. Ultrasensitive cloth-based microfluidic chemiluminescence detection of *Listeria monocytogenes* hlyA gene by hemin/G-quadruplex DNAzyme and hybridization chain reaction signal amplification. *Anal. Bioanal. Chem.* **2020**, *412*, 3787–3797. [CrossRef] [PubMed]
88. Baskaran, N.; Sakthivel, R.; Karthik, C.S.; Lin, Y.-C.; Liu, X.; Wen, H.-W.; Yang, W.; Chung, R.-J. Polydopamine-modified 3D flower-like ZnMoO4 integrated MXene-based label-free electrochemical immunosensor for the food-borne pathogen *Listeria monocytogenes* detection in milk and seafood. *Talanta* **2024**, *282*, 127008. [CrossRef]
89. Cheng, C.; Peng, Y.; Bai, J.; Zhang, X.; Liu, Y.; Fan, X.; Ning, B.; Gao, Z. Rapid detection of *Listeria monocytogenes* in milk by self-assembled electrochemical immunosensor. *Sens. Actuators B Chem.* **2014**, *190*, 900–906. [CrossRef]
90. Liu, H.; Zhou, X. Paper-based bipolar electrode electrochemiluminescence (pBPE-ECL) analysis system for sensitive detection of pathogenic bacteria. *Anal. Chem.* **2016**, *88*, 10191–10197. [CrossRef]
91. Ren, J.; He, F.; Yi, S.; Cui, X. A new MSPQC for rapid growth and detection of *Mycobacterium tuberculosis*. *Biosens. Bioelectron.* **2008**, *24*, 403–409. [CrossRef] [PubMed]
92. He, F.; Xiong, Y.; Liu, J.; Tong, F.; Yan, D. Construction of Au-IDE/CFP10-ESAT6 aptamer/DNA-AuNPsMSPQC for rapid detection of *Mycobacterium tuberculosis*. *Biosens. Bioelectron.* **2016**, *77*, 799–804. [CrossRef]
93. Mudgal, N.; Yupapin, P.; Ali, J.; Singh, G. $BaTiO_3$-Graphene-Affinity Layer–Based Surface Plasmon Resonance (SPR) Biosensor for Pseudomonas Bacterial Detection. *Plasmonics* **2020**, *15*, 1221–1229. [CrossRef]
94. Zhang, P.; Chen, Y.P.; Wang, W.; Shen, Y.; Guo, J.S. Surface plasmon resonance for water pollutant detection and water process analysis. *TrAC Trends Anal. Chem.* **2016**, *85*, 153–165. [CrossRef]
95. Kim, G.; Moon, J.; Moh, C.; Lim, J. A microfuidic nano-biosensor for the detection of pathogenic *Salmonella*. *Biosens. Bioelectron.* **2014**, *67*, 243–247. [CrossRef]
96. Duan, N.; Wu, S.; Ma, X.; Xia, Y.; Wang, Z. A universal fuorescent aptasensor based on AccuBlue dye for the detection of pathogenic bacteria. *Anal. Biochem.* **2014**, *454*, 1–6. [CrossRef] [PubMed]
97. Oh, S.Y.; Heo, N.S.; Shukla, S.; Cho, H.J.; Vilian, A.E.; Kim, J.; Huh, Y.S. Development of gold nanoparticle-aptamer-based LSPR sensing chips for the rapid detection of *Salmonella* Typhimurium in pork meat. *Sci. Rep.* **2017**, *7*, 10130. [CrossRef]
98. Sheikhzadeh, E.; CHamsaz, M.; Turner, A.P.F.; Jager, E.W.H.; Beni, V. Label-free impedimetric biosensor for *Salmonella* Typhimurium detection based on poly [pyrrole-co-3-carboxyl-pyrrole] copolymer supported aptamer. *Biosens. Bioelectron.* **2016**, *80*, 194–200. [CrossRef] [PubMed]
99. Bagheryan, Z.; Raoof, J.B.; Golabi, M.; Turner, A.P.; Beni, V. Diazonium-based impedimetric aptasensor for the rapid label-free detection of *Salmonella* Typhimurium in food sample. *Biosens. Bioelectron.* **2016**, *80*, 566–573. [CrossRef]
100. Ozalp, V.C.; Bayramoglu, G.; Erdem, Z.; Arica, M.Y. Pathogen detection in complex samples by quartz crystal microbalance sensor coupled to aptamer functionalized core–shell type magnetic separation. *Anal. Chim. Acta* **2015**, *853*, 533–540. [CrossRef] [PubMed]
101. Farka, Z.; Juřík, T.; Pastucha, M.; Skládal, P. Enzymatic precipitation enhanced surface plasmon resonance immunosensor for the detection of *Salmonella* in powdered milk. *Anal. Chem.* **2016**, *88*, 11830–11836. [CrossRef]

102. Zelada-Guillén, G.A.; Sebastián-Avila, J.L.; Blondeau, P.; Riu, J.; Rius, F.X. Label-free detection of *Staphylococcus aureus* in skin using real-time potentiometric biosensors based on carbon nanotubes and aptamers. *Biosens. Bioelectron.* **2012**, *31*, 226–232. [CrossRef]
103. Arora, S.; Ahmed, D.N.; Khubber, S.; Siddiqui, S. Detecting food borne pathogens using electrochemical biosensors: An overview. *IJCS* **2018**, *6*, 1031–1039.
104. Pohanka, M. QCM immunosensor for the determination of *Staphylococcus aureus* antigen. *Chem. Pap.* **2020**, *74*, 451–458. [CrossRef]
105. Noi, K.; Iijima, M.; Kuroda, S.I.; Ogi, H. Ultrahigh-sensitive wireless QCM with bio-nanocapsules. *Sens. Actuators B Chem.* **2019**, *293*, 59–62. [CrossRef]
106. Vásquez, G.; Rey, A.; Rivera, C.; Iregui, C.; Orozco, J. Amperometric biosensor based on a single antibody of dual function for rapid detection of *Streptococcus agalactiae*. *Biosens. Bioelectron.* **2017**, *87*, 453–458. [CrossRef] [PubMed]
107. Arachchillaya, B.P.A.P. *Development and Evaluation of a Paper Based Biochemical Sensor for Realtime Detection of Food Pathogen*; Bachelor Project; Asian Institute of Technology: Khlong Luang, Thailand, 2018.
108. Jiang, H.; Sun, Z.; Guo, Q.; Weng, X. Microfluidic Thread-Based Electrochemical Aptasensor for Rapid Detection of *Vibrio parahaemolyticus*. *Biosens. Bioelectron.* **2021**, *182*, 113191. [CrossRef] [PubMed]
109. Jin, X.; Gong, L.; Liang, J.; Wang, Z.; Wang, K.; Yang, T.; Zeng, H. Polydopamine-Enhanced Vertically-Ordered Mesoporous Silica Film Anti-Fouling Electrochemical Aptasensor for Indicator-Free Vibrio parahaemolyticus Discrimination Using Stable Inherent Au Signal. *Sens. Actuators B Chem.* **2024**, *407*, 135485. [CrossRef]
110. Tian, L.; Li, Y.; Wang, H.; Li, X.; Gao, Q.; Liu, Y.; Liu, Y.; Wang, Q.; Ma, C.; Shi, C. A pH Ultra-Sensitive Hydrated Iridium Oxyhydroxide Films Electrochemical Sensor for Label-Free Detection of *Vibrio parahaemolyticus*. *Anal. Biochem.* **2024**, *693*, 115597. [CrossRef] [PubMed]
111. Sha, Y.; Zhang, X.; Li, W.; Wu, W.; Wang, S.; Guo, Z.; Zhou, J.; Su, X. A label-free multi-functionalized graphene oxide based electrochemiluminscence immunosensor for ultrasensitive and rapid detection of *Vibrio parahaemolyticus* in seawater and seafood. *Talanta* **2016**, *147*, 220–225. [CrossRef]
112. Li, J.; Lin, X.; Wu, J.; Ying, D.; Duan, N.; Wang, Z.; Wu, S. Multifunctional Magnetic Composite Nanomaterial for Colorimetric-SERS Dual-Mode Detection and Photothermal Sterilization of Vibrio parahaemolyticus. *Chem. Eng. J.* **2023**, *477*, 147113. [CrossRef]
113. Xu, C.; Xie, J.; Yu, L.; Shu, B.; Liu, X.; Chen, S.; Li, Q.; Qi, S.; Zhao, S. Sensitive Colorimetric Detection of Vibrio vulnificus Based on Target-Induced Shielding against the Peroxidase-Mimicking Activity of CeO2@PtRu Nanozyme. *Food Chem.* **2024**, *454*, 139757. [CrossRef] [PubMed]
114. Baek, S.H.; Kim, M.W.; Park, C.Y.; Choi, C.S.; Kailasa, S.K.; Park, J.P.; Park, T.J. Development of a rapid and sensitive electrochemical biosensor for detection of human norovirus via novel specific binding peptides. *Biosens. Bioelectron.* **2019**, *123*, 223–229. [CrossRef] [PubMed]
115. Zhou, Q.; Tang, D. Recent advances in photoelectrochemical biosensors for analysis of mycotoxins in food. *TrAC Trends Anal. Chem.* **2020**, *124*, 115814. [CrossRef]
116. Oladoye, P.O.; Olowe, O.M.; Asemoloye, M.D. Phytoremediation technology and food security impacts of heavy metal contaminated soils: A review of literature. *Chemosphere* **2022**, *288*, 132555. [CrossRef] [PubMed]
117. Parker, G.H.; Gillie, C.E.; Miller, J.V.; Badger, D.E.; Kreider, M.L. Human health risk assessment of arsenic, cadmium, lead, and mercury ingestion from baby foods. *Toxicol. Rep.* **2022**, *9*, 238–249. [CrossRef]
118. Chailapakul, O.; Korsrisakul, S.; Siangproh, W.; Grudpan, K. Fast and simultaneous detection of heavy metals using a simple and reliable microchip-electrochemistry route: An alternative approach to food analysis. *Talanta* **2008**, *74*, 683–689. [CrossRef] [PubMed]
119. Yuan, M.; Qian, S.; Cao, H.; Yu, J.; Ye, T.; Wu, X.; Chen, L.; Xu, F. An ultra-sensitive electrochemical aptasensor for simultaneous quantitative detection of Pb2+ and Cd2+ in fruit and vegetable. *Food Chem.* **2022**, *382*, 132173. [CrossRef] [PubMed]
120. Wang, X.; Liu, M.; Wang, X.; Wu, Z.; Yang, L.; Xia, S.; Chen, L.; Zhao, J. p-Benzoquinone-mediated amperometric biosensor developed with *Psychrobacter* sp. for toxicity testing of heavy metals. *Biosens. Bioelectron.* **2013**, *41*, 557–562. [CrossRef]
121. Gammoudi, I.; Raimbault, V.; Tarbague, H.; Morote, F.; Grauby-Heywang, C.; Othmane, A.; Kalfat, R.; Moynet, D.; Rebiere, D.; Dejous, C.; et al. Enhanced bio-inspired microsensor based on microfluidic/ bacterial/love wave hybrid structure for continuous control of heavy metals toxicity in liquid medium. *Sens. Actuators B Chem.* **2014**, *198*, 278–284. [CrossRef]
122. Ghica, M.E.; Carvalho, R.C.; Amine, A.; Brett, C.M.A. Glucose oxidase enzyme inhibition sensors for heavy metals at carbon film electrodes modified with cobalt and copper hexacyanoferrate. *Sens. Actuators B Chem.* **2013**, *178*, 270–278. [CrossRef]
123. Magar, H.S.; Ghica, M.E.; Abbas, M.N.; Brett, C.M. Highly sensitive choline oxidase enzyme inhibition biosensor for lead ions based on multiwalled carbon nanotube modified glassy carbon electrodes. *Electroanalysis* **2017**, *29*, 1741–1748. [CrossRef]
124. Wang, N.; Lin, M.; Dai, H.; Ma, H. Functionalized gold nanoparticles/reduced graphene oxide nanocomposites for ultrasensitive electrochemical sensing of mercury ions based on thymine–mercury–thymine structure. *Biosens. Bioelectron.* **2016**, *79*, 320–326. [CrossRef] [PubMed]

125. Dai, X.; Wu, S.; Li, S. Progress on electrochemical sensors for the determination of heavy metal ions from contaminated water. *J. Chin. Adv. Mater. Soc.* **2018**, *6*, 91–111. [CrossRef]
126. Tao, H.C.; Peng, Z.W.; Li, P.S.; Yu, T.A.; Su, J. Optimizing cadmium and mercury specificity of CadRbased E-coli biosensors by redesign of CadR. *Biotechnol. Lett.* **2013**, *35*, 1253–1258. [CrossRef] [PubMed]
127. Amaro, F.; Turkewitz, A.P.; Martin-Gonzalez, A.; Gutierrez, J.C. Functional GFP metallothionein fusion protein from *Tetrahymena thermophila*: A potential whole-cell biosensor for monitoring heavy metal pollution and a cell model to study metallothionein overproduction effects. *Biometals* **2014**, *27*, 195–205. [CrossRef] [PubMed]
128. Kim, M.; Lim, J.W.; Kim, H.J.; Lee, S.K.; Lee, S.J.; Kim, T. Chemostat-like microfluidic platform for highly sensitive detection of heavy metal ions using microbial biosensors. *Biosens. Bioelectron.* **2015**, *65*, 257–264. [CrossRef] [PubMed]
129. Long, F.; Zhu, A.; Shi, H.; Wang, H.; Liu, J. Rapid on-site/in-situ detection of heavy metal ions in environmental water using a structure-switching DNA optical biosensor. *Sci. Rep.* **2013**, *3*, 2308. [CrossRef] [PubMed]
130. Zhou, X.; Pu, H.; Sun, D.W. DNA functionalized metal and metal oxide nanoparticles: Principles and recent advances in food safety detection. *Crit. Rev. Food Sci. Nutr.* **2021**, *61*, 2277–2296. [CrossRef] [PubMed]
131. Lake, R.J.; Yang, Z.; Zhang, J.; Lu, Y. DNAzymes as activity-based sensors for metal ions: Recent applications, demonstrated advantages, current challenges, and future directions. *Acc. Chem. Res.* **2019**, *52*, 3275–3286. [CrossRef]
132. Tang, S.; Tong, P.; Li, H.; Tang, J.; Zhang, L. Ultrasensitive electrochemical detection of Pb^{2+} based on rolling circle amplification and quantum dots tagging. *Biosens Bioelectron.* **2013**, *42*, 608–611. [CrossRef] [PubMed]
133. Miao, P.; Tang, Y.; Wang, L. DNA modified Fe3O4@ Au magnetic nanoparticles as selective probes for simultaneous detection of heavy metal ions. *ACS Appl. Mater. Interfaces* **2017**, *9*, 3940–3947. [CrossRef]
134. Wen, S.-H.; Wang, Y.; Yuan, Y.-H.; Liang, R.-P.; Qiu, J.-D. Electrochemical sensor for arsenite detection using graphene oxide assisted generation of prussian blue nanoparticles as enhanced signal label. *Anal. Chim. Acta* **2018**, *1002*, 82–89. [CrossRef]
135. Shi, J.-J.; Zhu, J.-C.; Zhao, M.; Wang, Y.; Yang, P.; He, J. Ultrasensitive photoelectrochemical aptasensor for lead ion detection based on sensitization effect of CdTe QDs on MoS2-CdS: Mn nanocomposites by the formation of G-quadruplex structure. *Talanta* **2018**, *183*, 237–244. [CrossRef]
136. Lee, C.-S.; Yu, S.H.; Kim, T.H. A "turn-on" electrochemical aptasensor for ultrasensitive detection of Cd2+ using duplexed aptamer switch on electrochemically reduced graphene oxide electrode. *Microchem. J.* **2020**, *159*, 105372. [CrossRef]
137. Gumpu, M.B.; Krishnan, U.M.; Rayappan, J.B.B. Design and development of amperometric biosensor for the detection of lead and mercury ions in water matrix—A permeability approach. *Anal. Bioanal. Chem.* **2017**, *409*, 4257–4266. [CrossRef] [PubMed]
138. European Food Safety Authority. Scientific Topic: Pesticides | European Food Safety Authority. Available online: https://www.efsa.europa.eu/en/topics/topic/pesticides (accessed on 2 January 2025).
139. Yadav, I.C.; Devi, N.L. Pesticides Classification and Its Impact on Environment. *Environ. Eng. Sci.* **2017**, *6*, 140–158.
140. Hu, H.; Yang, L. Development of enzymatic electrochemical biosensors for organophosphorus pesticide detection. *J. Environ. Sci. Health Part B* **2021**, *56*, 168–180. [CrossRef]
141. Tun, W.S.T.; Saenchoopa, A.; Daduang, S.; Daduang, J.; Kulchat, S.; Patramanon, R. Electrochemical biosensor based on cellulose nanofibers/graphene oxide and acetylcholinesterase for the detection of chlorpyrifos pesticide in water and fruit juice. *RSC Adv.* **2023**, *13*, 9603–9614. [CrossRef]
142. Guerrero-Esteban, T.; Gutiérrez-Sánchez, C.; Martínez-Periñán, E.; Revenga-Parra, M.; Pariente, F.; Lorenzo, E. Sensitive glyphosate electrochemiluminescence immunosensor based on electrografted carbon nanodots. *Sens. Actuators B Chem.* **2021**, *330*, 129389. [CrossRef]
143. Ba Hashwan, S.S.; Khir, M.H.B.M.; Al-Douri, Y.; Ahmed, A.Y. Recent progress in the development of biosensors for chemicals and pesticides detection. *IEEE Access* **2020**, *8*, 82514–82527. [CrossRef]
144. Bucur, B.; Munteanu, F.-D.; Marty, J.-L.; Vasilescu, A. Advances in enzyme-based biosensors for pesticide detection. *Biosensors* **2018**, *8*, 27. [CrossRef]
145. Mirres, A.C.d.M.; Silva, B.E.P.d.M.d.; Tessaro, L.; Galvan, D.; de Andrade, J.C.; Aquino, A.; Joshi, N.; Conte-Junior, C.A. Recent advances in nanomaterial-based biosensors for pesticide detection in foods. *Biosensors* **2022**, *12*, 572. [CrossRef] [PubMed]
146. Tsounidi, D.; Soulis, D.; Manoli, F.; Klinakis, A.; Tsekenis, G. AChE-based electrochemical biosensor for pesticide detection in vegetable oils: Matrix effects and synergistic inhibition of the immobilized enzyme. *Anal. Bioanal. Chem.* **2023**, *415*, 615–625. [CrossRef] [PubMed]
147. Surribas, A.; Barthelmebs, L.; Noguer, T. Monoclonal antibody-based immunosensor for the electrochemical detection of chlortoluron herbicide in groundwaters. *Biosensors* **2021**, *11*, 513. [CrossRef]
148. Liu, B.; Tang, Y.; Yang, Y.; Wu, Y. Design an aptamer-based sensitive lateral flow biosensor for rapid determination of isocarbophos pesticide in foods. *Food Control* **2021**, *129*, 108208. [CrossRef]
149. Taghizadeh-Behbahani, M.; Shamsipur, M.; Hemmateenejad, B. Detection and discrimination of antibiotics in food samples using a microfluidic paper-based optical tongue. *Talanta* **2022**, *241*, 123242. [CrossRef]

150. Li, H.; Huang, X.; Huang, J.; Bai, M.; Hu, M.; Guo, Y.; Sun, X. Fluorescence assay for detecting four organophosphorus pesticides using fluorescently labeled aptamer. *Sensors* **2022**, *22*, 5712. [CrossRef]
151. Dong, J.; Yang, H.; Li, Y.; Liu, A.; Wei, W.; Liu, S. Fluorescence sensor for organophosphorus pesticide detection based on the alkaline phosphatase-triggered reaction. *Anal. Chim. Acta* **2020**, *1131*, 102–108. [CrossRef] [PubMed]
152. Poudyal, D.C.; Dhamu, V.N.; Samson, M.; Muthukumar, S.; Prasad, S. Portable pesticide electrochem-sensor: A label-free detection of glyphosate in human urine. *Langmuir* **2022**, *38*, 1781–1790. [CrossRef]
153. Chen, C.; Zhou, J.; Li, Z.; Xu, Y.; Ran, T.; Gen, J. Wearable electrochemical biosensors for in situ pesticide analysis from crops. *J. Electrochem. Soc.* **2023**, *170*, 117512. [CrossRef]
154. Dhamu, V.N.; Poudyal, D.C.; Muthukumar, S.; Prasad, S. A highly sensitive electrochemical sensor system to detect and distinguis. *J. Electrochem. Soc.* **2021**, *168*, 057531. [CrossRef]
155. Verma, N.; Bhardwaj, A. Biosensor Technology for Pesticides—A review. *Appl. Biochem. Biotechnol.* **2015**, *175*, 3093–3119. [CrossRef]
156. Marrazza, G. Piezoelectric biosensors for organophosphate and carbamate pesticides: A review. *Biosensors* **2014**, *4*, 301–317. [CrossRef] [PubMed]
157. Arduini, F.; Cinti, S.; Caratelli, V.; Amendola, L.; Palleschi, G.; Moscone, D. Origami Multiple Paper-Based Electrochemical Biosensors for Pesticide Detection. *Biosens. Bioelectron.* **2019**, *126*, 346–354. [CrossRef]
158. Pérez-Fernández, B.; Costa-García, A.; Muñiz, A.d.l.E. Electrochemical (Bio)Sensors for Pesticides Detection Using Screen-Printed Electrodes. *Biosensors* **2020**, *10*, 32. [CrossRef]
159. Tran, H.; Yougnia, R.; Reisberg, S.; Piro, B.; Serradji, N.; Nguyen, T.; Tran, L.; Dong, C.; Pham, M. A label-free electrochemical immunosensor for direct, signal-on and sensitive pesticide detection. *Biosens. Bioelectron.* **2012**, *31*, 62–68. [CrossRef]
160. Liu, R.; Guan, G.; Wang, S.; Zhang, Z. Core-shell nanostructured molecular imprinting fluorescent chemosensor for selective detection of atrazine herbicide. *Analyst* **2011**, *136*, 184–190. [CrossRef] [PubMed]
161. Boro, R.C.; Kaushal, J.; Nangia, Y.; Wangoo, N.; Bhasin, A.; Suri, C.R. Gold nanoparticles catalyzed chemiluminescence immunoassay for detection of herbicide 2,4-dichlorophenoxyacetic acid. *Analyst* **2011**, *136*, 2125–2130. [CrossRef]
162. Shakhih, M.F.M.; Rosslan, A.S.; Noor, A.M. Review-enzymatic and non-enzymatic electrochemical sensor for lactate detection in human. *Biofluids J. Electrochem. Soc.* **2021**, *168*, 067502. [CrossRef]
163. Hassan, M.H.; Vyas, C.; Grieve, B. Recent advances in enzymatic and non-enzymatic electrochemical glucose sensing. *Sensors* **2021**, *21*, 4672. [CrossRef] [PubMed]
164. Sanati, A.; Jalali, M.; Raeissi, K. A review on recent advancements in electrochemical biosensing using carbonaceous nanomaterials. *Mikrochim. Acta* **2019**, *186*, 773. [CrossRef] [PubMed]
165. Wang, P.; Li, H.; Hassan, M.M. Fabricating an acetylcholinesterase modulated UCNPs-Cu^{2+} fluorescence biosensor for ultrasensitive detection of organophosphorus pesticides-diazinon in food. *J. Agric. Food Chem.* **2019**, *67*, 4071–4079. [CrossRef]
166. Itsoponpan, T.; Thanachayanont, C.; Hasin, P. Sponge-like CuInS2 microspheres on reduced graphene oxide as an electrocatalyst to construct an immobilized acetylcholinesterase electrochemical biosensor for chlorpyrifos detection in vegetables. *Sens. Actuators B Chem.* **2021**, *337*, 129775. [CrossRef]
167. Wang, X.; Lu, X.; Chen, J. Development of Biosensor Technologies for Analysis of Environmental Contaminants. *Trends Environ. Anal. Chem.* **2014**, *2*, 25–32. [CrossRef]
168. Liu, Z.; Xia, X.; Zhou, G.; Ge, L.; Li, F. Acetylcholinesterase-Catalyzed Silver Deposition for Ultrasensitive Electrochemical Biosensing of Organophosphorus Pesticides. *Analyst* **2020**, *145*, 2339–2344. [CrossRef] [PubMed]
169. Yao, Y.; Wang, G.; Chu, G.; An, X.; Guo, Y.; Sun, X. The Development of a Novel Biosensor Based on Gold Nanocages/Graphene Oxide-Chitosan Modified Acetylcholinesterase for Organophosphorus Pesticide Detection. *New J. Chem.* **2019**, *43*, 13816–13826. [CrossRef]
170. Aghoutane, Y.; Bari, N.E.; Laghrari, Z. Electrochemical detection of fenthion insecticide in olive oils by a sensitive non-enzymatic biomimetic sensor enhanced with Metal Nanoparticles. *Chem. Process.* **2021**, *5*, 64.
171. Silva, L.M.C.; Melo, A.F.; Salgado, A. Biosensors for environmental applications. In *Environmental Biosensors*; Somerset, V., Ed.; InTech: London, UK, 2011; ISBN ISBN 978-9-53307-486-3.
172. Hashemi Goradel, N.; Mirzaei, H.; Sahebkar, A.; Poursadeghiyan, M.; Masoudifar, A.; Malekshahi, Z.V.; Negahdari, B. Biosensors for the Detection of Environmental and Urban Pollutions. *J. Cell. Biochem.* **2018**, *119*, 207–212. [CrossRef] [PubMed]
173. Hondred, J.A.; Breger, J.C.; Alves, N.J.; Trammell, S.A.; Walper, S.A.; Medintz, I.L.; Claussen, J.C. Printed Graphene Electrochemical Biosensors Fabricated by Inkjet Maskless Lithography for Rapid and Sensitive Detection of Organophosphates. *ACS Appl. Mater. Interfaces* **2018**, *10*, 11125–11134. [CrossRef]
174. Borah, H.; Dutta, R.R.; Gogoi, S.; Medhi, T.; Puzari, P. Glutathione-S-Transferase-Catalyzed Reaction of Glutathione for Electrochemical Biosensing of Temephos, Fenobucarb and Dimethoate. *Anal. Methods* **2017**, *9*, 4044–4051. [CrossRef]
175. Borah, H.; Gogoi, S.; Kalita, S.; Puzari, P. A Broad Spectrum Amperometric Pesticide Biosensor Based on Glutathione S-Transferase Immobilized on Graphene Oxide-Gelatin Matrix. *J. Electroanal. Chem.* **2018**, *828*, 116–123. [CrossRef]

176. Prabhakaran, D.C.; Riotte, J.; Sivry, Y.; Subramanian, S. Electroanalytical Detection of Cr(VI) and Cr(III) Ions Using a Novel Microbial Sensor. *Electroanalysis* **2017**, *29*, 1222–1231. [CrossRef]
177. Pabbi, M.; Mittal, S.K. An Electrochemical Algal Biosensor Based on Silica Coated ZnO Quantum Dots for Selective Determination of Acephate. *Anal. Methods* **2017**, *9*, 1672–1680. [CrossRef]
178. Dasriya, V.; Joshi, R.; Ranveer, S.; Dhundale, V.; Kumar, N.; Raghu, H.V. Rapid detection of pesticide in milk, cereal and cereal based food and fruit juices using paper strip-based sensor. *Sci. Rep.* **2021**, *11*, 18855. [CrossRef] [PubMed]
179. Zamora-Sequeira, R.; Starbird-Pérez, R.; Rojas-Carillo, O.; Vargas-Villalobos, S. What are the main sensor methods for quantifying pesticides in agricultural activities? A review. *Molecules* **2019**, *24*, 2659. [CrossRef] [PubMed]
180. Bucur, B.; Purcarea, C.; Andreescu, S.; Vasilescu, A. Addressing the selectivity of enzyme biosensors: Solutions and perspectives. *Sensors* **2021**, *21*, 3038. [CrossRef]
181. NDA. Opinion of the scientific panel on dietetic products, nutrition and allergies (NDA) on a request from the Commission relating to the evaluation of allergenic foods for labelling purposes. *EFSA J.* **2004**, *32*, 1–197.
182. Wang, J.; Sampson, H.A. Treatments for food allergy: How close are we? *Immunol. Res.* **2012**, *54*, 83–94. [CrossRef]
183. Turner, P.J.; Bognanni, A.; Arasi, S.; Ansotegui, I.J.; Schnadt, S.; La Vieille, S.; O'B. Hourihane, J.; Zuberbier, T.; Eigenmann, P.; Ebisawa, M.; et al. Time to ACT-UP: Update on precautionary allergen labelling (PAL). *World Allergy Organ. J.* **2024**, *17*, 100972. [CrossRef] [PubMed]
184. Codex Alimentarius General Standard for Labelling of Pre-Packaged Foods (CODEX STAN 1-1985). 2010, pub WHO/FAO Rome. Available online: https://www.fao.org/fao-who-codexalimentarius/sh-proxy/es/?lnk=1&url=https%253A%252F%252Fworkspace.fao.org%252Fsites%252Fcodex%252FStandards%252FCXS+1-1985%252FCXS_001e.pdf (accessed on 20 January 2025).
185. European Commission. Commission Notice of 13 July 2017 Relating to the provision of information on substances or products causing allergies or intolerances as listed in Annex II to Regulation (EU) No 1169/2011 of the European Parliament and of the Council on the provision of food information to consumers. *Off. J. Eur. Union* **2017**, 5.C428/01. Available online: https://eur-lex.europa.eu/legal-content/EN/TXT/PDF/?uri=CELEX:52017XC1213(01) (accessed on 20 January 2025).
186. FALCPA. Food Allergen Labeling and Consumer Protection Act of 2004 (PublicLaw 108-282, Title II). 21 USC 301. Available online: https://public4.pagefreezer.com/browse/FDA/23-11-2021T07:28/https://www.fda.gov/food/allergens/food-allergen-labeling-and-consumer-protection-act-2004-falcpa (accessed on 20 January 2025).
187. US Congress. S.578—FASTER Act of 2021. Available online: https://www.congress.gov/bill/117th-congress/senate-bill/578 (accessed on 20 January 2025).
188. FDA Issues Guidances on Food Allergen Labeling Requirements. 2022. Available online: https://www.fda.gov/food/hfp-constituent-updates/fda-issues-guidances-food-allergen-labeling-requirements (accessed on 20 January 2025).
189. Food Allergen Labelling, Government of Canada. 13 April 2022. Available online: https://www.canada.ca/en/health-canada/services/food-nutrition/food-labelling/allergen-labelling.html (accessed on 20 January 2025).
190. FSANZ Australia New Zealand Food Standards Code Pub Commonwealth of Australia. 2023. Available online: http://www.foodstandards.gov.au/code/Pages/default.aspx (accessed on 20 January 2025).
191. Food Sanitation Act (Act No. 233, 24 February 1947). Available online: https://www.cas.go.jp/jp/seisaku/hourei/data/fsa.pdf (accessed on 20 January 2025).
192. MFDS Korea Food & Drug Administration: Foods Labeling System Pub Ministry of Food and Drug Safety, Seoul. 2003. Available online: https://www.mfds.go.kr/eng/wpge/m_14/de0110051001.do (accessed on 20 January 2025).
193. Tammineedi, C.V.; Choudhary, R. Recent advances in processing for reducing dairy and food allergenicity. *Int. J. Food Sci. Nutr. Eng.* **2014**, *4*, 36–42.
194. Nwaru, B.I.; Hickstein, L.; Panesar, S.S.; Roberts, G.; Muraro, A.; Sheikh, A. Prevalence of common food allergies in Europe: A systematic review and meta-analysis. *Allergy* **2014**, *69*, 992–1007. [CrossRef] [PubMed]
195. Radauer, C.; Bublin, M.; Wagner, S.; Mari, A.; Breiteneder, H. Allergens are distributed into few protein families and possess a restricted number of biochemical functions. *J. Allergy Clin. Immunol.* **2008**, *121*, 847–852. [CrossRef]
196. Hosu, O.; Selvolini, G.; Marrazza, G. Recent advances of immunosensors for detecting food allergens *Curr. Opin. Electrochem.* **2018**, *10*, 149–156. [CrossRef]
197. Boye, J.; Danquah, A.; Lam, C.; Thang Zhao, X. Food Allergens. In *Food Biochemistry and Food Processing*; John Wiley & Sons, Inc.: Hoboken, NJ, USA; Wiley-Backell: Hoboken, NJ, USA, 2012; pp. 798–819.
198. Zhang, M.; Wu, P.; Wu, J.; Ping, J.; Wu, J. Advanced DNA-based methods for the detection of peanut allergens in processed food. *TrAC Trends Anal. Chem.* **2019**, *114*, 278–292. [CrossRef]
199. Khanmohammadi, V.; Aghaie, G.; Qazvini, H.; Afkhami, B. Electrochemical biosensors for the detection of lung cancer biomarkers: A review. *Talanta* **2020**, *206*, 120251. [CrossRef] [PubMed]
200. Khedri, R.; Rafatpanah, A. Detection of food-born allergens with aptamer-based biosensors. *TRAC Trends Anal. Chem.* **2018**, *103*, 126–136. [CrossRef]

201. Gupta, B.; Raza, V.; Kim, B. Advances in nanomaterial-based electrochemical biosensors for the detection of microbial toxins, pathogenic bacteria in food matrices. *J. Hazard. Mater.* **2021**, *401*, 123379. [CrossRef] [PubMed]
202. Aydin, A.; Sezginturk, A. Advances in immunosensor technology. *Adv. Clin. Chem.* **2021**, *102*, 1–62.
203. Chinnappan, R.; AlZabn, K.; Lopata, A.-S.; Zourob, A. Aptameric biosensor for the sensitive detection of major shrimp allergen, tropomyosin. *Food Chem.* **2020**, *314*, 126133. [CrossRef] [PubMed]
204. Fang, L.; Jia, S.; Kang, Z. Recent progress in immunosensors for pesticides. *Biosens. Bioelectron* **2020**, *164*, 112255. [CrossRef] [PubMed]
205. Figueroa-Miranda, W.; Zhang, N.; Lo, T.; Elling, O. Mayer Polyethylene glycol-mediated blocking and monolayer morphology of an electrochemical aptasensor for malaria biomarker detection in human serum. *Bioelectrochemistry* **2020**, *136*, 107589. [CrossRef]
206. Zhang, Z.; Liu, Y.; Cao, X. Highly sensitive sandwich electrochemical sensor based on DNA-scaffolded bivalent split aptamer signal probe. *Sens. Actuators B Chem.* **2020**, *311*, 127920. [CrossRef]
207. Sundhoro, M.; Agnihotra, S.R.; Amberger, B.; Augustus, K.; Khan, N.D.; Barnes, A.; BelBruno, J.; Mendecki, L. An electrochemical molecularly imprinted polymer sensor for rapid and selective food allergen detection. *Food Chem.* **2021**, *344*, 128648. [CrossRef]
208. Freitas, M.; Neves, M.M.P.S.; Nouws, H.P.A.; Delerue-Matos, C. Electrochemical Immunosensor for the Simultaneous Determination of Two Main Peanut Allergenic Proteins (Ara h 1 and Ara h 6) in Food Matrices. *Foods* **2021**, *10*, 1718. [CrossRef] [PubMed]
209. Freire, F.D.C.O.; da Rocha, M.E. Impact of mycotoxins on human health. *Fungal Metab.* **2017**, *1*, 239–261. [CrossRef]
210. Hussein, H.S.; Brasel, J.M. Toxicity, metabolism, and impact of mycotoxins on humans and animals. *Toxicology* **2001**, *167*, 101–134. [CrossRef]
211. Nabok, A.; Al-Rubaye, A.; Al-Jawdah, A.; Tsargorodska, A.; Marty, J.-L.; Catanante, G.; Szekacs, A.; Takacs, E. Novel optical biosensing technologies for detection of mycotoxins. *Opt. Laser Technol.* **2019**, *109*, 212–221. [CrossRef]
212. Byrne, B.; Stack, E.; Gilmartin, N.; O'Kennedy, R. Antibody-based sensors: Principles, problems and potential for detection of pathogens and associated toxins. *Sensors* **2009**, *9*, 4407–4445. [CrossRef]
213. Krska, R.; Schubert-Ullrich, P.; Molinelli, A.; Sulyok, M.; MacDonald, S.; Crews, C. Mycotoxin analysis: An update. *Food Addit. Contam. Part A* **2008**, *25*, 152–163. [CrossRef]
214. Zhang, L.; Dou, X.W.; Zhang, C.; Logrieco, A.F.; Yang, M.H. A review of current methods for analysis of mycotoxins in herbal medicines. *Toxins* **2018**, *10*, 65. [CrossRef] [PubMed]
215. Yousefi, S.; Saraji, M. Optical aptasensor based on silver nanoparticles for the colorimetric detection of adenosine. *Spectrochim. Acta A Mol. Biomol. Spectrosc.* **2019**, *213*, 1–5. [CrossRef] [PubMed]
216. Dey, D.; Goswami, T. Optical biosensors: A revolution towards quantum nanoscale electronics device fabrication. *BioMed Res. Int.* **2011**, *2011*, 348218. [CrossRef] [PubMed]
217. Li, M.; Chen, L.; Zhang, W.; Chou, S.Y. Pattern transfer fidelity of nanoimprint lithography on six-inch wafers. *Nanotechnology* **2002**, *14*, 33. [CrossRef]
218. Turner, D.C.; Chang, C.; Fang, K.; Brandow, S.L.; Murphy, D.B. Selective adhesion of functional microtubules to patterned silane surfaces. *Biophys. J.* **1995**, *69*, 2782–2789. [CrossRef] [PubMed]
219. Damborský, P.; Švitel, J.; Katrlík, J. Optical biosensors. *Essays Biochem.* **2016**, *60*, 91–100. [PubMed]
220. Santos, A.; Vaz, A.; Rodrigues, P.; Veloso, A.; Venâncio, A.; Peres, A. Thin films sensor devices for mycotoxins detection in foods: Applications and challenges. *Chemosensors* **2019**, *7*, 3. [CrossRef]
221. Alahi, M.; Eshrat, E.; Mukhopadhyay, S.C. Detection methodologies for pathogen and toxins: A review. *Sensors* **2017**, *17*, 1885. [CrossRef] [PubMed]
222. Kaminiaris, M.D.; Mavrikou, S.; Georgiadou, M.; Paivana, G.; Tsitsigiannis, D.I.; Kintzios, S. An impedance based electrochemical immunosensor for aflatoxin B1 monitoring in pistachio matrices. *Chemosensors* **2020**, *8*, 121. [CrossRef]
223. Jusoh, N.S.; Awaludin, N.; Salam, F.; Kadir, A.; Said, N. Label-free electrochemical Immunosensor development for mycotoxins detection in grain corn. *Malays. J. Anal. Sci.* **2022**, *26*, 1205–1215.
224. Goryacheva, I.Y.; Saeger, S.D.; Eremin, S.A.; Peteghem, C.V. Immunochemical methods for rapid mycotoxin detection: Evolution from single to multiple analyte screening: A review. *Food Addit Contam.* **2007**, *24*, 1169–1183. [CrossRef] [PubMed]
225. Garden, S.R.; Strachan, N.J. Novel colorimetric immunoassay for the detection of aflatoxin B1. *Anal. Chim. Acta* **2001**, *444*, 187–191. [CrossRef]
226. Puiu, M.; Istrate, O.; Rotariu, L.; Bala, C. Kinetic approach of aflatoxin B1–acetylcholinesterase interaction: A tool for developing surface plasmon resonance biosensors. *Anal. Biochem.* **2012**, *421*, 587–594. [CrossRef]
227. Moscone, D.; Arduini, F.; Amine, A.; Arduini, F.; Amine, A. A rapid enzymatic method for aflatoxin B detection. In *Microbial Toxins*; Humana Press: Totowa, NJ, USA, 2011; pp. 217–235. [CrossRef]
228. Stepurska, K.V.; Soldatkin, O.O.; Kucherenko, I.S.; Arkhypova, V.M.; Dzyadevych, S.V.; Soldatkin, A.P. Feasibility of application of conductometric biosensor based on acetylcholinesterase for the inhibitory analysis of toxic compounds of different nature. *Anal. Chim. Acta* **2015**, *854*, 161–168. [CrossRef] [PubMed]

229. Soldatkin, O.O.; Burdak, O.S.; Sergeyeva, T.A.; Arkhypova, V.M.; Dzyadevych, S.V.; Soldatkin, A.P. Acetylcholinesterase-based conductometric biosensor for determination of aflatoxin B1. *Sens. Actuators B Chem.* **2013**, *188*, 999–1003. [CrossRef]
230. Egbunike, G.N.; Ikegwuonu, F.I. Effect of aflatoxicosis on acetylcholinesterase activity in the brain and adenohypophysis of the male rat. *Neurosci. Lett.* **1984**, *52*, 171–174. [CrossRef] [PubMed]
231. Moon, J.; Byun, J.; Kim, H.; Lim, E.-K.; Jeong, J.; Jung, J.; Kang, T. On-site detection of aflatoxin B1 in grains by a palm-sized surface plasmon resonance sensor. *Sensors* **2018**, *18*, 598. [CrossRef] [PubMed]
232. Liu, T.; Zhao, Y.; Zhang, Z.; Zhang, P.; Li, J.; Yang, R.; Yang, C.; Zho, L. A fiber-optic biosensor for specific identification of dead Escherichia coli O157:H7. *Sens. Actuators B Chem.* **2014**, *196*, 161–167. [CrossRef]
233. Panini, N.V.; Salinas, E.; Messina, G.A.; Raba, J. Modified paramagnetic beads in a microfluidic system for the determination of zearalenone in feedstuffs samples. *Food Chem.* **2011**, *125*, 791–796. [CrossRef]
234. Mirasoli, M.; Buragina, A.; Dolci, L.S.; Simoni, P.; Anfossi, L.; Giraudi, G.; Roda, A. Chemiluminescence-based biosensor for fumonisins quantitative detection in maize samples. *Biosens Bioelectron.* **2012**, *32*, 283–287. [CrossRef]
235. Lu, L.; Gunasekaran, S. Dual-channel ITO-microfluidic electrochemical immunosensor for simultaneous detection of two mycotoxins. *Talanta* **2019**, *194*, 709–716. [CrossRef] [PubMed]
236. Xu, Y.; Huang, Z.; He, Q.; Deng, S.; Li, L.; Li, Y. Development of an immunochromatographic strip test for the rapid detection of deoxynivalenol in wheat and maize. *Food Chem.* **2010**, *119*, 834–839. [CrossRef]
237. Romanazzo, D.; Ricci, F.; Volpe, G.; Elliott, C.T.; Vesco, S.; Kroeger, K.; Moscone, D.; Stroka, J.; Egmond, H.V.; Vehniäinene, M.; et al. Development of a recombinant Fab-fragment based electrochemical immunosensor for deoxynivalenol detection in food samples. *Biosens. Bioelectron.* **2010**, *25*, 2615–2621. [CrossRef]
238. Pennacchio, A.; Ruggiero, G.; Staiano, M.; Piccialli, G.; Oliviero, G.; Lewkowicz, A.; Synak, A.; Bojarski, P.; D'Auria, S. A surface plasmon resonance based biochip for the detection of patulin toxin. *Opt. Mater.* **2014**, *36*, 1670–1675. [CrossRef]
239. Funari, R.; Ventura, B.D.; Carrieri, R.; Morra, L.; Lahoz, E.; Gesuele, F.; Altucci, C.; Velotta, R. Detection of parathion and patulin by quartz-crystal microbalance functionalized by the photonics immobilization technique. *Biosens. Bioelectron.* **2014**, *67*, 224–229. [CrossRef] [PubMed]
240. Ivnitski, D.; Abdel-Hamid, I.; Atanasov, P.; Wilkins, E. Biosensors for detection of pathogenic bacteria. *Biosens. Bioelectron* **1999**, *14*, 599–624. [CrossRef]
241. Banerjee, P.; Bhunia, A.K. Cell-based biosensor for rapid screening of pathogens and toxins. *Biosens. Bioelectron.* **2010**, *26*, 99–106. [CrossRef] [PubMed]
242. Panwar, S.; Duggirala, K.S.; Yadav, P.; Debnath, N.; Yadav, A.K.; Kumar, A. Advanced diagnostic methods for identification of bacterial foodborne pathogens: Contemporary and upcoming challenges. *Crit. Rev. Biotechnol.* **2023**, *43*, 982–1000. [CrossRef]
243. Korsak, D.; Mackiw, E.; Rozynek, E.; Zylowska, M. Prevalence of *Campylobacter* spp. in retail chicken, turkey, pork, and beef meat in Poland between 2009 and 2013. *J. Food Prot.* **2015**, *78*, 1024–1028. [CrossRef]
244. Torso, L.M.; Voorhees, R.E.; Forest, S.A.; Gordon, A.Z.; Silvestri, S.A.; Kissler, B.; Schlackman, J.; Sandt, C.H.; Toma, P.; Bachert, J.; et al. Escherichia coli O157:H7 outbreak associated with restaurant beef grinding. *J. Food Prot.* **2015**, *78*, 1272–1279. [CrossRef] [PubMed]
245. Zhao, X.; Lin, C.-W.; Wang, J.; Oh, D.H. Advances in rapid detection methods for foodborne pathogens. *J. Microbiol. Biotechnol.* **2014**, *24*, 297–312. [CrossRef] [PubMed]
246. Senturk, E.; Aktop, S.; Sanlibaba, P.; Tezel, B.U. Biosensors: A novel approach to detect food-borne pathogens. *Appl. Microbiol. Open Access* **2018**, *4*, 1–8. [CrossRef]
247. Arora, P.; Sindhu, A.; Dilbaghi, N.; Chaudhury, A. Biosensors as in novative tools for the detection of food borne pathogens. *Biosens. Bioelectron.* **2011**, *28*, 1–12. [CrossRef]
248. Chai, Y.; Horikawa, S.; Li, S.; Wikle, H.C.; Chin, B.A. A surface-scanning coil detector for real-time, in-situ detection of bacteria on fresh food surfaces. *Biosens. Bioelectron.* **2013**, *50*, 311–317. [CrossRef] [PubMed]
249. Pilevar, M.; Kim, K.T.; Lee, W.H. Recent advances in biosensors for detecting viruses in water and wastewater. *J. Hazard. Mater.* **2021**, *410*, 124656. [CrossRef] [PubMed]
250. Cossettini, A.; Vidic, J.; Maifreni, M.; Marino, M.; Pinamonti, D.; Manzano, M. Rapid detection of Listeria monocytogenes, Salmonella, *Campylobacter* spp., and *Escherichia coli* in food using biosensors. *Food Control* **2022**, *137*, 108962. [CrossRef]
251. Servarayan, K.L.; Krishnamoorthy, G.; Sundaram, E.; Karuppusamy, M.; Murugan, M.; Piraman, S.; Vasantha, V.S. Optical immunosensor for the detection of *Listeria monocytogenes* in food matrixes. *ACS Omega* **2023**, *8*, 15979–15989. [CrossRef]
252. Pathirana, S.T.; Barbaree, J.; Chin, B.A.; Hartell, M.G.; Neely, W.C.; Vodyanoy, V. Rapid and sensitive biosensor for Salmonella. *Biosens. Bioelectron.* **2000**, *15*, 135–141. [CrossRef]
253. Gertie, C.A.M.; Bokken, R.J.; Corbee, F.; van Knapen, A.; Bergwerff, A. Immunochemical detection of Salmonella group B, D and E using an optical surface plasmon resonance biosensor. *FEMS Microbiol. Lett.* **2003**, *222*, 75–82. [CrossRef]
254. Ko, S.; Sheila, A.; Grant, A. A novel FRET-based optical fiber biosensor for rapid detection of *Salmonella typhimurium*. *Biosens. Bioelectron.* **2006**, *21*, 1283–1290. [CrossRef]

255. Liu, J.; Jasim, I.; Shen, Z.; Zhao, L.; Dweik, M.; Zhang, S.; Almasri, M. A microfluidic based bio-sensor for rapid detection of Salmonella in food products. *PLoS ONE* **2019**, *14*, e0216873.
256. Mahari, S.; Roberts, A.; Gandhi, S. Probe-free nanosensor for the detection of Salmonella using gold nanorods as an electroactive modulator. *Food Chem.* **2022**, *390*, 133219. [CrossRef]
257. Eissa, S.; Zourob, M. Ultrasensitive peptide-based multiplexed electrochemical biosensor for the simultaneous detection of Listeria monocytogenes and Staphylococcus aureus. *Microchim. Acta* **2020**, *187*, 1. [CrossRef]
258. Saini, K.; Kaushal, A.; Gupta, S.; Kumar, D. PlcA-based nanofabricated electrochemical DNA biosensor for the detection of *Listeria monocytogenes* in raw milk samples. *3Biotech* **2020**, *10*, 1. [CrossRef]
259. Kaushal, S.; Priyadarshi, N.; Pinnaka, A.K.; Soni, S.; Deep, A.; Singhal, N.K. Glycoconjugates coated gold nanorods based novel biosensor for optical detection and photothermal ablation of food borne bacteria. *Sens. Actuators B Chem.* **2019**, *289*, 207–215. [CrossRef]
260. Du, J.; Yu, Z.; Hu, Z.; Chen, J.; Zhao, J.; Bai, Y. A low pH-based rapid and direct colorimetric sensing of bacteria using unmodified gold nanoparticles. *J. Microbiol. Methods* **2021**, *180*, 106110. [CrossRef]
261. Jin, S.; Dai, M.; Ye, B.-c.; Nugen, S.R. Development of a capillary flow microfluidic Escherichia coli biosensor with on-chip reagent delivery using water-soluble nanofibers. *Microsyst. Technol.* **2013**, *19*, 2011–2015. [CrossRef]
262. Zhou, H.; Guo, W.; Hao, T.; Xie, J.; Wu, Y.; Jiang, X.; Hu, Y.; Wang, S.; Guo, Z. Electrochemical sensor for single-cell determination of bacteria based on target-triggered click chemistry and fast scan voltammetry. *Food Chem.* **2023**, *417*, 135906. [CrossRef]
263. Morant-Miñana, M.C.; Elizalde, J. Microscale electrodes integrated on COP for real sample Campylobacter spp. Detection. *Biosens. Bioelectron.* **2015**, *70*, 491–497. [CrossRef]
264. Jiang, D.; Liu, F.; Liu, C.; Liu, L.; Li, Y.; Pu, X. Induction of an electrochemiluminescence sensor for DNA detection of Clostridium perfringens based on rolling circle amplification. *Anal. Methods* **2014**, *6*, 1558–1562. [CrossRef]
265. Ghadeer, A.R.Y.S.; Alhogail, S.; Zourob, M. Rapid and low-cost biosensor for the detection of *Staphylococcus aureus*. *Biosens. Bioelectron.* **2017**, *90*, 230–237. [CrossRef]
266. Jiang, S.; Hua, E.; Liang, M.; Liu, B.; Xie, G. A novel immunosensor for detecting toxoplasma gondii-specific IgM based on goldmag nanoparticles and graphene sheets. *Colloids Surf. B Biointerfaces* **2013**, *101*, 481–486. [CrossRef]
267. Bacchu, M.S.; Ali, M.R.; Das, S.; Akter, S.; Sakamoto, H.; Suye, S.-I.; Rahman, M.M.; Campbell, K.; Khan, M.Z.H. A DNA functionalized advanced electrochemical biosensor for identification of the foodborne pathogen *Salmonella enterica* serovar Typhi in real samples. *Anal. Chim. Acta* **2022**, *1192*, 339332. [CrossRef] [PubMed]
268. Angelopoulou, M.; Petrou, P.; Misiakos, K.; Raptis, I.; Kakabakos, S. Simultaneous Detection of *Salmonella* Typhimurium and Escherichia coli O157:H7 in Drinking Water and Milk with Mach–Zehnder Interferometers Monolithically Integrated on Silicon Chips. *Biosensors* **2022**, *12*, 507. [CrossRef] [PubMed]
269. da Silva, A.D.; Paschoalino, W.J.; Neto, R.C.; Kubota, L.T. Electrochemical Point-Of-Care Devices for Monitoring Waterborne Pathogens: Protozoa, Bacteria, and Viruses—An Overview. *Case Stud. Chem. Environ. Eng.* **2022**, *5*, 100182. [CrossRef]
270. Roy, E.; Maity, S.K.; Patra, S.; Madhuri, R.; Sharma, P.K. A metronidazole-probe sensor based on imprinted biocompatible nanofilm for rapid and sensitive detection of anaerobic protozoan. *RSC Adv.* **2014**, *4*, 32881. [CrossRef]
271. Ilkhani, H.; Zhang, H.; Zhou, A. A novel three-dimensional microTAS chip for ultra-selective single base mismatched Cryptosporidium DNA biosensor. *Sens. Actuator. B Chem.* **2019**, *282*, 675–683. [CrossRef]
272. Manzano, M.; Viezzi, S.; Mazerat, S.; Marks, R.; VIDIC, J. Rapid and label-free electrochemical DNA biosensor for detecting hepatitis A virus. *Biosens. Bioelectron.* **2018**, *100*, 89–95. [CrossRef]
273. Escobar, V.; Scaramozzino, N.; Vidic, J.; Buhot, A.; Mathey, R.; Chaix, C.; Hou, Y. Recent Advances on Peptide-Based Biosensors and Electronic Noses for Foodborne Pathogen Detection. *Biosensors* **2023**, *13*, 258. [CrossRef]
274. Gomes, N.O.; Teixeira, S.C.; Calegaro, M.L.; Machado, S.A.S.; Ferreira Soares, N.F.; de Oliveira, T.V.; Raymundo, P.A. Pereira Flexible and sustainable printed sensor strips for on-site, fast decentralized self-testing of urinary biomarkers integrated with a portable wireless analyzer. *Chem. Eng. J.* **2023**, *472*, 144775. [CrossRef]
275. Sakthivel, K.; Balasubramanian, S.; Chang-Chien, G.-P.; Wang, S.-F.; Ahammad Billey, W.; Platero, J.; Soundappan, T.; Sekhar, P. Editors' Choice—Review—Advances in Electrochemical Sensors: Improving Food Safety, Quality, and Traceability. *ECS Sens. Plus* **2024**, *3*, 020605. [CrossRef]

processes

MDPI

Article

Predicting and Understanding Emergency Shutdown Durations Level of Pipeline Incidents Using Machine Learning Models and Explainable AI

Lemlem Asaye [1], Chau Le [2,*], Ying Huang [1,*], Trung Q. Le [3], Om Prakash Yadav [4] and Tuyen Le [5]

1 Department of Civil, Construction and Environmental Engineering, North Dakota State University, Fargo, ND 58102, USA; lemlem.asaye@ndsu.edu
2 Department of Engineering Technology and Construction Management, University of North Carolina at Charlotte, Charlotte, NC 28223, USA
3 Department of Industrial and Management Systems Engineering, University of South Florida, Tampa, FL 33620, USA; tqle@usf.edu
4 Department of Industrial and Systems Engineering, North Carolina Agricultural and Technical State University, Greensboro, NC 27411, USA; oyadav@ncat.edu
5 Glenn Department of Civil Engineering, Clemson University, Clemson, SC 29634, USA; tuyenl@clemson.edu
* Correspondence: chau.le@charlotte.edu (C.L.); ying.huang@ndsu.edu (Y.H.)

Abstract: Pipeline incidents pose significant concerns due to their potential environmental, economic, and safety risks, emphasizing the critical need to understand and manage this vital infrastructure. While existing studies predominantly focus on the causes of pipeline incidents and failures, few have investigated the consequences, such as shutdown duration, and most lack comprehensive models capable of accurately predicting and providing actionable insights into the risk factors. This study bridges this gap by employing machine learning (ML) techniques, including Random Forest and Light Gradient Boosting Machine (LightGBM), for classifying pipeline incidents' emergency shutdown duration levels. These techniques are specifically designed to capture complex, nonlinear patterns and interdependencies within the data, addressing the limitations of traditional linear approaches. The proposed model has further enhanced with Explainable AI (XAI) techniques, such as Shapley Additive exPlanations (SHAP) values, to improve interpretability and provide insights into the factors influencing shutdown durations. Historical incident data, collected from the Pipeline and Hazardous Materials Safety Administration (PHMSA) from 2010 to 2022, were utilized to examine the risk factors. K-Fold Cross-Validation with 5 folds was employed to ensure the model's robustness. The results demonstrate that the LightGBM model achieved the highest accuracy of 75.0%, closely followed by Random Forest at 74.8%. The integration of XAI techniques provides actionable insights into key factors such as pipeline material, age, installation layout, and commodity type, which significantly influence shutdown durations. These findings underscore the practical implications of the proposed approach, enabling pipeline operators, emergency responders, and regulatory authorities to make informed decisions that optimize resource allocation and mitigate risks effectively.

Keywords: emergency shutdown duration; pipeline incidents; Explainable AI; infrastructure resilience; risk management

1. Introduction

Pipelines are integral to global energy infrastructure, enabling the efficient transportation of critical resources such as oil and natural gas, which underpin industrial processes and social functions [1]. Despite their operational efficiency, pipelines are vulnerable to risks, and incidents often lead to shutdowns that can have severe environmental, economic, and social impacts [2]. These incidents often necessitate emergency shutdowns, disrupting normal operations, and requiring swift responses to mitigate cascading effects. Among the various consequences of pipeline incidents, Emergency Shutdowns Duration (ESDs) stand out as a crucial factor, as they directly impact the mitigation of damage, downtime, and recovery efforts [3]. ESDs in pipeline operations are critical safety measures designed to immediately halt operations during severe hazards or imminent dangers, such as gas leaks, fires, or equipment failures [4]. Managing these incidents often requires emergency shutdowns, whose duration can vary greatly based on the incident's nature and context. Understanding and classifying these durations at different levels is essential to optimize emergency response decision-making processes and minimize cascading impacts on safety, the environment, and the economy.

Despite significant advances, research on the classification and prediction of emergency shutdown durations remains limited. Many studies prioritize causal factors and risk assessments over operational consequences, specifically in classifying and understanding the levels of shutdown durations [5–8]. For example, Lam and Zhou [9] employed statistical methods to identify the frequent causes of pipeline failures, such as corrosion and third-party damage. Similarly, Halim et al. [10] analyzed causal factors and background conditions influencing pipeline integrity. However, few studies have investigated the shutdown duration [9,11–16]. For instance, Zhu et al. [16] have investigated how data and information quality impact emergency shutdown systems decision-making. They highlighted the importance of leveraging robust data frameworks to enhance safety-critical decisions. However, this study did not address predictive modeling or classification of shutdown durations, leaving a gap in applying such frameworks to duration-specific contexts. In the same line of research, Vitali et al. [17] assessed the locations of incidents, along with the corresponding number of shutdown incidents for CO_2 pipelines, and calculated the average shutdown durations. However, these studies focus on the frequency and duration of shutdown incidents without considering the broader context of predicting and understanding the levels of shutdown duration incidents. Moreover, Hainen et al. [18] utilized hazard-based duration analysis with Weibull models to explore shutdown durations for hazardous liquid pipelines. They identified critical factors like incident location, maintenance practices, and pipeline material, offering actionable recommendations for reducing durations. Nonetheless, the investigation considered only limited causal and background factors. In addition, their scope was limited to binary outcomes, neglecting the multifaceted nature of real-world incidents. These studies provide valuable insights but often lack predictive frameworks to classify shutdown durations effectively based on incident characteristics.

Current practices in pipeline incident management largely rely on simplistic statistical models, historical trends, and rule-based systems for ESD durations [19–21]. While these approaches are cost-effective and accessible, they are significantly limited in both accuracy and adaptability. The reliance on historical averages, for instance, assumes that similar incidents will yield comparable outcomes. However, this overlooks the complex and dynamic factors influencing incidents, such as environmental conditions, equipment age, and response efficiency [22]. Additionally, rule-based systems operate based on predefined protocols, which lack the flexibility to adapt to evolving scenarios or incorporate nuanced data, often leading to generic and inaccurate predictions [23,24]. Moreover, traditional practices emphasize post-incident analysis and reactive decision-making rather than proactive prediction and mitigation, contributing to delayed responses and prolonged shutdowns. These limitations underscore the critical need for advanced, explainable predictive models that integrate diverse data sources to improve incident management and reduce ESD durations. The absence of a systematic, data-driven methodology to predict ESD durations can lead to several challenges [25–27]. Firstly, it restricts the ability to optimize resource allocation during emergencies, potentially delaying response times and increasing the risk to public safety and environmental health. Secondly, it complicates the scheduling of maintenance and inspection activities, as uncertainties in shutdown durations can disrupt planned operations and lead to increased operational costs. Lastly, the lack of predictive accuracy can undermine stakeholder confidence, including that of regulatory bodies and the public, in the operator's ability to manage emergencies effectively. Existing frameworks lack predictive tools for categorizing shutdown durations into actionable levels, which are crucial for tailored response strategies. Moreover, advanced data-driven methods, such as ML and natural language processing, remain underutilized in this domain [28]. While duration models have identified key variables, their static nature fails to capture the dynamic interdependencies of operational and environmental factors. Furthermore, traditional predictive models operate as "black boxes", limiting their applicability in safety-critical contexts. XAI frameworks, which can provide transparency and actionable insights, are yet to be integrated into pipeline incident management.

This study aims to bridge these gaps by leveraging ML techniques enhanced with XAI tools to develop a robust classification system for pipeline shutdown durations. ML models can uncover complex, nonlinear relationships among variables [29,30], making them ideal for analyzing pipeline incident data. XAI bridges this black box gap by providing transparent insights into model predictions [31]. Techniques such as SHapley Additive exPlanations (SHAP) show the contribution of factors like pipeline material, installation layout, and incident timing to predicted shutdown durations and providing actionable insights into the factors influencing shutdown durations to support informed decision-making during emergencies. The integration of predictive modeling and explainability offers a dual benefit. First, it improves the accuracy of shutdown duration predictions, enabling more precise resource allocation and intervention planning. Second, it equips stakeholders with a deeper understanding of incident dynamics, facilitating the design of tailored strategies to mitigate risks and enhance safety. The study leverages historical incident data from the Pipeline and Hazardous Materials Safety Administration (PHMSA), ensuring a robust analysis of diverse variables. By prioritizing both predictive performance and interpretability, this research bridges a critical gap in the literature, advancing the field of pipeline incident management through innovative applications of ML and XAI. Classifying shutdown durations is essential for optimizing emergency responses and minimizing disruptions. Short-term shutdowns typically require rapid but less intensive interventions. In contrast, medium-term or long-term shutdowns require comprehensive strategies to address severe impacts such as environmental remediation and supply chain interruptions. A structured classification system enables decision makers to allocate resources proportional to the severity of an incident, ensuring a balanced approach to immediate mitigation and long-term resilience. Furthermore, identifying patterns and predictors of shutdown durations facilitates proactive planning and preventive measures, reducing the frequency and severity of future incidents. The findings aim to improve operational practices, inform policy decisions, and contribute to infrastructure resilience.

This study's novelty lies in addressing these limitations through the application of machine learning (ML) models enhanced with Explainable AI (XAI) techniques to classify pipeline emergency shutdown durations. Specifically, the study has novelty in (1) integrating SHAP values to make machine learning predictions transparent and interpretable, addressing the traditional "black-box" limitation of ML models; (2) targeting shutdown durations as a key focus, an underexplored but critical consequence of pipeline incidents, to provide actionable insights for emergency management; and (3) identifying key influencing factors, such as pipeline material, installation layout, and commodity type, to provide a nuanced understanding and support data-driven decisions. Furthermore, this study presents innovations by advancing operational decision-making through predictive modeling and explainability. Notably, the study has innovations in (1) developing a predictive framework that integrates machine learning and Gaussian Mixture Models (GMM) to classify shutdown durations into actionable categories (short, medium, and long); (2) employing XAI techniques, particularly SHAP values, to provide actionable and explainable insights into the factors influencing shutdown durations; and (3) enabling proactive emergency management by anticipating the severity of shutdowns and facilitating resource allocation to mitigate risks. This comprehensive approach not only addresses an underexplored aspect of pipeline incident management but also sets the stage for proactive risk mitigation strategies, ensuring safer and more reliable pipeline operations.

Related Studies and Research Gap

Previous studies have investigated various aspects of pipeline incidents, as shown in Table 1. The existing literature on pipeline shutdown durations and incident management can be classified into four main categories: pipeline failure causes and risk factors, pipeline failure prediction, emergency response and risk management, and pipeline incident consequences.

The first approach examines pipeline failure causes and risk factors using statistical methods and causal analysis techniques. Studies such as Hameed et al. [12] developed a risk-based methodology for optimizing shutdown intervals by considering system availability and risk, using the Markov process to calculate critical equipment's risk profiles. These methods offer valuable insights into the causes of pipeline incidents, helping improve inspection and maintenance practices. However, these approaches primarily focus on identifying the causal factors of pipeline failures rather than understanding and predicting the duration of shutdowns caused by these incidents. As a result, while they help mitigate pipeline failures, they do not address the crucial aspect of shutdown duration, which is essential for optimizing emergency response strategies and minimizing operational disruptions.

The second approach involves the use of predictive models to estimate pipeline failure or shutdown durations, particularly using machine learning algorithms. Capshaw and Padgett [32] introduced a predictive model for estimating refinery shutdown durations during hurricanes, using logistic regression and Poisson distribution to analyze resilience impacts. Though this study provided a valuable predictive framework, it was confined to hurricane-related incidents and did not address generalizable models for pipeline incidents under diverse conditions. The model does not offer a generalizable solution for predicting shutdown durations in diverse operational and environmental contexts.

The third approach involves emergency response and risk management. Studies like those by Vitali et al. [17] and Zhu et al. [16] explore methods such as Markov processes and Monte Carlo simulations to optimize emergency response times, mitigate risks during pipeline incidents, and improve real-time monitoring. While these studies provide a comprehensive framework for responding to pipeline incidents, they primarily focus on mitigation strategies, without directly addressing the classification of shutdown durations.

Finally, the fourth approach focuses on evaluating the broader consequences of pipeline incidents, such as the economic, environmental, and social impacts [33,34]. For instance, Al-Douri et al. [35] employed statistical analyses to investigate causal factors in shutdown incidents, identifying differences between hazardous liquid pipelines and natural gas transmission but did not examine shutdown durations. While these studies contribute valuable insights into the systemic effects of pipeline disruptions, they do not specifically address the duration of shutdowns, which are essential to improving decision-making during emergency responses.

While several studies have investigated the causes and impacts of pipeline failures and shutdowns, a clear gap remains in research focused on classifying and understanding emergency shutdown durations. Current approaches typically rely on general models that predict incidents or analyze risk factors but do not specifically address the duration of these incidents or classify them into actionable categories. The proposed study aims to bridge this gap by utilizing machine learning models enhanced with XAI techniques to classify pipeline shutdown durations into meaningful categories (short, medium, and long-term). By leveraging advanced predictive models and XAI techniques, this study offers actionable insights into the factors influencing shutdown durations and supports more efficient decision-making during pipeline incidents.

Table 1. Overview of pipeline incident studies: approaches and findings.

Study	Research Focus	Key Techniques/Methods	Main Findings
Halim et al. [10], Lam and Zhou [9], Hameed et al. [12]	Pipeline Failure Causes and Risk Factors	Statistical analysis, Causal factor identification, Fault Tree Analysis	Identify frequent failure causes like corrosion, damage, and equipment failure, highlighting areas for improved inspection
Capshaw and Padett [32], Hassan et al. [6]	Pipeline Failure Prediction	Statistical models, Poisson distribution, Markov processes, Bayesian networks	Predict pipeline incidents, helping to anticipate future failures.
Vitali et al. [17], Zhu et al. [16], Yu et al. [14]	Emergency Response and Risk Management	Markov Processes, GIS for Pipeline Monitoring, Monte Carlo Simulation, Time Series Analysis	Addresses emergency response times, optimized shutdown strategies, mitigated risk during pipeline incidents
Al-Douri et al. [35], Ramírez-Camacho et al. [33], Xiao et al. [2], Aalirezaei et al. [34]	Pipeline Incidents Consequences	Cost–benefit analysis, Multi-Criteria Decision Analysis, Environmental Impact Assessment	Evaluated the broader impacts of pipeline shutdowns, emphasizing economic, environmental, and social factors to aid in decision-making regarding pipeline maintenance and risk management.

2. Methodology

Figure 1 describes the methodology employed in this study for classifying emergency shutdown duration levels in pipeline incidents. The process begins with the collection of incident reports from the PHMSA. To ensure the robustness of the analysis, data irregularities, including missing values, are addressed. Missing values in continuous variables are handled through imputation with their respective means, while categorical variables are treated using one-hot encoding to enable effective inclusion in machine learning models. Additionally, Min–Max scaling is applied to normalize continuous variables, ensuring that the relative importance of each feature is treated equitably during modeling. Influential features are identified using the wrapper-based backward elimination method, which iteratively refines the feature set by evaluating their impact on model performance. This study employed Python (version 3.11.5), a widely used programming language, for data preprocessing and modeling. Specifically, the Pandas library was used for data manipulation and analysis. Following data preprocessing, shutdown durations are categorized into sub-groups based on their severity using the Gaussian Mixture Clustering Algorithm (GMM), implemented through Scikit-learn (version 1.5.2). GMM is selected for its ability to model complex, nonlinear patterns in the data and probabilistically assign data points to clusters [36]. The optimal number of clusters is determined using the Akaike Information Criterion (AIC), resulting in three distinct categories: short-term, medium-term, and long-term shutdown durations. This classification provides a structured framework for assessing incident severity and tailoring responses accordingly. To develop predictive models, a range of machine learning algorithms is employed, including Multilayer Perceptron Neural Networks (MLPNN), Random Forest (RF), eXtreme Gradient Boosting (XGBoost), Categorical Boosting (CatBoost), and Light Gradient Boosting Machine (LightGBM). These algorithms are chosen for their ability to capture intricate patterns and dependencies within the dataset. Rigorous hyperparameter tuning is performed using randomized search techniques to optimize model configurations and enhance performance. The models are evaluated using metrics such as accuracy, precision, recall, and F1-score to ensure comprehensive performance assessment. K-Fold Cross-Validation, with five folds, is employed to validate the robustness and generalizability of the models across diverse data samples. To enhance the transparency and interpretability of the predictive models, XAI techniques are integrated. SHAP values are employed to interpret model predictions [37], providing detailed insights into the influence of various factors on shutdown duration classifications. Key features such as pipeline material, installation layout, timing of incidents, and commodity type are identified as critical determinants of shutdown durations. By bridging the gap between predictive accuracy and interpretability, the integration of XAI ensures actionable insights for pipeline operators, emergency responders, and regulatory authorities. Overall, this methodological framework represents a robust approach to addressing the challenges associated with classifying shutdown durations in pipeline incidents. By combining advanced machine learning techniques with explainability, the study not only achieves high predictive performance but also provides valuable insights to support effective decision-making in emergency management scenarios.

2.1. Data Collection and Preprocessing

This study obtained oil pipeline incident data from PHMSA for the years 2010 to 2022. Earlier incident data prior to 2010 was excluded from the analysis due to challenges in extracting and analyzing records from a different format that significantly differed from the 2010–2022 data. Moreover, reporting criteria and forms have evolved over time, leading to variations in the attributes provided across years, which could influence the dataset's consistency. Additionally, the geographical coverage of the dataset includes incidents from pipelines across the United States, with data reflecting various pipeline types, operational environments, and geographical regions. However, the distribution of incidents is not uniform across regions, as certain areas may have a higher frequency of pipeline infrastructure, such as densely populated or industrial regions, while others may have fewer incidents due to less infrastructure or differing operational conditions. Additionally, operational and geographic factors like weather conditions, terrain, and access to emergency response resources may vary significantly across regions, potentially affecting shutdown durations. The incident report includes causal factors, background factors, and shutdown duration. A causal factor directly triggers an incident [10]. PHMSA categorizes these factors into eight major groups: excavation damage, corrosion failure, and equipment failure. Background factors, such as pipeline material, type, installation year, diameter, and transported commodity, influence the occurrence of a causal factor without directly impacting pipeline failure [10]. Furthermore, shutdown duration refers to the period during which a pipeline remains inactive following an incident. Among the 4974 oil pipeline incidents reported to PHMSA, 2472 instances led to shutdowns. To ensure data integrity, missing values for continuous variables were imputed by using their respective means, and for categorical variables, missing values were addressed by employing the modes specific to each category. Additionally, the respective variable was excluded from the analysis in cases of high missing values. The GMM was employed to categorize the numerical value of shutdown duration into sub-groups due to its flexibility in handling complex, nonlinear patterns in the data and its ability to model probabilistic relationships. GMM assumes that all data points are generated from a mixture of normal distributions, and this assumption is reasonable for this study due to the large dataset and the diverse factors influencing pipeline shutdown durations [38]. By incorporating variables such as pipeline material, incident location, and commodity type, the dataset approximates distributions that can be effectively modeled with Gaussian components. The validation of this assumption was supported by the GMM's ability to identify distinct clusters with meaningful interpretations, aligning well with the real-world characteristics of shutdown incidents. GMM assigns probabilities to data points within clusters rather than strictly classifying them, aligning well with potentially varied patterns and distributions of shutdown durations in pipeline incidents. Unlike hard clustering methods such as K-means, GMM's probabilistic approach allows it to capture overlapping distributions and varied patterns in the data [39], which are likely influenced by diverse factors such as pipeline material, location, and commodity type. To determine the optimal number of clusters, the Akaike Information Criterion (AIC) score was used. AIC is a robust model selection criterion that balances goodness-of-fit with model complexity, ensuring that the number of clusters selected avoids overfitting while accurately representing the data [40]. This guided the categorization of shutdown duration into three groups: short (less than 12 h and 28 min), medium (12 h and 28 min to 213 h and 31 min), and long-term (over 213 h and 31 min) (see Figure 2). However, it is acknowledged that clustering durations into these discrete categories might oversimplify the inherent variability of pipeline shutdown incidents, particularly when durations at the boundaries of these clusters are grouped together. An alternative approach would involve applying regression techniques to predict shutdown durations as a continuous variable, which could provide finer-grained insights into incident severity and resource allocation needs. While this approach could allow for more precise predictions, the three-class system was selected to strike a balance between model interpretability and operational relevance, ensuring practical applicability for stakeholders. Short-term shutdowns typically represent minor incidents requiring minimal intervention, medium-term shutdowns involve moderate complexities demanding coordinated responses, and long-term shutdowns reflect significant disruptions requiring extensive planning and resource allocation. This approach ensures practical applicability for stakeholders while effectively capturing the inherent variance in the data.

2.2. Feature Selection

This study employs wrapper methods, specifically backward elimination, to identify influential features related to the target variable, shutdown duration. The feature selection process begins with the complete set of features, systematically assessing each feature's contribution to the model's performance using a chosen machine learning algorithm and a performance metric, such as accuracy. In each iteration, one feature is removed from the set, and the model is re-evaluated to determine the impact of its exclusion. The feature whose removal has the least adverse effect on model performance is eliminated. This iterative approach continues until further removals no longer improve performance significantly, ensuring an optimized and parsimonious feature set. This method identified several critical features, as illustrated in Figure 3, which ranks their relative importance based on a Random Forest model. Among the top predictors are unintentional material release, pipeline age, and estimated pressure during the incident. For example, incidents involving higher unintentional material release volumes and older pipelines are more likely to result in extended shutdown durations, reflecting the increased complexity and scale of required responses. Operational and structural variables, such as maximum allowable pressure and pipeline installation layout, also emerged as significant factors influencing shutdown durations. These features affect the accessibility and intricacy of repairs, directly shaping the time needed to restore operations. The feature selection process underscores the importance of understanding both causal and contextual factors in categorizing and predicting shutdown durations. Table 2 provides a comprehensive overview of the key variables identified as crucial for this analysis, offering actionable insights into emergency pipeline management strategies.

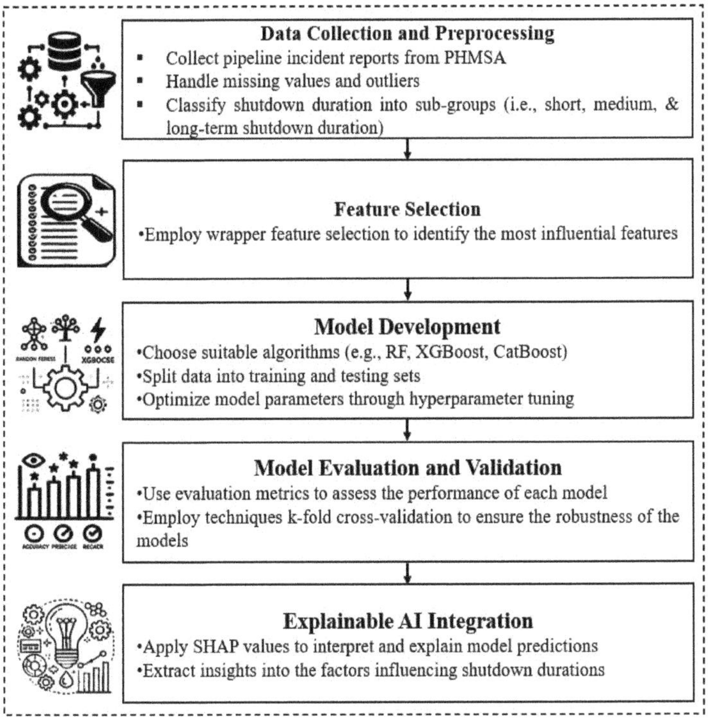

Figure 1. Framework for classifying shutdown duration of a pipeline incident.

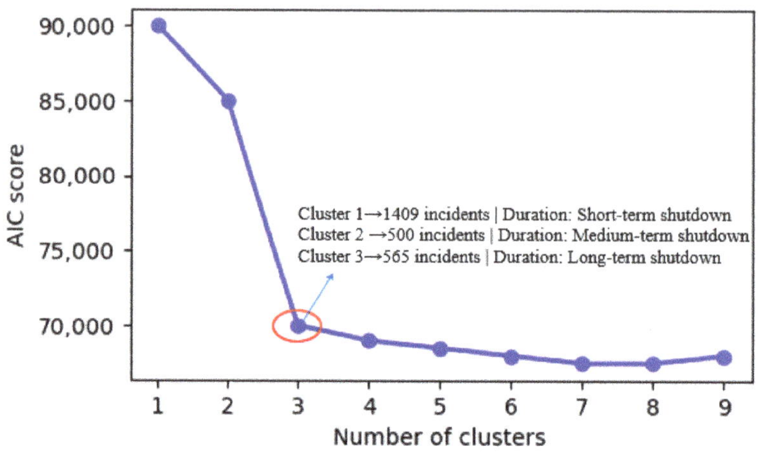

Figure 2. Optimal number of clusters for shutdown durations using GMM.

Table 2. Input and output variables.

Variable		**Subcategories**	**Data Type**
Input	Causal factor	Excavation damage, corrosion failure, equipment failure, natural force damage, incorrect operation, material failure of pipe or weld, other outside force damage, and other	Nominal
	Pipeline installation layout	Aboveground, underground, and transition area	Nominal
	Part of item involved	Pipe, tank, meter, tubing, and other	Nominal
	Commodity released type	Biofuel, CO_2, crude oil, High-Vapor Pressure Liquids (HVLs), and petroleum product	Nominal
	Leak type	Pinhole, seal, crack, connection failure, and other	Nominal
	Failure mode	Leakage, rupture, mechanical puncture, overflow, and other	Nominal
	Incident location	Off-site migration from operator-controlled property, on operator-controlled property, and Pipeline right-of-way	Nominal
	Pipeline material type	Carbon steel, and other than carbon steel	Nominal
	Pipe facility type	Intrastate and interstate	Nominal
	On/off shore	Onshore and offshore	Nominal
	Incident occurred week	Weekday and weekend	Nominal
	Crossing	Yes and no	Binary
	Pipeline age	Ranging from 1 year to 103 years	Discrete
	Recovered material	Ranging from 0.01 to 18,245 bbls	Continuous
	Unintentional release material	Ranging from 0.01 to 48,400 bbls	Continuous
	Pressure during the incident	Ranging from 0.25 to 2940 psig	Continuous
	Maximum allowable pressure	Ranging from 0.25 to 5000 psig	Continuous
	Pressure during the incident	Pressure did not exceed mop, exceeded 110% of mop, and exceeded mop, but did not exceed 110% of mop	Ordinal
	Incident occurred day	Morning, afternoon, evening, and night	Ordinal
Output	Shutdown duration	Short, medium, and long-term shutdown	Ordinal

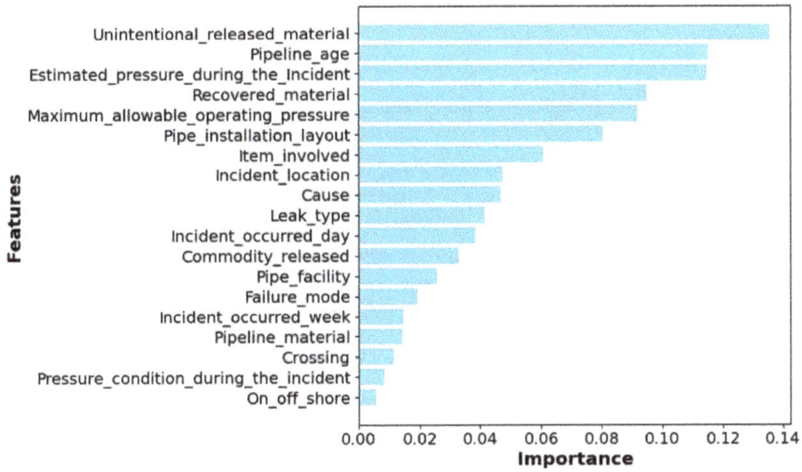

Figure 3. Feature importance for predicting shutdown duration levels.

2.3. Model Development

The development of predictive models is a critical component of this study, aimed at accurately classifying emergency shutdown durations in pipeline incidents. A range of machine learning algorithms, including Random Forest (RF), eXtreme Gradient Boosting (XGBoost), Categorical Boosting (CatBoost), Light Gradient Boosting Machine (LightGBM), and Multilayer Perceptron (MLP), are employed due to their ability to capture intricate patterns and relationships in the data. To ensure the effectiveness and generalizability of these models, the dataset is divided into training and testing subsets. This split allows for objective evaluation of model performance on unseen data, ensuring that the models are robust and not overfitted. Hyperparameter tuning is performed to optimize the performance of each model. A randomized search approach is utilized to explore a wide range of hyperparameter combinations systematically. This process identifies the optimal configuration for each algorithm, improving their predictive capabilities while maintaining computational efficiency. Table 3 summarizes the selected hyperparameters for each model, reflecting the best configurations determined through this tuning process.

Table 3. Hyperparameters for the proposed models.

Model	Hyperparameters
RF	Estimators: 100, Max Depth: 5, Min Samples Split: 2, Min Samples Leaf: 1, Bootstrap: True
XGBoost	Estimators: 100, Max Depth: 3, Learning Rate: 0.1, Subsample: 0.8
CatBoost	Iterations: 100, Depth: 4
LightGBM	Estimators: 100, Max Depth: 5, Learning Rate: 0.1
MLP	eurons per Layer: 100, Activation: ReLU, Learning Rate: 0.01

2.4. Model Evaluation and Validation

The evaluation and validation of the predictive models are conducted using comprehensive performance metrics to ensure their reliability and robustness. Key metrics, including accuracy, precision, recall, and F1-score, are employed to compare the performance of the different models. Accuracy provides an overall measure of the models' correctness, while precision, recall, and F1-score offer deeper insights into their ability to handle class imbalances and specific categories within the shutdown duration classification. To further enhance the robustness of the evaluation, K-Fold Cross-Validation is utilized. This approach divides the dataset into five subsets, or folds, to iteratively train and test the models across different partitions of the data. During each iteration, one fold is held out as the testing set, while the remaining folds are used for training. This process is repeated five times, ensuring that every data point is used for both training and validation. By averaging the performance metrics across all folds, this method provides a comprehensive assessment of the models' generalizability and prevents overfitting. This rigorous evaluation framework ensures that the models are not only accurate but also capable of maintaining consistent performance across diverse datasets, making them suitable for real-world applications in classifying pipeline shutdown durations.

2.5. Explainable AI Integration

XAI techniques are integrated into this study to improve the transparency and interpretability of the machine learning models used for predicting pipeline shutdown durations. Among these techniques, SHAP values are utilized to quantify the contribution of each feature to the model's predictions. SHAP values are a prominent method in XAI that assigns a value to each feature based on its impact on predictions, allowing users to understand the rationale behind individual predictions [41]. By assigning the importance scores to individual features, SHAP values help users understand how and why specific predictions are made. These insights provide actionable information for stakeholders, enabling them to identify factors most significantly affecting shutdown durations. The integration of XAI ensures that the predictive models are not only accurate but also interpretable, fostering trust and practical usability among pipeline operators and emergency responders.

3. Results and Discussion

This section provides an overview of the performance and implications of the predictive models developed to classify emergency shutdown durations in pipeline incidents. The models demonstrate strong predictive capabilities and highlight important factors influencing shutdown durations. Through the integration of machine learning and XAI, this study not only achieves accurate classification but also enhances understanding of the underlying variables driving predictions. These insights support better decision-making for pipeline operators and emergency responders, offering practical applications for mitigating risks and improving response strategies. The findings underscore the potential of combining advanced analytics and interpretability tools to address challenges in pipeline incident management.

3.1. Model Performance Comparison

Table 4 illustrates the comparative performance metrics across various models employed in this study. LightGBM demonstrated the highest accuracy at 75.00% on the testing data, closely followed by Random Forest at 74.8%. Additionally, these models exhibited competitive precision, recall, and F1-score metrics, indicating their efficacy in classifying shutdown duration levels. To ensure the reliability of the findings, a robust K-Fold Cross-Validation method with K set to 5 was employed. The results indicated that the LightGBM had an average accuracy of 74.74% on the validation data. While the differences in performance across models were relatively small, this consistency underscores the robustness of the feature set and preprocessing pipeline employed, as the models effectively captured the primary patterns in the data. However, the observed prediction accuracy, while moderate, highlights certain inherent challenges. These include the complexity of pipeline shutdown incidents, where numerous factors, such as regulatory delays or environmental conditions, could influence durations but were not captured in the dataset. Moreover, features related to the broader operational context, such as workforce availability, material stockpiles, or proximity to repair facilities, might also play a significant role in shaping shutdown durations and should be explored further. Additionally, potential class imbalances in the dataset may have contributed to reduced generalizability across short-, medium-, and long-term shutdowns. These imbalances could skew the model's ability to accurately predict minority classes, leading to biases in predictions. The feature set, while comprehensive, may also lack certain explanatory variables, such as real-time operational constraints or detailed pipeline configurations, which could enhance model precision.

Table 4. Comparison of the performance of proposed algorithms.

Metrics	RF	MLP	XGBoost	CatBoost	LightGBM
Accuracy (%)	74.8	74.60	73.99	73.19	75.00
Precision (%)	70.92	70.49	69.96	68.96	70.74
Recall (%)	71.28	72.94	70.92	70.32	71.68
F1-Score (%)	70.20	71.41	69.80	69.17	70.45

Despite these challenges, the models provide a foundational framework for understanding and predicting shutdown durations. The 75% accuracy, while moderate, represents a reasonable balance between predictive capability and operational feasibility, particularly in complex real-world scenarios. To enhance the framework's predictive accuracy and address current limitations, future research could incorporate additional explanatory variables, such as geographic data, real-time operational constraints, or detailed pipeline configurations, which may offer a more detailed understanding of the factors driving shutdown durations. Moreover, the integration of temporal data (such as incident seasonality or historical response times) could provide a more detailed context for improving predictions. Further, the nature of pipeline shutdowns, being affected by a large number of interacting dynamic factors, suggests that the models may benefit from more complex model architectures such as deep learning-based recurrent neural networks (RNN) or long short-term memory networks (LSTM) to capture temporal dependencies and long-range patterns. Advanced modeling approaches, such as ensemble learning methods that combine traditional machine learning with deep learning models, could also be explored to capture more complex interactions within the data. Additionally, applying resampling techniques, such as Synthetic Minority Over-sampling Technique (SMOTE), or cost-sensitive learning, could mitigate the effects of class imbalance and improve the performance of the predictive framework for underrepresented categories. Incorporating continuous prediction models alongside classification models could help in making more precise predictions for shutdown durations and enhance real-time decision-making. Furthermore, integrating regression models to predict continuous durations alongside classification models could provide more precise insights into shutdown durations, significantly enhancing the practical utility of the framework. Another potential avenue for improvement involves incorporating external factors, such as economic or environmental variables, which could influence the shutdown duration indirectly and further refine model predictions.

3.2. Explainable AI: Feature-Based Insights into Shutdown Durations

XAI techniques provide a detailed understanding of the model's predictions by elucidating the importance of features and their impacts on individual outcomes. SHAP values were integrated for model interpretability, offering critical insights that influence practical applications such as resource allocation and scheduling decisions. By providing transparency into the contribution of each feature to the model's predictions, SHAP values enable stakeholders to make informed decisions about pipeline incident management based on predicted shutdown durations. Understanding the key factors—such as pipeline material, location, and timing—that influence shutdown durations allows emergency response teams to prioritize resources more effectively and optimize their response strategies. Figure 4 displays the SHAP summary plots by explaining for a LightGBM model, offering a comprehensive view of variable importance across all instances through mean absolute SHAP values. These values quantify the average impact of each feature on the model's predictions, allowing for the identification of critical factors that significantly influence the classification of shutdown durations. This analysis highlights key variables, including pipe installation layout, time of occurrence, and the type of transported commodities, which collectively shape the duration of pipeline shutdowns. For instance, incidents involving underground pipelines or occurring at night can be flagged as high-priority scenarios, requiring specialized equipment and additional personnel due to the complexity of repairs or limited resource availability during off-hours. Similarly, incidents involving highly volatile liquids (HVLs), as highlighted by SHAP analysis, necessitate more intensive safety protocols, such as containment measures, due to the higher risks they pose. These insights empower operators to allocate the appropriate resources in advance, minimizing delays and improving response efficiency.

3.2.1. Pipe Installation Layout

As depicted in Figure 4a, failures in underground pipelines are the predominant contributors to long-term shutdown durations. This is due to accessibility challenges, excavation requirements, and the intricate repairs needed for underground pipelines. Such repairs often involve specialized equipment and labor-intensive procedures, leading to significant delays in restoring operations [15]. On the other hand, incidents involving aboveground pipelines, shown in Figure 4c, are associated with short-term shutdown durations. The accessibility of these pipelines simplifies inspection and repair efforts, facilitating faster resolutions. Aboveground pipelines are particularly advantageous in emergency scenarios, where immediate access to the damaged infrastructure is critical. Reducing shutdown durations in these cases leads to lower operational downtime costs for operators and greater economic stability by minimizing disruptions in the supply of essential commodities.

3.2.2. Timing of Incidents

The time of occurrence is a decisive factor influencing shutdown durations. Incidents occurring during weekdays, as highlighted in Figure 4a, benefit from the availability of emergency response teams and operational resources, resulting in shorter durations. In contrast, weekend incidents experience delays due to reduced staffing levels and limited resource accessibility. Similarly, nighttime incidents pose unique challenges, as shown in Figure 4a. Limited visibility and reduced resource availability during the night necessitate additional safety measures and extended response times. Conversely, incidents occurring during the afternoon, as depicted in Figure 4c, are resolved more quickly due to optimal visibility, resource availability, and heightened staff engagement. These temporal dynamics underscore the importance of aligning emergency response strategies with the timing of incidents to optimize efficiency. By reducing shutdown durations during less optimal times (like weekends and nights), operators can lower emergency response costs, while society benefits from fewer interruptions in essential services.

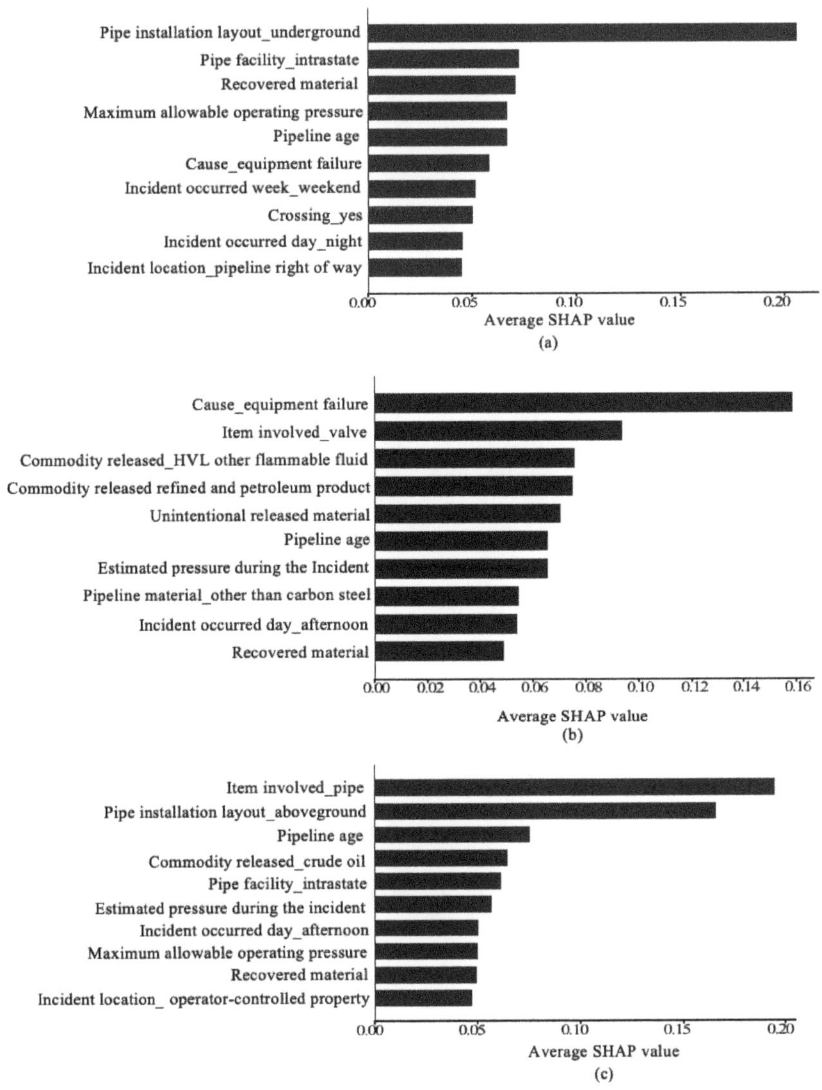

Figure 4. Global feature importance: (**a**) long-, (**b**) medium-, and (**c**) short-term shutdown duration.

3.2.3. Commodity Type

The type of commodity transported also plays a critical role in determining shutdown durations. As illustrated in Figure 4b, incidents involving highly volatile liquids (HVLs), such as refined petroleum products, often lead to medium-term shutdowns. The handling of HVLs necessitates comprehensive safety measures, including containment protocols and rigorous inspections, to mitigate secondary hazards. In contrast, incidents involving crude oil, as shown in Figure 4c, are typically resolved more swiftly due to well-documented repair procedures and standardized containment strategies. This highlights the need for tailored response approaches based on the commodity involved in an incident.

3.2.4. Incident Location

The location of pipeline incidents significantly impacts shutdown durations. As observed in Figure 4a, incidents occurring within the pipeline right-of-way are associated with prolonged durations. These locations often involve regulatory compliance requirements and the coordination of multiple stakeholders, which can delay response efforts. Conversely, incidents occurring within operator-controlled properties, as depicted in Figure 4c, are resolved more efficiently. Operator-controlled locations allow for streamlined decision-making and the quicker implementation of containment and repair measures, minimizing external dependencies and expediting recovery processes. Faster resolution of incidents on operator-controlled properties leads to lower operational disruption and financial losses, while also reducing the societal costs related to service interruptions.

3.2.5. Material and Facility Characteristics

The material of the pipeline and its facility type also influence shutdown durations. Failures in pipelines made of materials other than carbon steel, as shown in Figure 4b, require specialized repair techniques and materials, contributing to longer shutdown durations. Similarly, intrastate pipeline failures, as indicated in Figure 4a, often result in extended durations due to localized resource constraints and logistical challenges. In contrast, interstate pipelines benefit from centralized resource allocation and well-established repair protocols, enabling shorter resolution times.

3.2.6. Pressure Conditions and Pipeline Age

Maintaining pressure within allowable operating limits is a critical factor in minimizing shutdown durations. Pipelines operating within these limits, as highlighted in Figure 4c, experience reduced stress on their infrastructure, lowering the risk of critical failures and extensive repairs. Conversely, older pipelines, as depicted in Figure 4a,b, are strongly associated with prolonged shutdown durations. Aging infrastructure often requires comprehensive inspections and meticulous repairs to ensure safety and functionality, resulting in longer recovery times. Upgrading aging infrastructure can significantly reduce shutdown durations, leading to lower maintenance costs, enhanced pipeline reliability, and reduced societal costs such as energy shortages and delayed services.

The insights derived from this XAI analysis offer valuable guidance for optimizing emergency response strategies and resource allocation during pipeline incidents. By identifying specific scenarios that contribute to prolonged shutdowns, stakeholders such as pipeline operators, emergency responders, and regulatory authorities can prioritize resources more effectively. For instance, incidents involving underground pipelines or occurring at night can be flagged as high-priority scenarios requiring immediate attention. Additionally, understanding the influence of commodity type and incident location can inform the development of tailored response plans, ensuring timely and efficient resolution. Incorporating economic assessments into the predictive model could guide future investment decisions for pipeline infrastructure, ensuring that resources are allocated effectively to minimize both operational and societal costs associated with prolonged shutdowns. Future research could further enhance the predictive framework by incorporating additional contextual features, such as geographic and demographic data, to capture the broader impact of pipeline incidents. Integrating real-time data on material availability and emergency resource distribution could also improve the model's accuracy and practical applicability. By leveraging these insights, the overall management of pipeline incidents can be significantly improved, minimizing economic, environmental, and safety risks associated with prolonged shutdown durations.

4. Conclusions

The primary contribution of this study to the body of knowledge lies in the development of an innovative framework that integrates ML techniques and XAI to classify emergency shutdown durations and identify influential factors in pipeline incidents. This approach provides a structured methodology for classifying shutdown durations into different levels based on causal factors and pipeline characteristics, achieving an accuracy of 75% with the LightGBM. By addressing the complexities of pipeline incidents, this study introduces a proactive mechanism for emergency response and resource optimization, contributing to both academic research and practical applications. Furthermore, the integration of XAI techniques, particularly SHAP values, ensures transparency and interpretability, bridging the gap between predictive analytics and domain expertise. This allows pipeline operators, emergency responders, and regulatory authorities to better understand the factors influencing shutdown durations and make informed decisions during critical incidents. Moreover, the study's methodological rigor, combining robust preprocessing, advanced modeling, and interpretability, sets a foundation for future research in infrastructure management. The practical implications of this work are substantial, as the framework facilitates efficient resource allocation, risk mitigation, and emergency planning. By providing actionable insights into the key factors influencing pipeline shutdowns, it helps optimize decision-making and emergency response strategies. The study highlights the potential of ML and XAI to address complex infrastructure challenges, paving the way for future advancements in predictive modeling and emergency management strategies. Through its contributions to both academic and practical domains, this research enhances the understanding of pipeline incidents, offering a roadmap for integrating data-driven insights into real-world applications.

Despite the contributions made, this study acknowledges certain limitations. The dataset, while comprehensive, may lack important contextual variables, such as geographic constraints, workforce availability, or real-time operational factors, that could further refine model precision. Moreover, incorporating more diverse data sources, such as incident response times and external weather conditions, could capture additional complexity that influences shutdown durations. Additionally, potential class imbalances in the dataset may have impacted the predictive performance for underrepresented categories, such as long-term shutdown durations. Addressing these limitations could significantly enhance the model's applicability and reliability in real-world scenarios.

Future research will focus on incorporating these missing variables, such as geographic data, workforce factors, and real-time operational constraints, which will help refine the model and improve its predictive capabilities. Integrating continuous regression models alongside the current classification techniques to provide more granular insights into shutdown durations. Additionally, exploring ensemble approaches that combine regression and classification models could increase the robustness and accuracy of the predictions. Additionally, addressing class imbalances using techniques like SMOTE or incorporating more granular pipeline data, such as detailed configurations or material-specific properties, will likely improve the model's generalizability. Future work will explore economic impact assessments to evaluate the financial implications of reduced shutdown durations, providing stakeholders with a dual perspective on both the operational and financial aspects of pipeline incident management. This will help quantify cost savings, operational efficiencies, and the broader societal benefits of faster recovery times.

Author Contributions: Conceptualization, C.L. and T.Q.L.; methodology, C.L. and T.Q.L.; software, L.A.; validation, C.L., O.P.Y. and T.L.; formal analysis, L.A.; investigation, C.L.; resources, Y.H. and O.P.Y.; data curation, C.L. and L.A.; writing—original draft preparation, L.A.; writing—review and editing, C.L., Y.H. and T.L.; visualization, T.L.; supervision, C.L., O.P.Y. and Y.H.; project administration, Y.H. and O.P.Y.; funding acquisition, Y.H., O.P.Y. and T.Q.L. All authors have read and agreed to the published version of the manuscript.

Funding: The authors express their gratitude for the funding provided to support this study from the National Science Foundation (NSF) EPSCoR RII Track-2 Program under grant number OIA-2119691. The findings and opinions expressed in this article are those of the authors only and do not necessarily reflect the views of the sponsors.

Data Availability Statement: Data are contained within the article.

Conflicts of Interest: The authors declare no conflicts of interest.

Abbreviations

The following abbreviations are used in this manuscript:

ESD	Emergency Shutdowns Duration
SMOTE	Synthetic Minority Over-sampling Technique
GMM	Gaussian Mixture Mode
AIC	Akaike Information Criterion
SHAP	Shapley Additive exPlanations
ML	Machine Learning
XAI	Explainable AI
PHMSA	Pipeline and Hazardous Materials Safety Administration
LightGBM	Light Gradient Boosting Machine
RF	Random Fores
XGBoost	eXtreme Gradient Boosting
CatBoost	Categorical Boosting
MLP	Multilayer Perceptron
HVL	Highly Volatile Liquids

References

1. Chen, C.; Li, C.; Reniers, G.; Yang, F. Safety and security of oil and gas pipeline transportation: A systematic analysis of research trends and future needs using WoS. *J. Clean. Prod.* **2021**, *279*, 123583. [CrossRef]
2. Xiao, R.; Zayed, T.; Meguid, M.A.; Sushama, L. Understanding the factors and consequences of pipeline incidents: An analysis of gas transmission pipelines in the US. *Eng. Fail. Anal.* **2023**, *152*, 107498. [CrossRef]
3. Rusin, A.; Stolecka-Antczak, K.; Kapusta, K.; Rogoziński, K.; Rusin, K. Analysis of the effects of failure of a gas pipeline caused by a mechanical damage. *Energies* **2021**, *14*, 7686. [CrossRef]
4. Sotoodeh, K. Review of the role of safety engineers in the prevention and mitigation of fires during oil and gas plant design. *Life Cycle Reliab. Saf. Eng.* **2023**, *12*, 83–92. [CrossRef]
5. Liu, G.; Boyd, M.; Yu, M.; Halim, S.Z.; Quddus, N. Identifying causality and contributory factors of pipeline incidents by employing natural language processing and text mining techniques. *Process Saf. Environ. Prot.* **2021**, *152*, 37–46. [CrossRef]
6. Hassan, S.; Wang, J.; Kontovas, C.; Bashir, M. An assessment of causes and failure likelihood of cross-country pipelines under uncertainty using bayesian networks. *Reliab. Eng. Syst. Saf.* **2022**, *218*, 108171. [CrossRef]
7. Abdoul Nasser, A.H.; Ndalila, P.D.; Mawugbe, E.A.; Emmanuel Kouame, M.; Arthur Paterne, M.; Li, Y. Mitigation of risks associated with gas pipeline failure by using quantitative risk management approach: A descriptive study on gas industry. *J. Mar. Sci. Eng.* **2021**, *9*, 1098. [CrossRef]
8. Kumari, P.; Wang, Q.; Khan, F.; Kwon, J.S.I. A unified causation prediction model for above-ground onshore oil and refined product pipeline incidents using artificial neural network. *Chem. Eng. Res. Des.* **2022**, *187*, 529–540. [CrossRef]

9. Lam, C.; Zhou, W. Statistical analyses of incidents on onshore gas transmission pipelines based on PHMSA database. *Int. J. Press. Vessel. Pip.* **2016**, *145*, 29–40. [CrossRef]
10. Halim, S.Z.; Yu, M.; Escobar, H.; Quddus, N. Towards a causal model from pipeline incident data analysis. *Process Saf. Environ. Prot.* **2020**, *143*, 348–360. [CrossRef]
11. Asaye, L.; Ali Moriyani, M.; Le, C.; Le, T.; Prakash Yadav, O. Insights from Applying Association Rule Mining to Pipeline Incident Report Data. In Proceedings of the ASCE International Conference on Computing in Civil Engineering, Corvallis, OR, USA, 25–28 June 2023; pp. 763–771. [CrossRef]
12. Hameed, A.; Khan, F.; Ahmed, S. A risk-based methodology to estimate shutdown interval considering system availability. *Process Saf. Prog.* **2015**, *34*, 267–279. [CrossRef]
13. Hameed, A.; Khan, F.; Ahmed, S. A risk-based shutdown inspection and maintenance interval estimation considering human error. *Process Saf. Environ. Prot.* **2016**, *100*, 9–21. [CrossRef]
14. Yu, J.; Yi, J.; Mahgerefteh, H. Optimal emergency shutdown valve configuration for pressurised pipelines. *Process Saf. Environ. Prot.* **2022**, *159*, 768–778. [CrossRef]
15. Asaye, L.; Le, C.; Le, T.; Yadav, O.P.; Le, T. Insights into the Interactions of Pipeline Risk Factors and Consequences Using Association Rule Mining. *J. Perform. Constr. Facil.* **2025**, *39*, 04024059. [CrossRef]
16. Zhu, P.; Liyanage, J.P.; Kumar, R.; Panesar, S.S. Decision quality related to emergency shutdown system in the oil and gas industry: Influences from data and information. *Int. J. Decis. Sci. Risk Manag.* **2021**, *10*, 131–159. [CrossRef]
17. Vitali, M.; Zuliani, C.; Corvaro, F.; Marchetti, B.; Tallone, F. Statistical analysis of incidents on onshore CO_2 pipelines based on PHMSA database. *J. Loss Prev. Process Ind.* **2022**, *77*, 104799. [CrossRef]
18. Hainen, A.M.; Harbin, K.B.; Dye, D.; Lindly, J.K. Duration analysis of emergency shutdown incidents regarding hazardous liquid pipelines. *J. Perform. Constr. Facil.* **2020**, *34*, 04020040. [CrossRef]
19. Remil, Y.; Bendimerad, A.; Mathonat, R.; Kaytoue, M. Aiops solutions for incident management: Technical guidelines and a comprehensive literature review. *arXiv* **2024**, arXiv:2404.01363.
20. Khan, F.; Rathnayaka, S.; Ahmed, S. Methods and models in process safety and risk management: Past, present and future. *Process Saf. Environ. Prot.* **2015**, *98*, 116–147. [CrossRef]
21. Christodoulou, S.; Deligianni, A.; Aslani, P.; Agathokleous, A. Risk-based asset management of water piping networks using neurofuzzy systems. *Comput. Environ. Urban Syst.* **2009**, *33*, 138–149. [CrossRef]
22. Chen, Y.; Zhang, L.; Hu, J.; Liu, Z.; Xu, K. Emergency response recommendation for long-distance oil and gas pipeline based on an accident case representation model. *J. Loss Prev. Process Ind.* **2022**, *77*, 104779. [CrossRef]
23. Amare, M.D.; Gedafa, D.S. Assessing Traffic Safety in Cold Regions for Sustainable and Resilient Infrastructure: A Spatial Analysis and Association Rule Mining Approach. In *Cold Regions Engineering 2024: Sustainable and Resilient Engineering Solutions for Changing Cold Regions*; ASCE: Anchorage, AK, 2024; pp. 174–185.
24. Amare, M.D.; Gedafa, D.S. Comparison of the Effect of In-Crosswalk Traffic Signs on Pedestrian Safety. In Proceedings of the International Conference on Transportation and Development 2024, Atlanta, Georgia, 15–18 June 2024; pp. 281–293.
25. Xiang, Q.; Yang, Z.; He, Y.; Fan, L.; Su, H.; Zhang, J. Enhanced method for emergency scheduling of natural gas pipeline networks based on heuristic optimization. *Sustainability* **2023**, *15*, 14383. [CrossRef]
26. Mukhopadhyay, A.; Pettet, G.; Vazirizade, S.M.; Lu, D.; Jaimes, A.; El Said, S.; Baroud, H.; Vorob-eychik, Y.; Kochenderfer, M.; Dubey, A. A review of incident prediction, resource allocation, and dispatch models for emergency management. *Accid. Anal. Prev.* **2022**, *165*, 106501. [CrossRef]
27. Mukhopadhyay, A.; Pettet, G.; Vazirizade, S.; Lu, D.; Said, S.E.; Jaimes, A.; Baroud, H.; Vorobey-chik, Y.; Kochenderfer, M.; Dubey, A. A Review of Incident Prediction, Resource Allocation, and Dispatch Models for Emergency Management. *arXiv* **2020**, arXiv:2006.04200.
28. Moriyani, M.A.; Asaye, L.; Le, C.; Le, T.; Le, T. Natural Language Processing for Infrastructure Resilience to Natural Disasters: A Scientometric Review. In Proceedings of the International Conference series on Geotechnics, Civil Engineering and Structures, Ho Chi Minh City, Vietnam, 4–5 April 2024; Springer: Berlin/Heidelberg, Germany, 2024; pp. 1506–1513.
29. Maulud, D.; Abdulazeez, A.M. A review on linear regression comprehensive in machine learning. *J. Appl. Sci. Technol. Trends* **2020**, *1*, 140–147. [CrossRef]
30. Qiu, W.; Chen, H.; Dincer, A.B.; Lundberg, S.; Kaeberlein, M.; Lee, S.I. Interpretable machine learning prediction of all-cause mortality. *Commun. Med.* **2022**, *2*, 125. [CrossRef]
31. Mersha, M.; Lam, K.; Wood, J.; AlShami, A.; Kalita, J. Explainable artificial intelligence: A survey of needs, techniques, applications, and future direction. *Neurocomputing* **2024**, *599*, 128111. [CrossRef]

32. Capshaw, K.M.; Padgett, J.E. Development and Application of a Predictive Model for Estimating Refinery Shutdown Duration and Resilience Impacts Due to Hurricane Hazards. *ASCE-ASME J. Risk Uncertain. Eng. Syst. Part B Mech. Eng.* **2023**, *9*, 031101. [CrossRef]
33. Ramírez-Camacho, J.G.; Carbone, F.; Pastor, E.; Bubbico, R.; Casal, J. Assessing the consequences of pipeline accidents to support land-use planning. *Saf. Sci.* **2017**, *97*, 34–42. [CrossRef]
34. Aalirezaei, A.; Kabir, D.G.; Khan, M.S.A. Dynamic predictive analysis of the consequences of gas pipeline failures using a Bayesian network. *Int. J. Crit. Infrastruct. Prot.* **2023**, *43*, 100638. [CrossRef]
35. Al-Douri, A.; Halim, S.Z.; Quddus, N.; Kazantzi, V.; El-Halwagi, M.M. A stochastic approach to evaluating the economic impact of disruptions in feedstock pipelines on downstream production. *Process Saf. Environ. Prot.* **2022**, *162*, 187–199. [CrossRef]
36. Liu, B.; Liu, C.; Zhou, Y.; Wang, D.; Dun, Y. An unsupervised chatter detection method based on AE and merging GMM and K-means. *Mech. Syst. Signal Process.* **2023**, *186*, 109861. [CrossRef]
37. Ekanayake, I.; Meddage, D.; Rathnayake, U. A novel approach to explain the black-box nature of machine learning in compressive strength predictions of concrete using Shapley additive explanations (SHAP). *Case Stud. Constr. Mater.* **2022**, *16*, e01059. [CrossRef]
38. Fan, C.; Zhang, N.; Jiang, B.; Liu, W.V. Preprocessing large datasets using Gaussian mixture modelling to improve prediction accuracy of truck productivity at mine sites. *Arch. Min. Sci.* **2022**, *67*, 661–680.
39. Patel, E.; Kushwaha, D.S. Clustering cloud workloads: K-means vs gaussian mixture model. *Procedia Comput. Sci.* **2020**, *171*, 158–167. [CrossRef]
40. Chan, H.M.T. On Evaluating the Watanabe-Akaike Information Criteria for Bayesian Modelling of Point Referenced Spatial Data. Ph.D. Thesis, University of Southampton, Southampton, UK, 2024.
41. Rodríguez-Pérez, R.; Bajorath, J. Interpretation of machine learning models using shapley values: Application to compound potency and multi-target activity predictions. *J. Comput.-Aided Mol. Des.* **2020**, *34*, 1013–1026. [CrossRef]

processes

MDPI

Article

Machine Learning Predictions for the Comparative Mechanical Analysis of Composite Laminates with Various Fibers

Baha Eddine Ben Brayek [1], Sirine Sayed [2,3], Viorel Mînzu [4,*] and Mostapha Tarfaoui [1,5]

[1] Mohammed VI Polytechnic University, GEP/GSMI, Ben Guerir 43150, Morocco; bahaeddine.benbrayek@um6p.ma (B.E.B.B.); mostapha.tarfaoui@um6p.ma (M.T.)
[2] ENSTA Bretagne, IRDL–UMR CNRS 6027, F-29200 Brest, France; sirine.sayed@ensta.fr
[3] International University of Rabat, LERMA, Rabat 11103, Morocco
[4] Control and Electrical Engineering Department, "Dunarea de Jos" University of Galati, 800008 Galati, Romania
[5] Mohammed VI Polytechnic University, Green Energy Park, Ben Guerir 43150, Morocco
* Correspondence: viorel.minzu@ugal.ro

Abstract: This article addresses the complex behavior of composite laminates under varied layer orientations during tensile tests, focusing on carbon fiber and epoxy matrix composites. Data characterizing the mechanical load behavior are obtained using twelve composite laminates with different layer orientations and the DIGIMAT-VA software (version 2023.3). First, these data were used to elaborate a complex comparative analysis of composite laminates from the perspective of materials science. Composite laminates belong to three classes: unidirectional, off-axis oriented, and symmetrically balanced laminates, each having a specific behavior. From the perspective of designing a new material, a prediction model that is faster than the finite element analysis is needed to apply this comparative analysis's conclusions. As a novelty, this paper introduces several machine learning prediction models for composite laminates with 16 layers arranged in different orientations. The Regression Neural Network model performs best, effectively replacing expensive tensile test simulations and ensuring good statistics (RMSE = 34.385, R^2 = 1, MAE = 19.829). The simulation time decreases from 34.5 s (in the case of finite element) to 0.6 s. The prediction model returns the stress–strain characteristic of the elastic zone given the new layer orientations. These models were implemented in the MATLAB system 2024, and their running proved good models' generalization power and accuracy. Even specimens with randomly oriented layers were successfully tested.

Keywords: composite materials; machine learning; regression neuronal network

1. Introduction

Composite materials stand at the intersection of innovation and practical engineering, blending diverse constituents to unlock superior properties and address the limitations of conventional materials. As Nachtane and Tarfaoui [1,2] insightfully observe, these materials ingeniously merge fibers with matrices, crafting systems surpassing traditional materials in mechanical robustness and offering significant weight reduction and durability advantages, fulfilling critical roles in demanding applications. El Moumen et al. [3] elaborate that this precise integration allows for the strategic engineering of fiber arrangements within the matrix, enhancing the adaptability of composites to meet exacting performance standards across various industrial uses. The versatility of composite materials is particularly evident in sectors requiring high-performance materials. Daly et al. [4] note that composites are ideally suited for critical aerospace, automotive, and military applications due to their ability to be molded into complex shapes and intrinsic properties like corrosion and fatigue resistance. This adaptability also extends to infrastructure projects and medical implants, where composites provide innovative solutions to traditional challenges, tailoring mechanical properties to meet precise engineering needs. Technological advancements are at the heart of the evolution of composite materials. Using machine learning (ML) techniques, Yang et al. [5] propose stress–strain models beyond the elastic limit for a category of composite materials; these models establish a link between their composite microstructures and their mechanical properties.

Ongoing innovations in fiber technology and matrix improvements have broadened the capabilities of composites, with developments in nanotechnology and enhanced interfacial bonding techniques allowing these materials to withstand more extreme conditions. These advancements are pivotal for industries like renewable energy, where long-term performance under harsh environmental conditions is crucial. Further driving the capabilities of composite materials, Tarfaoui et al. [2] and Rajak et al. [6] discuss how recent innovations in fiber technology and matrix formulations have enabled the creation of materials that can endure increasingly harsh conditions. These technological signs of progress are essential for meeting the stringent demands of sectors such as the renewable energy and automotive industries, where materials must perform reliably over extended periods. The science of composite materials harnesses the unique properties of diverse constituents, innovatively combining them to meet exacting performance standards. Khammassi et al. [7] demonstrate how adding vermiculite, silver, and graphene oxide to PLA-based nanocomposites significantly boosts their thermal stability and mechanical properties, enabling their use in high-demand environments. Gan [8] explores the crucial role of interface structures, revealing that enhancing interfacial bond strength can lead to a 40% increase in overall material durability and performance, proving pivotal in applications that demand high strength and longevity.

Additionally, Rajak et al. [6] cover expanding composites' capabilities through novel reinforcement materials that boost mechanical attributes, enhance thermal resistance and environmental adaptability, and expand potential uses across cutting-edge engineering fields. The matrix's selection directly influences a composite's performance by defining its response to environmental and mechanical stresses. Polymer matrices such as epoxies provide excellent adhesion and flexibility, with typical tensile strengths up to 100 MPa and moduli ranging from 3 to 4 GPa, making them versatile for aerospace and consumer applications [9,10]. Metal matrices incorporate materials like aluminum and titanium alloys, offering superior thermal conductivity and mechanical properties with 200 to 400 MPa yield strengths, tailored for high-load aerospace applications [11,12].

In the strategic synthesis of composite materials, selecting fibers and particles is critical for balancing strength, stiffness, and environmental resistance. Carbon fibers, for instance, are known for their high tensile strength, reaching up to 7 GPa. A modulus of elasticity of around 230 GPa, making them ideal for applications demanding minimal weight and maximum strength [13–16]. Glass fibers offer a more economical reinforcement option with a tensile strength of around 3.5 GPa and a modulus of 85 GPa, well suited for less critical structural applications [17–19]. Aramid fibers, noted for their impact resistance, provide tensile strengths up to 3.6 GPa and a modulus of 70 GPa, which are ideal for military and aerospace applications where durability is paramount [20–23].

Integrating artificial intelligence (AI) into composite material science has catalyzed significant advancements in predictive modeling and optimization. For instance, Vahed et al. [24] demonstrate how neural networks can enhance the prediction of dynamic mechanical properties, achieving a notable accuracy improvement of up to 30% over traditional methods. This precision significantly reduces the experimental workload by optimizing material properties efficiently.

Similarly, Ho et al. [25] use ML to predict the Young's modulus of polymer composites reinforced with carbon nanotubes, achieving a predictive accuracy that surpasses traditional testing methods by approximately 25%, thus modernizing the development process for new materials. Yang et al. [5] further contribute to this field by employing convolutional neural networks combined with principal component analysis to accurately predict stress–strain behaviors from microstructural images of composites. Their method has reduced computational time by over 40% compared to conventional finite element analysis, providing a rapid and scalable tool for material design. Moreover, as explored by Dotoli et al. [26], virtual testing allows for the simulation of material performance under real-world conditions with a precision of up to 95% correlation with physical testing results. This approach saves substantial time and resources and ensures that aerospace composites meet rigorous safety standards before implementation.

In conjunction with traditional composite constituents (such as fibers, particles, and matrices), the application of AI in optimizing the interface and interaction between these components has shown promising results. Research indicates that AI-driven optimization of fiber orientation and matrix bonding can enhance composites' mechanical strength and thermal stability by up to 50%, depending on the material specifications and environmental conditions [4,14]. Campbell [27] emphasizes AI's role in predicting and replicating composites' mechanical behavior under varied stress conditions. This capability allows for developing materials tailored to specific industrial needs with improved reliability and performance, further pushing the boundaries of what can be achieved with composite technologies.

Our work has developed along two axes from distinct scientific fields: composite material science and AI, more precisely, machine learning. The connection between the two axes lies in that machine learning aims to address specific challenges in using composite materials: optimal design and precise behavior prediction.

Considering the first axis, this article delves into the intricate behavior of composite laminates under varied orientations during tensile testing [28–32], focusing on carbon fiber and epoxy matrix composites. Employing the advanced DIGIMAT-VA software, we analyzed the stress–strain relationships and assessed rupture criteria to gain profound insights into the material's response under mechanical loading. This analysis forms the basis for the subsequent explorations into material science and engineering [33,34]; it refers to both zones of the stress–strain curve obtained during tensile testing, i.e., undamaged (elastic) and damaged zones.

The second axis is an innovative method that utilizes ML models to predict complex behaviors for known or new stratification combinations. These models, developed using ML algorithms [35–39], enable the simulation and prediction of composite performance with unmatched accuracy and efficiency, significantly reducing the time and cost associated with experimental testing. Integrating ML into research methodologies [40,41] transcends traditional boundaries, leading to a new predictive and adaptive materials science stage. The potential of this approach is significant, leading to optimizing material design to enhance performance. However, this paper does not address the optimization of composite laminates.

Currently, our ML prediction method only focuses on the undamaged zone. We have postponed developing models for the damaged zone as future work. Therefore, the proposed ML models are based solely on data from the undamaged zone. Nevertheless, this article also presents aspects of comparative mechanical analysis, including the damaged zone, for twelve specimens of laminate composites with multiple layers. Besides the other interesting aspects this analysis reveals, knowing where the damaged zone begins is crucial.

This article's structure is briefly presented. The material used in our work is described in Section 2. It is about twelve specimens used in tensile tests made of the unidirectional composite material AS4/8552-UD that combines AS4 carbon fibers with an 8552-epoxy resin matrix. Each specimen has 16 layers of different orientations, which are given for all 12 specimens. Section 3 presents how the DIGIMAT-VA software is leveraged to produce the stress–strain curves. These curves provide the data for constructing the ML models, with preparation details described in Section 3.2.

The implementation of ML models is the goal of Section 4. Special attention was paid to the implementation aspects, illustrated using specific datasets, to help the interested reader understand and eventually construct their model. This section explains how ML algorithms can be harnessed to train and test the dataset characterizing a specific specimen. Among the algorithms we implemented, we present three along with their models: Multiple Linear Regression, Support Vector Machine, and Regression Neural Network. The details in this section (and Section 5) relate to the models' implementation using the MATLAB system 2024 [38,39]. The predicted and actual values are compared for both the training and test processes. Statistics for the training and test process results are provided for all five constructed ML models.

Section 5 first aims to prove the generalization power of the prediction models constructed in the previous section context. The generalization power of an ML prediction model (RNN2, a Regression Neural Network [39] developed in Section 4) is evaluated using data points the algorithm has never encountered; these points are not part of the training or test datasets.

Section 5.1 addresses the cases where new combinations share the same structure as a specimen used in the model construction, differing only in a few layer orientations. A variable parametrizes the set of new stratification combinations. We can determine the optimal characteristic of stress versus strain by predicting all new stratification combinations and conducting a mechanical analysis, allowing us to find the best parameter value. This procedure can be regarded as an optimization tool for a design problem.

In Section 5.2, the generalization power of RNN2 is evaluated under a tougher context: specimens with randomly generated layer orientations. To stay within the generalization area of the ML model, we first selected four base specimens (to make this presentation easy to follow) subjected to tensile tests, contributing to the dataset used for training the RNN2 model. These specimens exhibit different behaviors in the stress–strain space.

Each layer orientation is independently modified using a uniformly distributed perturbation. The resulting specimens, with randomly generated orientations, are significantly different from the basic ones but remain within the representation area of the model. Then, we compare the predictions for these specimens with the results of DIGIMAT simulations. The comparison shows that the prediction accuracy is greatly satisfactory for the new randomly generated specimens; RNN2 has very good generalization power. However, the ML model must be properly employed under some constraints. For example, in this stage, the strain values must be inside the range corresponding to the elasticity zone.

Section 6 mainly discusses the results of tensile tests (42–49) and provides a comparative mechanical analysis of the twelve specimens to explain their load behavior. The analysis brings to light the fact that the twelve specimens belong to three classes: unidirectional laminates, off-axis oriented laminates, and symmetrically balanced laminates, each of them having a specific behavior.

The section devoted to conclusions briefly reviews our paper's findings and establishes the subject of future research.

Our presentation focused on practical implementation [38,39] to help interested readers understand and potentially replicate or apply parts of this work to their projects. To this end, all algorithms used in our work are fully implemented, with accompanying scripts provided as Supplementary Materials. Additionally, all necessary details are included in the appendices.

2. Materials

2.1. Material Properties

AS4/8552-UD is a meticulously engineered unidirectional composite material that combines AS4 carbon fibers with an 8552-epoxy resin matrix. It is specifically designed for applications requiring exceptional mechanical properties. The unidirectional fiber orientation optimizes mechanical strength along the axis of fiber alignment. It is particularly suited for components subjected to uniaxial stresses, such as those found in aerospace structures and high-performance automotive parts. For this study, AS4/8552-UD has been selected due to its significant advantages over conventional materials like aluminum or steel, particularly regarding strength-to-weight efficiency. The high fiber volume fraction of approximately 59%, combined with the exceptional mechanical properties of the fibers and the matrix, allows the composite to maintain a lightweight structure while offering substantial structural integrity. As summarized in Table 1, the AS4 carbon fibers exhibit outstanding axial tensile properties, with an axial Young's modulus of approximately 217,687 MPa and a tensile strength of around 3413 MPa. These properties significantly enhance the composite's tensile load-bearing capacity along the primary fiber axis. The fibers also display high compressive strength and favorable transverse mechanical properties, contributing to the composite's overall performance under multidirectional stresses.

Table 1. DIGIMAT material model for AS4 carbon fibers.

Property	Tension	Compression
Axial Young's modulus (MPa)	217,687	184,220
In-plane Young's modulus (MPa)	15,236	16,595
Transverse shear modulus (MPa)	15,818	15,818
In-plane Poisson's ratio	0.22	0.26
Transverse Poisson's ratio	0.27	0.32
Tensile strength (MPa)	3413	3413
Compressive strength (MPa)	2366	2366

Similarly, the 8552-epoxy resin matrix, detailed in Table 2, provides a Young's modulus of about 4668 MPa and tensile and compressive strengths of approximately 56 MPa and 232 MPa, respectively. The matrix's mechanical properties ensure effective stress transfer between fibers and enhance the composite's ability to absorb energy during deformation, improving toughness and durability. The matrix also contributes to the composite's resistance to environmental factors, including corrosion and fatigue, which are common issues with traditional metallic materials.

The synergy between the high-strength fibers and the resilient matrix results in a composite material that outperforms traditional materials regarding strength-to-weight ratio, fatigue resistance, and environmental durability. This fact makes AS4/8552-UD an ideal choice for high-performance aerospace and automotive applications, where reducing weight is directly tied to economic and ecological benefits. For instance, utilizing AS4/8552 composites in modern aircraft can enhance fuel efficiency by up to 12%, aligning with industry priorities for sustainability and reducing carbon emissions (see [1,40]). By incorporating the superior mechanical properties highlighted in Tables 1 and 2, this study leverages the advanced characteristics of the AS4/8552-UD composite. This alignment with contemporary advancements in composite material technology enables us to meet the rigorous demands of modern engineering applications, providing tailored solutions that enhance both performance and sustainability.

Table 2. Digimat material model for 8552 Epoxy matrix.

Property	Tension	Compression
Young's modulus (MPa)	4668	4668
Poisson's ratio	0.35	0.35
Tensile strength (MPa)	56	56
Compressive strength (MPa)	232	232
Shear strength (MPa)	62	62

2.2. Orientation and Stress Application

In composite materials, the strategic selection of fiber orientations plays a pivotal role in tailoring their mechanical properties to meet specific operational demands. In this study, we focus on a primary type of fiber orientation: unidirectional, which is integral to optimizing the performance and application of the AS4/8552-UD composites. Unidirectional orientations, where fibers are aligned along a single axis, are essential for applications requiring high strength and stiffness in one direction. This option is particularly beneficial in aerospace and automotive applications where aligning fibers along the load path can significantly enhance the load-bearing capacity while minimizing weight, a critical factor in fuel efficiency and performance enhancements [1,37]. The practical implementation of these fiber orientations leverages advanced modeling and simulation tools to predict and optimize the behavior of composites under various stress conditions. Using DIGIMAT-VA software (2023.3 version), we apply virtual stress tests to explore the responses of different fiber orientations under simulated mechanical loads. This approach helps refine the composite design by providing insights into how each orientation influences the material's overall structural integrity and performance. Simulating these conditions is invaluable, significantly reducing the reliance on costly and time-consuming physical testing while accelerating the development cycle of new composite applications ([2,41]). The AS4/8552-UD composite material investigation was characterized by a unidirectional load and a fiber volume fraction of 59%. The capabilities of DIGIMAT-VA software were harnessed to conduct tensile tests in alignment with the actual standard, as illustrated in Figure 1. We carefully prepared twelve distinct specimen configurations, detailed in Table 3, each featuring unique fiber orientations to extensively assess the material's behavior under varied stress scenarios.

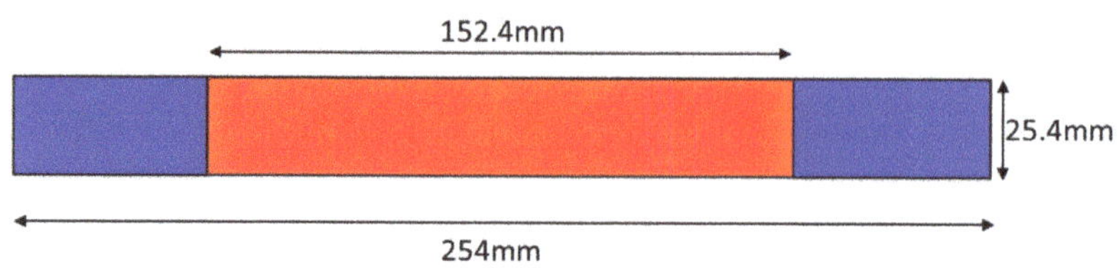

Figure 1. Tensile test specimen (blue color–seizing ends).

Table 3. Layup configurations and their designations.

Layups Configuration	Designation
$[0]_{16}$	S-1
$[\pm 20]_8$	S-2
$[\pm 30]_8$	S-3
$[\pm 45]_8$	S-4
$[\pm 60]_8$	S-5
$[\pm 70]_8$	S-6
$[90]_{16}$	S-7
$[0/45/0/90/0/-45/0/45]_s$	S-8
$[45/0/-45/90]_{2s}$	S-9
$[45/-45/0/45/-45/90/45/-45]_{2s}$	S-10
$[0/30/0/90/0/-30/0/30]_{2s}$	S-11
$[60/0/-60/90]_{2s}$	S-12

These configurations were segmented into three categories, as illustrated in Figure 2. Unidirectional specimens included S-1 with a $[0]_{16}$ orientation and S-7 with a $[90]_{16}$ orientation to test the response along parallel and perpendicular fiber alignments relative to the load. Off-axis oriented specimens comprised S-2 ($[\pm 20]_8$), S-3 ($[\pm 30]_8$), S-4 ($[\pm 45]_8$), S-5 ($[\pm 60]_8$), and S-6 ($[\pm 70]_8$), exploring the effects of fibers oriented at angles diverging from the principal stress directions. Lastly, symmetric balanced specimens, such as S-8 ($[0/45/0/90/0/-45/0/45]_s$), S-9 ($[45/0/-45/90]_{2s}$), S-10 ($[45/-45/0/45/-45/90/45/-45]_s$), S-11 ($[0/30/0/90/0/-30/0/30]_s$), and S-12 ($[60/0/-60/90]_{2s}$), were evaluated to determine their performance under symmetrically balanced loads, simulating complex operational conditions. This elaborate testing methodology highlights the versatility of the DIGIMAT-VA software. It provides essential insights into how fiber orientation impacts the mechanical performance of the composite, informing its potential applications across diverse industrial sectors.

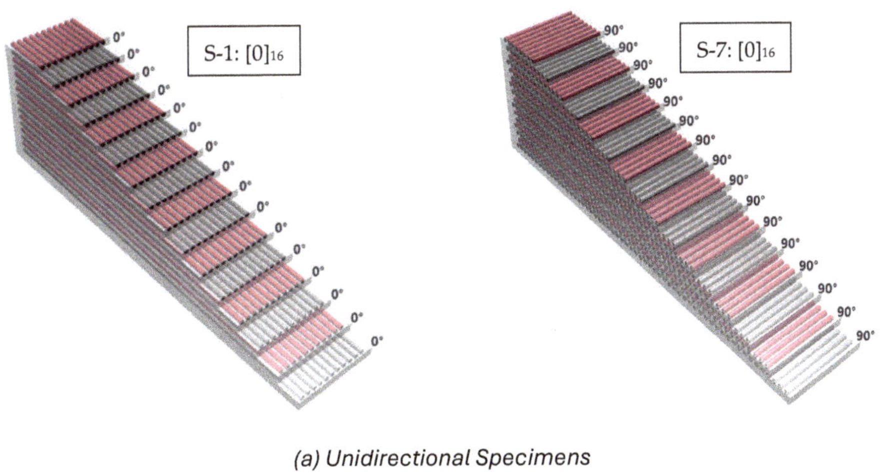

Figure 2. *Cont.*

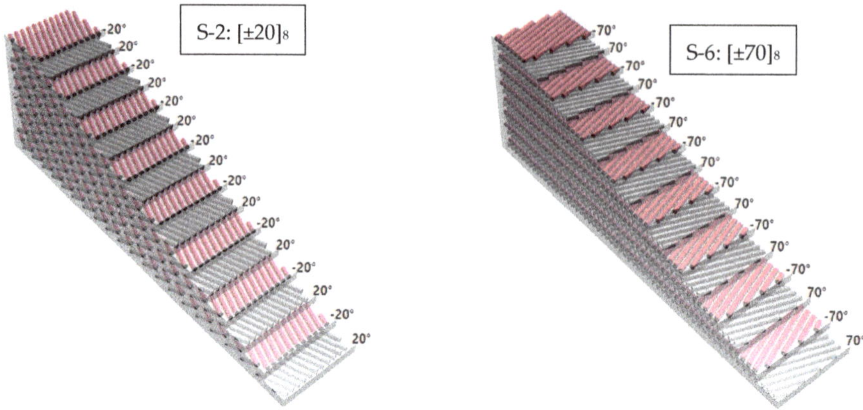

(b) Off-Axis Oriented Specimens

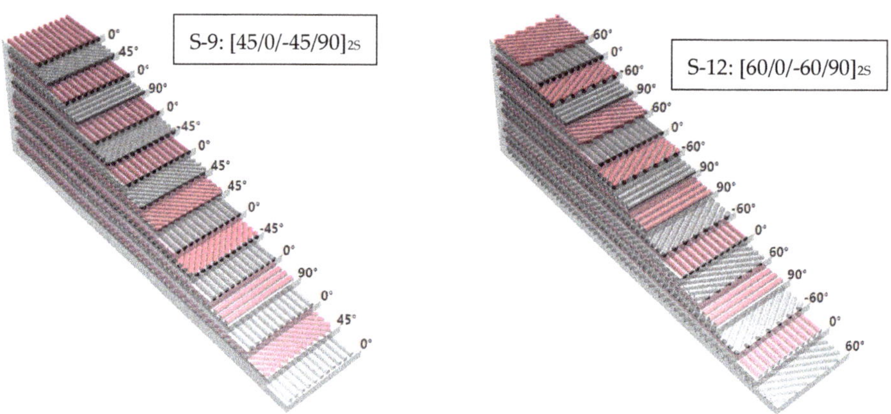

(c) Symmetric Balanced Specimens

Figure 2. Specimen classification.

Digimat-VA was employed in this study to input the essential experimental data needed to simulate composite materials accurately. These data include comprehensive mechanical properties for the matrix (Epoxy 8552-UD) and the reinforcement (AS4 fibers), which were incorporated into the software to ensure that the virtual material model closely mirrors real-world behavior under varying stress conditions. The fiber properties, detailed in Table 1, include axial and in-plane Young's modulus (217,687 MPa and 15,236 MPa, respectively), which ensure that the fibber's contribution to stiffness and load distribution is adequately accounted for. The tensile strength (3413.08 MPa) and compressive strength (2366.15 MPa) further solidify the robustness of the AS4 fibers under uniaxial and multiaxial stresses.

As illustrated in Table 2, key parameters such as Young's modulus (4667.7 MPa for tension and compression) and Poisson's ratio (0.35) were used to define the material's elastic and plastic behavior. Additionally, the matrix's tensile and compressive strength was set at 56.12 MPa and 231.67 MPa, respectively, providing a foundation for how the epoxy responds under applied loads. As highlighted in Tables 4–6, the experimental data used for the ply properties showcase essential metrics such as tensile and compressive moduli, strengths, and shear properties. These inputs (nomenclature given in Appendix A), including Young's modulus, shear modulus, and Poisson's ratio, help simulate the composite's behavior under different loading conditions, ensuring the material response is precisely captured. Incorporating these experimental values into Digimat VA streamlines the simulation process, eliminates the need for repeated physical testing, and creates a highly accurate virtual model. This data-driven approach enhances the fidelity of simulations and ensures reliable predictions of the composite's performance in structural applications.

Table 4. Experimental data (tension/compression).

Ply Property	Value (MPa)	Volume Fraction
E_1^t	131,550 $\pm$ 12	0.5956
F_1^t	2063 $\pm$ 10	0.5956
E_2^t	9239 $\pm$ 12	0.5872
F_2^t	64 $\pm$ 4	0.5872
ν_{12}^t	0.302	0.5956
E_1^c	115,560 $\pm$ 12	0.6176
F_1^c	1484 $\pm$ 7	0.6176
E_2^c	9860 $\pm$ 10	0.6148
F_2^c	268 $\pm$ 12	0.6148
ν_{12}^c	0.335	0.6148

Table 5. Experimental data (shear).

Ply Property	Value
G_{12}	4826 $\pm$ 14 MPa
F_{12} (0.2% offset)	55 MPa
F_{12}	92 MPa
Shear properties volume fraction	0.5885

Table 6. Experimental data (matrix).

Ply Property	Value
Density t/mm^3	1.301×10^{-9}
Young's modulus (tension) (MPa)	4668 $\pm$ 11
Young's modulus (compression) (MPa)	4668 $\pm$ 16
Poisson's ratio	0.35

3. Method

3.1. Stress vs. Strain Analysis

DIGIMAT-VA (version 2023.3) software yielded the function stress vs. strain for each specimen, including the tensile test's elastic and damaged zones. In our work concerning ML predictions, only the elastic zone of these curves is considered because, so far, our interest covers only this zone. We have highly accurate simulation results equivalent to physical tensile tests. Therefore, the stress value for a given strain will be regarded as the actual stress value when compared to the ML predicted value to assess the quality of the ML model. In other words, we work with two curves, stress vs. strain type, resulting from the DIGIMAT-VA simulation and the ML model's prediction. Both curves are considered "experimental" results.

The simulation results for the 12 specimens allow us to perform a quite complex mechanical analysis covering the elastic and damaged zones (details in Section 6). Our main objective is to propose ML models that can produce accurate predictions.

By analyzing these behaviors, we establish the foundation for the next phase of this study, which utilizes artificial intelligence (AI) techniques to predict these complex material responses. Prediction models for the behavior of composite laminates will be constructed using various machine learning methods. Once validated against experimental data, the ML approach extends the analysis beyond the initial composite set. The predictive capability validates the models' reliability and enables us to explore new combinations of materials and orientations, optimizing their mechanical properties without requiring extensive physical testing. This integration of ML models marks a significant advancement, enabling the discovery of advanced composite configurations with customized performance characteristics.

3.2. Machine Learning Approach

The ML approach generates models for the stress–strain behavior observed during tensile tests of various specimens. Preparing the dataset for the learning process is the initial step, followed by constructing different parametric and nonparametric ML models. As mentioned, the main objective is to construct ML models that can comprehend all the measurements described earlier and predict the stress value for any pattern (combination of orientations) at a given strain value. We established specific objectives to achieve our ultimate aim, which will be accomplished through the following steps:

- Generate a dataset of significant size. This dataset will be used to construct the ML model, enabling it to provide a generalized response for any pattern and an appropriate strain value.
- Construct a parametric model (e.g., multiple linear regression) that is easy to understand and apply and can be compared with the following models.
- Construct some nonparametric models (SVM, decision trees, Gaussian process regression, and neural networks), analyze their accuracy, and compare them to the parametric model. Out of many ML models investigated, we chose to present two nonparametric models (SVM and Regression NN). The last ones yielded four trained and tested models, the most effective and appropriate to the considered dataset.
- Select the most accurate parametric models that could be used in further research, providing a solid foundation for future studies.

Remark 1. *For our problem, many SVM and Regression NN models could be constructed, with some of them having potentially superior capabilities to predict the behavior of the stencils. Initially, our objective was not to find the best but to validate our approach: to prove that ML models can accurately predict stress–strain behavior.*

3.3. Data Preparation

As presented before, the data obtained using the synergistic combination of numerical/experimental investigations are processed to generate the dataset needed for the ML models' construction. This dataset is used for the training and testing phases of the considered ML model.

To describe our approach, we consider the data collected from twelve specimens' tests, which were loaded within different stress and strain ranges. Linear segments approximated the results, as shown in Figure 3.

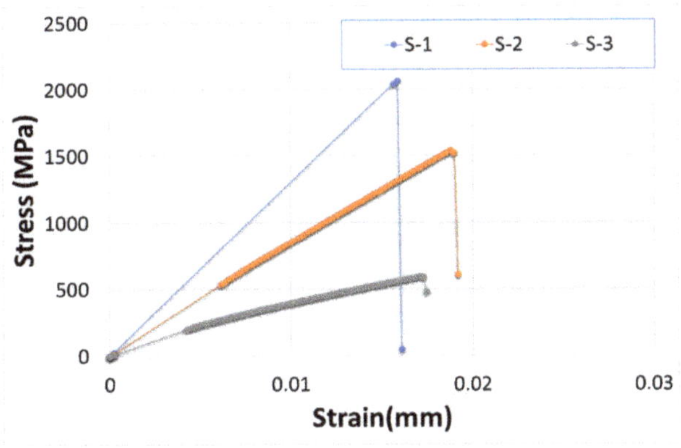

(a) The stress–strain dependence during the tensile test of S1 to S3.

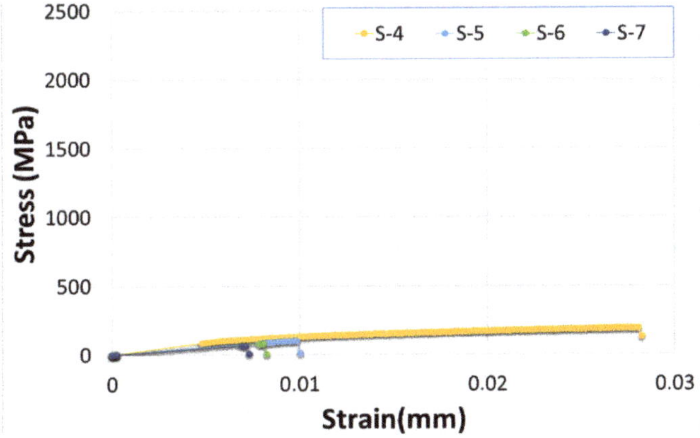

(b) The stress–strain dependence during the tensile test of S4 to S7.

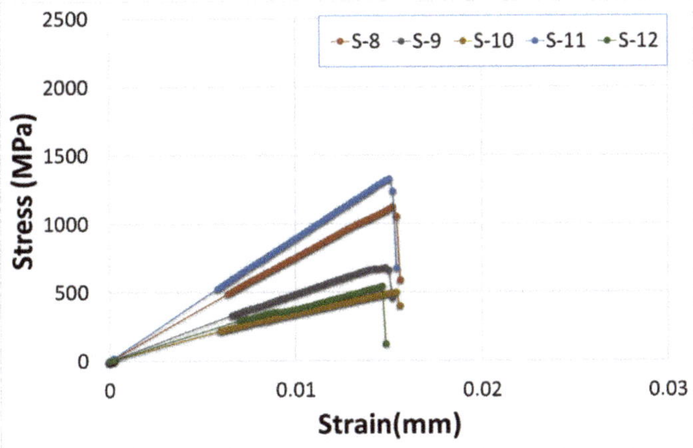

(c) The stress–strain dependence during the tensile test of S8 to S12.

Figure 3. The stress–strain dependence during the tensile test of the twelve specimens.

Table 7 shows the minimum and maximum limits of the stress and strain parameters for the twelve specimens.

Table 7. Limits of strain and stress values.

	Strain (Max)	Stress (Max)
S-1	0.01587015	2064.24019
S-2	0.01874477	1540.16043
S-3	0.01705347	593.258184
S-4	0.02798728	192.457154
S-5	0.00981914	101.786008
S-6	0.00803761	77.3136431
S-7	0.00709987	63.9780266
S-8	0.01537992	1121.76279
S-9	0.01478241	680.072911
S-10	0.01536462	499.329696
S-11	0.01499824	1327.03122
S-12	0.0146055	540.079359

The stress values obtained through physical measurements can be expressed with an accuracy of ± 1 MPa, which is already highly accurate. This fact contradicts the values indicated in Table 7 and throughout this paper because we have considered stress values expressed in MPa with a few decimal points. In our study, the stress values of the composite laminates during the tensile tests are obtained from the high-precision simulator DIGIMAX—VA rather than from actual measurements.

Each of the twelve tested specimens has a specific angle combination α_i, $i = 1, \ldots, 16$, which will be called a *pattern*. For example, the specimen denoted S-8 has the following pattern:

$$[0.\ 45.\ 0.\ 90.\ 0.\ -45.\ 0.\ 45.\ 45.\ 0.\ -45.\ 0.\ 90.\ 0.\ 45.\ 0.].$$

For each specimen Sk, Sk = 1, . . ., 12, we dispose of M pairs of values

$$(\text{strain}(Sk, j), \text{stress}(Sk, j)), j = 1, \ldots, M,$$

obtained by simulations or tensile tests. The M data points corresponding to the specimen Sk have the following structure:

$$\alpha_1, \alpha_2, \ldots, \alpha_{16}, (\text{strain}(Sk, j), \text{stress}(Sk, j)), j = 1, \ldots, M.$$

This sequence is related to our objective, which is to predict the stress value for the specimen having the sixteen orientations and loaded with the given strain value. Our approach is based on supervised learning algorithms, so the stress value is the label of each data point.

The data-generating process must use a probability distribution to meet the assumption of independent and identically distributed (i.i.d.) samples. The training and test sets will be generated independently using the same probability distribution. We have considered that a uniformly distributed noise perturbs the patterns; that is, it affects each orientation angle with a value belonging to $[-d, d]$ (e.g., $d = 2$ grads). This perturbation could model the imprecision in achieving the layer's orientation (in this work, it does not) but certainly diversifies the orientation values to make the generalization possible to some extent. For example, one of the M data points generated by specimen S8 is the following:

[1.391 45.331 0.34471 91.703 0.30031 −46.96 1.2375 45.435
44.92 0.92624 −45.968 −0.075 88.909 −1.8056 43.677 −0.96621
0.0051266 393.11].

The last elements are the strain and stress values, 0.0051266 and 393.11 MPa, respectively.

Finally, our dataset would have 12·M data points. In our tests, we have considered M equals 30, which means that the data-generating process yielded 360 data points used for both the training and testing phases. This dataset, called in our programs BigData, supplied the data for the tables TableTrain and TableTest, devoted to the training and test phases. The M data points corresponding to each specimen are split into two parts, included in TableTrain and TableTest, containing p·M (e.g., $p = 80\%$) and $(1 - p)$·M data points, respectively.

Validation sets and k-fold cross-validation were also used in the learning process when obtaining specific ML models that use hyperparameter optimization.

4. Implementation of Machine Learning Models

4.1. A Multiple Linear Regression Model

The first ML model constructed to fit the dataset is parametric: a multiple linear regression model [35–37] that allows the possibility of including nonlinear terms as the interactions, that is, the product of predictor variables. The model preserves its linear relationship concerning its coefficients.

Out of the linear regression models developed in our work, we present only that based on the step-wise regression strategy. The latter consists of adding or removing features from a constant model. In the MATLAB system used in our implementation, this strategy is implemented by a specialized function step-wise (T), which returns a model that fits the dataset in T [38].

The features of this model are named $x_1, \ldots, x_{16}$ for the orientations $\alpha_1, \alpha_2, \ldots, \alpha_{16}$, and St and Ss for the strain and stress, respectively. Appendix B describes the results obtained using the step-wise regression strategy. One of the best linear regression models for stress has the following structure:

$$\text{Ss} \sim 1 + \text{x1} + \text{x2} + \text{x9} + \text{x12} + \text{x13} + \text{St} +$$

$$+ \text{x1} \times \text{x9} + \text{x1} \times \text{St} + \text{x2} \times \text{x13} + \text{x2} \times \text{St} + \text{x9} \times \text{St} + \text{x12} \times \text{St} + \text{x13} \times \text{St}$$

Besides the intercept and terms corresponding to six predictors, all the other terms are interactions of the predictors. The regression coefficients are given in Table A1. Figure 4 presents the predicted and training values for all 300 training records (data points). Figure 5 shows a global image of the model's generalization efficiency using the 60 data points.

Usually, the training and test process results can be characterized by statistics, allowing the ML models constructed using the same dataset to be compared. The statistics given in Table 8 for different ML models are the root mean squared error (RMSE), R-squared, and mean absolute error (MAE) values (see [37,38]). This table provides data that characterizes the later-developed models and is shown here for comparative analysis.

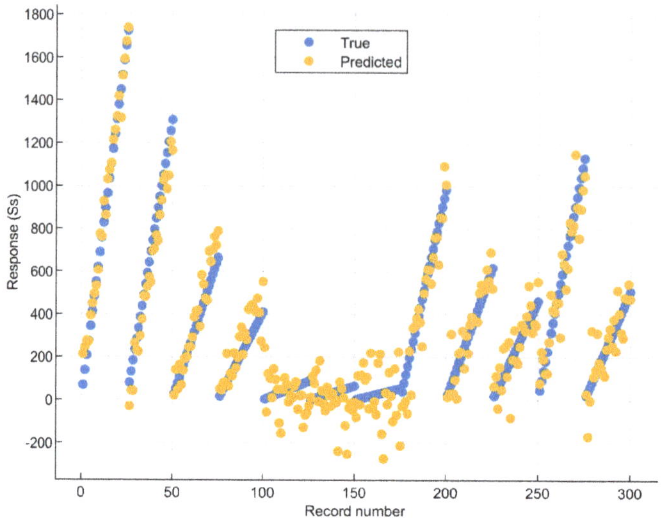

Figure 4. Predicted versus real values for the training dataset.

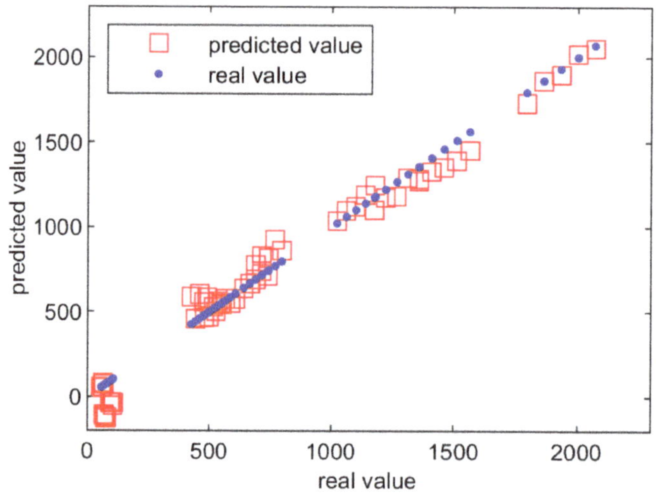

Figure 5. Predicted versus real values for the test dataset.

Let us notice the statistics characterizing the Step-Wise Linear Regression model presented in this sub-section, given in the column **SW Linear Regression**. These will allow us to ascertain the superiority of the next proposed ML models. Besides the first column, Table 8 presents the statistics of the four trained and tested ML models described in the next subsections. The model size is also given.

Although the regression model has good predictions for most specimens, it does not give good predictions for those with a small stress range; their behavior is not accurately "learned". Figures 4 and 5 show certain data points for which the predicted stress value is negative. There are ten such data points, which we shall call critical, corresponding to the specimens whose stress range is very narrow. These bad predictions are given in Table 9.

Table 8. Statistics of the training and test process results and model size for different ML models.

	Statistics	SW Linear Regress.	SVM Regress. 1	SVM Regress. 2	RNN1	RNN2
Training results	RMSE	52.045	34.93	46.903	41.135	6.375
	R-Squared	0.98	0.99	0.98	0.99	1.
	MAE	37.42	28.324	31.44	19.132	3.9465
Test results	RMSE	91.091	52.108	86.383	99.206	34.385
	R-Squared	0.97	0.99	0.98	0.97	1.
	MAE	66.667	43.38	66.255	67.875	19.829
Model size		22 kB	16 kB	40 kB	8 kB	1 MB

Table 9. The set of the worst predictions made by the SW Linear Regression model.

Stress True Value	90.999	94.498	97.997	101.5	105.	67.043	69.621	72.2	74.779	77.357
Stress Predicted Value	−46.32	−23.93	−29.72	−18.4	−27.3	−83.92	−118.7	−102.	−121.2	−129.2

This fact led to investigating *nonparametric* ML models capable of overpassing this drawback.

Remark 2. *Besides ameliorating the statistics, improving the critical data points' predictions was challenging for the new ML models. The following subsections present only models with better prediction capabilities, including the critical data points.*

4.2. Support Vector Machine Models

The Support Vector Machine generated good ML models for our problem. Out of the SVM models constructed in our work, we present, in the sequel, only two SVM models responding to our objectives. The first SVM model, called **SVM Regression 1**, has a cubic Kernel function and a set of intern hyperparameters (as defined within MATLAB system—see [39]): PolynomialOrder, Standardize, KernelScale, BoxConstraint, Epsilon. Appendix B gives details concerning the hyperparameters of SVM Regression 1.

This model was trained and tested using the Regression Learner application (see [39]), which led to the results presented in Table 8, column **SVM Regression 1**. The RMSE values are smaller than those of the **SW Linear Regression** model, proving that the SVM works better. Figures 6 and 7 support this statement compared to Figures 4 and 5.

The predictions for the critical data points are given in Table 10. Although they are not very good, they are better than those in Table 9, at least because they are positive.

Table 10. The set of the critical predictions made by the SVM Model 1.

Stress True Value (MPa)	90.999	94.498	97.997	101.5	105.	67.043	69.621	72.2	74.779	77.357
Stress Predicted Value (MPa)	47.679	35.616	39.292	42.684	34.877	82.007	89.711	81.055	81.242	56.943

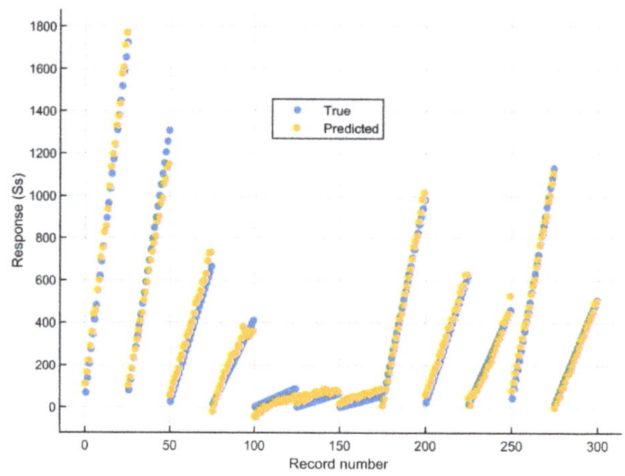

Figure 6. Predicted versus real values for the training dataset—SVM Regression 1.

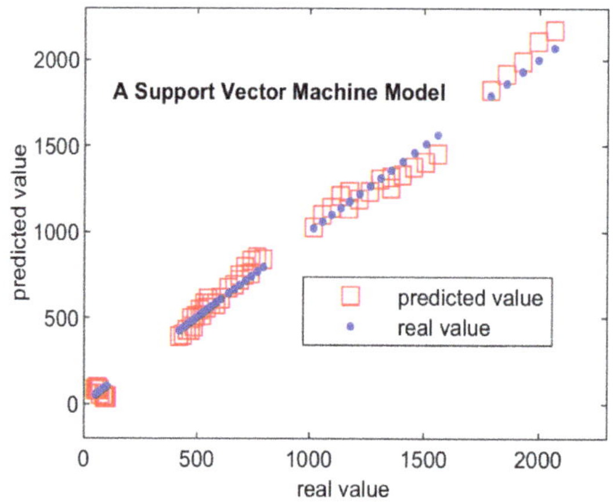

Figure 7. Predicted versus real values for the test dataset—SVM Regression 1.

We also constructed another SVM model that predicts the critical data points very well, whose statistics are given in a column called **SVM Regression 2**. The Hyperparameter Model is the following:

Preset: Optimizable SVM;

Kernel function: Quadratic;

Kernel scale: Automatic.

The training process uses Bayesian optimization to optimize the combination of hyperparameters.

Table 11 shows excellent predictions for the critical data points, but the statistics corresponding to this new SVM model are inferior to those of **SVM Regression 1.**

Table 11. The set of the critical predictions made by the SVM Model 2.

Stress True Value (MPa)	90.999	94.498	97.997	101.5	105.	67.043	69.621	72.2	74.779	77.357
Stress Predicted Value (MPa)	93.074	95.691	100.18	105.26	110.17	68.82	71.944	74.032	77.312	80.144

Moreover, the new model size, 40 kB, is larger than the first. In conclusion, owing to most of its characteristics, the **SVM Regression 1** model can be considered better than the second one.

4.3. Regression Neural Network Models

This subsection presents two other nonparametric ML models using Regression Neural Networks [39]. The first one uses a Narrow Neural Network in MATLAB system terminology. Its statistics are shown in column **RNN 1** of Table 8. Appendix B provides details about the model RNN 1. The hyperparameters are optimized using heuristic procedures. The RMSE and MAE values are better than those of the previous models in Table 8, showing better predicting accuracy. Moreover, the model's size is the smallest of all presented ML models, having 8 kB.

The predicted stress values for the critical data points are very good, like those presented in Table 11, proving that this problem is also solved.

The more accurate prediction is obtained using another Regression NN, the **RNN 2** model, whose statistics are displayed in the last column of Table 8. Details concerning the model **RNN 2** are given in Appendix B. It is an RNN with three layers whose hyperparameters are found using Bayesian optimization.

The RNN2 model has hyperparameters, and it is initialized with the following options:

Preset: Optimizable Neural Network;

Iteration limit: 1000;

Optimizer: Bayesian optimization.

The constraints for the hyperparameters search range are given below:

Number of fully connected layers: 1–3;

Activation: ReLU, Tanh, Sigmoid;

Standardization data: Yes, No;

Lambda: 3.33×10^{-8}–333.3;

Layers size: 1–300.

The application Regression Learner from MATLAB harnessing the Bayesian optimizer found the best hyperparameters' values:

Number of fully connected layers: 3;

Activation: ReLU;

Regularization strength (lambda): 3.6315×10^{-8};

Standardization data: Yes;

First layer size: 166;

Second layer size: 280;

Third layer size: 298.

The reader can construct all the ML models and redraw all the figures in this article using the folder ART_Matlb extracted from the Supplementary Materials. In the current state of the folder, the model RNN2 can be trained and tested directly using the script "H2_modelNN2" because it can load the necessary workspaces describing the problem context. The training takes approximately 5 min because of the Bayesian optimization of hyperparameters. Minimum programming details can also be found in Appendix B.

Table 12 shows that the predictions for the critical data points are very good, the best in comparison with the previous models.

Table 12. The set of the critical predictions made by the RNN 2.

Stress True Value	90.999	94.498	97.997	101.5	105.	67.043	69.621	72.2	74.779	77.357
Stress Predicted Value	92.233	96.111	102.73	107.33	113.38	67.583	70.387	73.247	76.896	80.282

The price to pay for the very good accuracy of RNN 2 is the larger size of the model, which is 1 MB.

Figures 8 and 9 show the efficiency of the training and test processes, respectively, and prove that RNN2 is the more accurate ML model for the given dataset. According to how the ML model is used, RNN1 can replace RNN2 and be a good solution for our prediction problem due to its small model size and good accuracy.

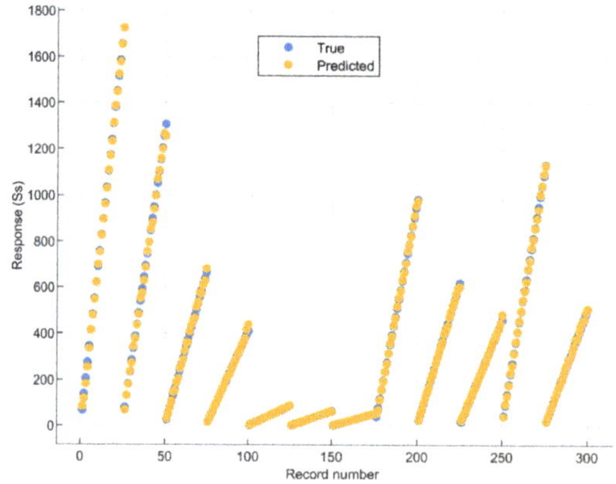

Figure 8. Predicted versus real values—the training of RNN2.

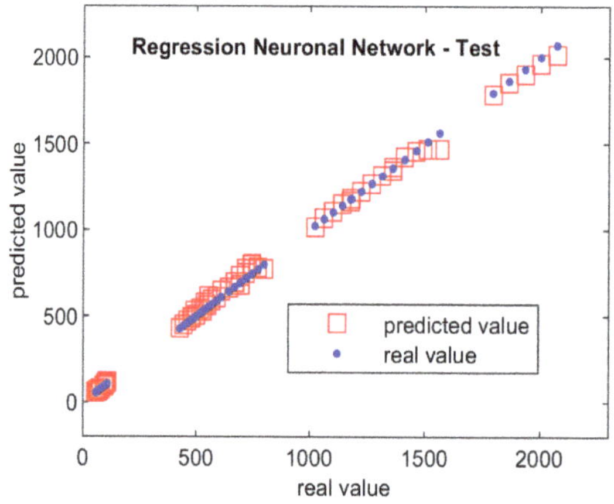

Figure 9. Predicted versus real values—the test of RNN2.

5. Forecasting New Stratification Combinations

This section will present how to exploit the ML models presented in Section 4, that is, to replicate and predict the behavior of carbon fiber–epoxy composites for different orientations, including novel stratification combinations.

5.1. Stress–Strain Predictions for New Combinations

So far, the generalization accuracy of ML models has been tested using the testing datasets reserved for this objective. The testing dataset comes from the same initial traction tests; they have the same real physical support. The generalization power of the ML model would be proven for data points that the model has never "seen"; that is, they belong neither to the training data nor the test data.

The RNN2 model was used as the most performant ML model for stress prediction in our tests.

This subsection considers the case when a new combination has the same structure as one that already contributed to the ML model construction, differing from this in only a few layer orientations. For example, we can generate novel stratification combinations derived from the S8's pattern as "neighbors" of this one: the middle sequence 45/45 is replaced by α/α. The new pattern, denoted Snew, is given below:

$$\text{Snew} = [0/45/0/90/0/-45/0/\alpha/\alpha/0/-45/0/90/0/45/0].$$

Under this hypothesis, the ML model can cover and predict the new combination's behavior.

Four values (denoted α) have been considered that generated the four stratifications presented below for the composite materials. Figure 10 presents the predicted and real stress values for $\alpha = 45°$ and, for comparison, the predicted and simulated stress values for $\alpha = 40°$. The blue curve is shorter because it stops at the beginning of the damaged zone; the Digimat VA simulation program can determine the latter. Currently, the predictions do not consider the damaged zones.

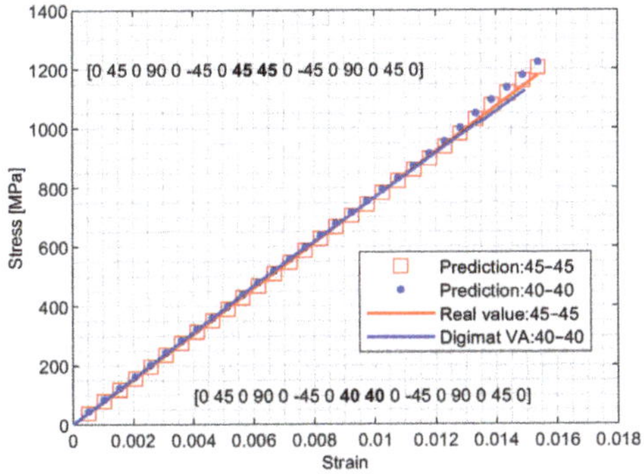

Figure 10. Predicted and real/simulated stress values for $\alpha = 45°$ and $\alpha = 40°$.

Figure 11 presents the predicted and simulated stress values for $\alpha = 20°$ and, for comparison, $\alpha = 30°$. The continuous blue and red curves are shorter because they stop at the beginning of the damaged zone.

Figures 10 and 11 suggest the following observations.

Remark 3. *Two aspects can be underlined:*

- The predictions made by the ML model are very good inside the considered elasticity zones.
- Predictions give more significant errors at the end of the elasticity zones while remaining within acceptable limits.

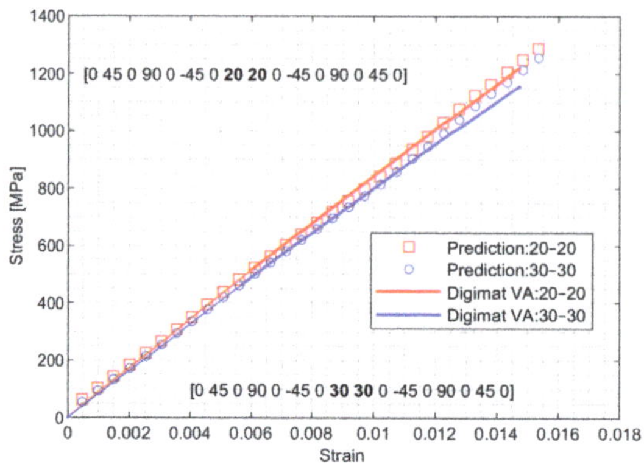

Figure 11. Predicted and simulated stress values for $\alpha = 20°$ and $\alpha = 30°$.

Beyond the opportunity to compare predictions to real/simulated values, these examples based on the Snew pattern suggest how to solve a possible peculiar problem that seeks the most resistant stratification with a given pattern.

5.2. Stress–Strain Predictions for New Random Combinations

In this part of our work, we considered specimens with randomly generated layer orientations, which, in other words, did not contribute to the dataset used to construct the ML model. Then, we compared the predictions for these specimens made by the ML model with the DIGIMAT simulation results. Only four specimens with randomly generated combinations are considered to make this presentation easy to follow.

To remain inside the ML model's generalization area, we first chose four base specimens submitted to tensile tests, PAT2, PAT6, PAT9, and PAT11, which contributed to the dataset used to train the ML model. They have different behaviors in the space stress–strain. Each layer orientation of these specimens was modified independently using a uniformly distributed perturbation in the range [−4°, +4°]. The resulting specimens with randomly generated orientations are NEW2, NEW6, NEW9, and NEW11, which are quite different from the initial ones but remain in the model representation area. For example, Figure 12 shows the sixteen orientation values of NEW2 and base specimen PAT2.

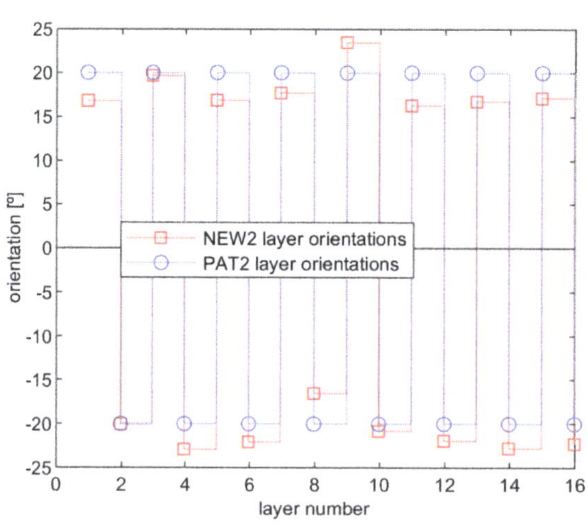

Figure 12. Layer orientations of NEW2 and PAT2.

Table A2 in Appendix C gives the layer orientations for all the base and perturbated specimens and the difference between them (DIFF2, DIFF6, DIFF9, DIFF11). Because the base specimens contributed to the dataset used to train and test the ML model, their stress predictions given strain values are already accurate. The accuracy of the stress prediction must be verified for the new specimens by comparing them with the values given by the DIGIMAT simulations.

Figure 13 presents all the curves obtained through simulation and prediction and ascertains the accuracy of the prediction of the ML model made for the four new specimens. The pairs of curves having the same color prove that there is a small prediction error for all strain values.

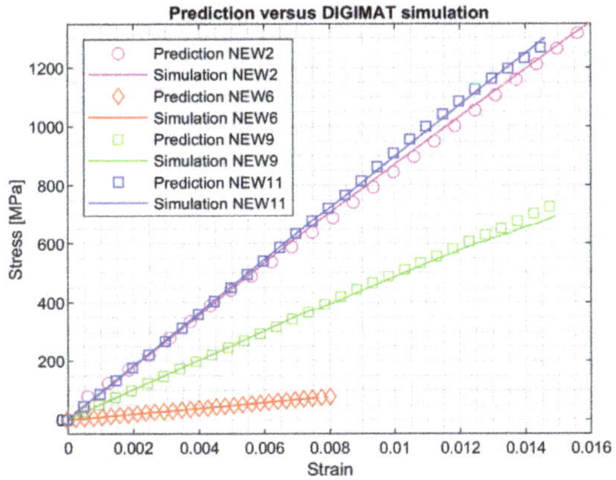

Figure 13. Comparison between the predicted and DIGIMAT values for the four randomly generated specimens.

To zoom in on the prediction error, Figure 14 shows the prediction relative error in a certain number of points situated in the strain range considered in Figure 13.

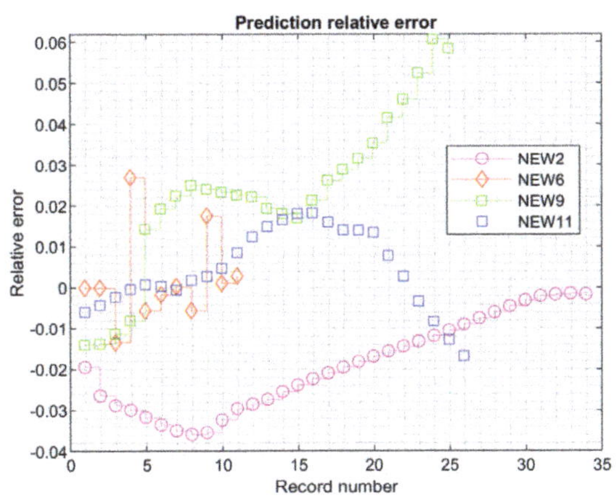

Figure 14. Relative prediction errors for the four randomly generated specimens.

The following equation gives the prediction relative error:

$$\text{prediction relative error} = \frac{\text{predicted stress value} - \text{simulated stress value}}{\text{simulated stress value}}.$$

The relative error is placed in the interval [−0.04, 0.06], which means the prediction accuracy is greatly satisfactory. The proposed ML model has a good generalization power if it is appropriately employed:

- The specimens are inside the ML model representation domain;
- The strain values are inside the range corresponding to the elasticity zone.

In our context, we can imprecisely define the representation domain of an ML model—not the model's capacity—as being all the curves in the strain–stress plane that resemble the curves the model has learned. When a data point is far from the actual examples, the model's predictions may be inaccurate. The second constraint is obvious because the ML predictor was trained using only the elastic zone.

Generally, there is no procedure to verify if the first constraint is rigorously met. Depending on the dataset and practical application, the user can consider a priori a certain ML model representation domain and estimate if this constraint is met. After that, reliable predictions can be made. Hybrid approaches combining physics-based simulations with ML have been suggested as potential solutions to address these limitations [42,43].

6. Discussion

6.1. Importance and Complexity of Comparative Mechanical Analysis of Composite Laminates with Various Fiber Orientations

As discussed in previous sections, the first axis of our work is to apprehend the complex behavior of carbon fiber and epoxy matrix composite laminates under various orientations during tensile testing. All twelve specimens underwent a minute analysis to reveal their behavior during tensile tests. In the sequel, we only present the key aspects of the mechanical analysis of the specimens' behavior, including the damaged zone.

The DIGIMAT-VA simulation's results of tensile tests are graphically represented in Figure 15. The latter describes the stress–strain behavior of the twelve composite layup configurations, considering both the elastic and damaged zones.

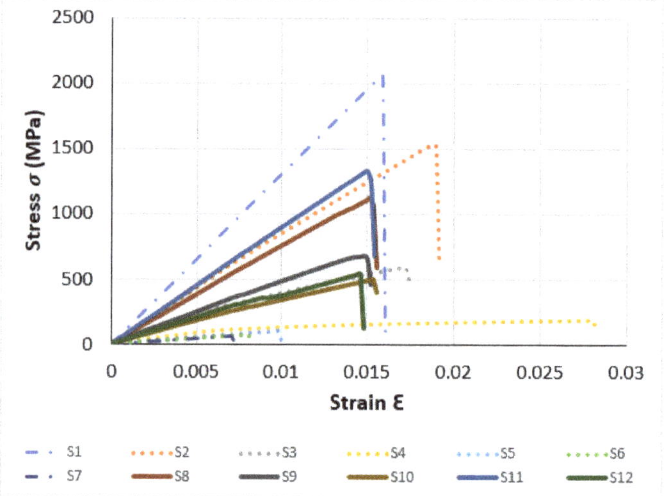

Figure 15. Comparative stress–strain behavior of twelve different composite layup configurations.

Following individual analyses, a comparative evaluation will highlight the key differences in performance, emphasizing the trade-offs between strength, stiffness, and strain to failure for each type of laminate.

6.1.1. The Unidirectional Laminates

The unidirectional laminates (S-1 $[0]_{16}$ and S-7 $[90]_{16}$) exhibit distinctly different mechanical behaviors based on fiber alignment relative to the tensile load. These laminates demonstrate how fiber orientation directly influences the composite materials' stiffness, strength, and failure mechanisms. In S-1 $[0]_{16}$, where the fibers are fully aligned with the load, the laminate displays an explicit linear elastic behavior up to 0.01567 strain, with a high tensile stress of 2064.24 MPa, as depicted in Figures 16 and 17. The steep slope of the stress–strain curve reflects a Young's modulus of 130,393.14 MPa, as represented in Figure 18, indicative of the laminate's high stiffness. This stiffness results from the fibers being in the direct path of the applied load, allowing them to bear most of the tensile stress with minimal deformation. As the laminate reaches its maximum strength at 2064.24 MPa, catastrophic failure occurs abruptly due to fiber breakage. This fiber-dominated failure mode is typical for unidirectional laminates where fibers are the primary load-bearing components. The First Ply Failure (FPF) in S-1 occurs at 2044.30 MPa, as outlined in Figure 19, almost simultaneous with ultimate failure, indicating minimal load redistribution once fiber breakage initiates. The failure is sudden and brittle, with the laminate unable to carry further load post-rupture, making it ideal for applications requiring high stiffness and strength, such as aerospace structures. In stark contrast, S-7 $[90]_{16}$, where the fibers are oriented perpendicular to the load, exhibits drastically different mechanical behavior.

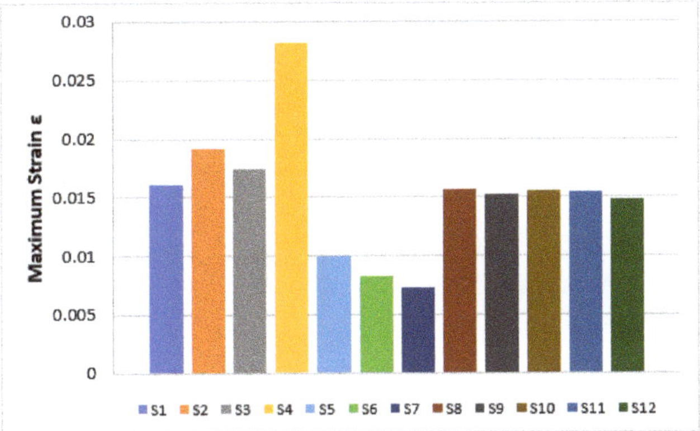

Figure 16. Strain to failure for various layup configurations.

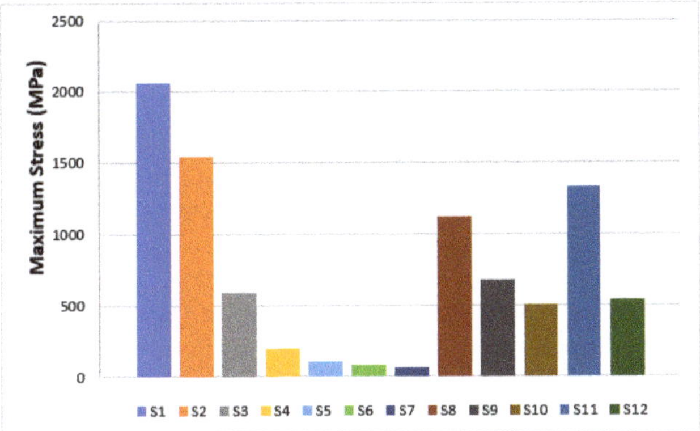

Figure 17. Comparison of maximum tensile strength across laminate configurations.

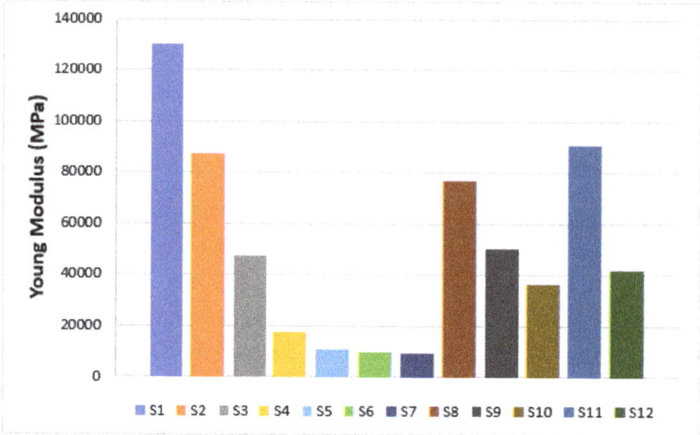

Figure 18. Young's modulus for various layup configurations.

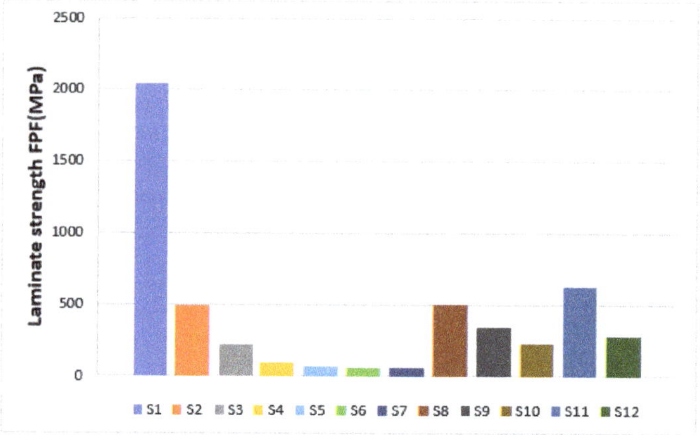

Figure 19. First Ply Failure (FPF) strength comparison across various composite laminate configurations.

The stress–strain curve remains linear only up to around 0.006 strain, and the maximum stress achieved is just 63.98 MPa, significantly lower than S-1's 2064.24 MPa, as illustrated in Figure 17. The Young's modulus in S-7 is correspondingly low, at 9267.99 MPa, since the fibers contribute minimally to axial load resistance in this configuration. Here, the matrix is the primary load-bearing component, and once matrix cracking begins, the laminate rapidly loses its load-bearing capacity. The First Ply Failure occurs at 63.95 MPa, almost identical to the ultimate failure, confirming that matrix failure dominates the behavior of this laminate. Unlike the fiber-driven failure of S-1, the failure mechanism in S-7 is matrix-driven, leading to a brittle, catastrophic rupture once the matrix can no longer carry the load. This low strength and stiffness render the S-7 laminate unsuitable for tensile strength applications. Still, they might be appropriate for secondary structures where minimal tensile loads and transverse reinforcement are more critical. The stark difference between S-1 and S-7 highlights the importance of fiber alignment in determining the mechanical properties of composite laminates. In S-1, where the fibers are perfectly aligned with the tensile load, the laminate demonstrates high stiffness, strength, and abrupt failure due to fiber rupture. In contrast, S-7 shows very low stiffness and strength, with early matrix failure as the dominant failure mode, emphasizing the need for fiber alignment when designing laminates for tensile applications.

6.1.2. The Off-Axis Oriented Laminates

The off-axis oriented laminates (S-2 to S-6) exhibit various mechanical behaviors as the fiber orientations progressively deviate from the load axis. These laminates demonstrate a clear trend of decreasing stiffness and tensile strength as the fiber alignment moves further off-axis, directly impacting the overall mechanical properties. In the elastic region, the stress–strain curve shows a linear relationship, although the stiffness significantly decreases with increasing fiber angles. For S-2 $[\pm 20°]_8$, the Young's modulus is 87155.07 MPa, a marked reduction from S-1's 130393.14 MPa. This reduction is expected, as the fibers are not fully aligned with the tensile load, requiring the matrix to bear more. As the fiber orientation increases, the stiffness continues to decrease. For instance, S-3 $[\pm 30°]_8$ has a Young's modulus of 47161.48 MPa, reflecting the further reduced effectiveness of the fibers in carrying axial loads. For S-4 $[\pm 45°]_8$ and S-5 $[\pm 60°]_8$, the stiffness drops significantly to 17579.89 MPa and 10707.33 MPa, respectively, as the fibers are now oriented primarily for shear load resistance rather than axial tension. The lowest stiffness is observed in S-6 $[\pm 70°]_8$, at 9624.43 MPa, where fibers are nearly perpendicular to the load axis. This position leads to the early onset of matrix-dominated deformation, where the matrix takes on most of the load until the fibers gradually engage as shear stresses increase. The tensile strength of each laminate follows a similar decreasing trend. S-2 $[\pm 20°]_8$ reaches a maximum stress of 1540.16 MPa, significantly lower than S-1's 2064.24 MPa, due to the fiber misalignment with the load. S-3 $[\pm 30°]_8$ further reduces the strength to 593.26 MPa, and for S-4 $[\pm 45°]_8$ and S-5 $[\pm 60°]_8$, the maximum stresses drop to 192.46 MPa and 101.79 MPa, respectively. S-6 $[\pm 70°]_8$ exhibits the lowest tensile strength at 77.31 MPa, as the fibers are almost perpendicular to the tensile load and contribute minimally to resisting axial stresses.

First Ply Failure (FPF) occurs earlier in these off-axis laminates than S-1, highlighting the early damage initiation. The strain to failure increases as the fiber angle deviates from the load axis. Regarding damage mechanisms, off-axis laminates exhibit more progressive damage than unidirectional laminates like S-1. In S-2 $[\pm 20°]_8$, damage initiates early with matrix cracking or fiber–matrix debonding, but the laminate can carry additional load due to stress redistribution among the plies. This progressive damage is also observed in S-3 $[\pm 30°]_8$, where fiber–matrix debonding occurs earlier, but ultimate failure is delayed. In S-4 $[\pm 45°]_8$, S-5 $[\pm 60°]_8$, and S-6 $[\pm 70°]_8$, shear stresses dominate the failure mechanisms.

Matrix cracking and debonding occur early, with fibers failing under shear stresses. These laminates exhibit abrupt failure after reaching maximum stress, with little ability to redistribute load, leading to catastrophic failure. The overall performance of these off-axis laminates suggests that they are less suited for tensile load-bearing applications than unidirectional laminates. However, their higher strain tolerance and more progressive failure mechanisms make them suitable for applications requiring shear resistance and flexibility. Laminates like S-4 $[\pm 45°]_8$, with higher strain to failure and more gradual damage progression, are ideal for structures subjected to shear loading. On the other hand, S-5 $[\pm 60°]_8$ and S-6 $[\pm 70°]_8$, with lower stiffness and early failure, may be used in applications where shear resistance is prioritized over tensile strength.

6.1.3. The Symmetrically Balanced Laminates

The symmetrically balanced laminates (S-8 to S-12) demonstrate a complex interaction between stiffness, strength, and strain tolerance, as these configurations include multiple fiber orientations that provide balanced load-bearing capabilities across various directions. These laminates are designed to offer a compromise between the high stiffness of unidirectional laminates and the flexibility of off-axis configurations, making them well suited for multidirectional loading.

These laminates (S-8 to S-12) demonstrate a precise balance between stiffness, strength, and strain tolerance compared to both the unidirectional laminates (S-1 and S-7) and the off-axis laminates (S-2 to S-6). While unidirectional laminates excel in stiffness and tensile strength along the fiber direction, they fail abruptly with little strain tolerance. Though lower in tensile strength, off-axis laminates provide greater flexibility and progressive failure. The symmetrically balanced laminates offer an intermediate solution, combining the tensile strength of 0° fibers with the flexibility and load-distributing capabilities of off-axis fibers; they are ideal for applications with expected multidirectional loading and gradual failure, such as aerospace, automotive, and energy sector components that need stiffness and flexibility to perform under complex stress conditions.

Table A3 from Appendix C summarizes the mechanical behavior of the twelve stratifications under tensile testing. It is noted that all parameters remain consistent across the stratifications, with the only variable being the fiber orientation.

Investigating the mechanical behavior of different composite materials has provided valuable insights into how various fiber orientations impact their performance under tensile loading.

6.2. ML Prediction Models—Justification of the Addressed Problem

The specialist in composite materials who wants to conceive a composite laminate that is adequate for their application needs to conduct this complex mechanical analysis for laminates with various fiber orientations, as in the previous subsection. This challenge has been addressed in recent studies emphasizing the importance of advanced modeling techniques to handle the complexity of composite structures [44,45]. Thus, many stress–strain characteristics must be determined quickly and precisely, avoiding physical tensile tests or simulations that are expensive and time-consuming. The response to this desideratum is to harness the generalization power of the ML prediction models [44,45].

Section 4 addressed how to construct a prediction model using machine learning algorithms. To do this, we presented only the models constructed by Multiple Linear Regression, Support Vector Machine, and Regression Neural Network algorithms, although we tested more algorithms. These algorithms gave good models for our problem concerning the data described in Section 3.3. Similar approaches have been validated in other studies where supervised machine learning methods were used effectively to predict the mechanical properties of unidirectional fiber composites [42,46]. The predictions are accurate except for the so-called critical data points, a situation presented in Tables 9–12.

Undoubtedly, the statistics presented in Table 8 indicate that the RNN2 model is the most accurate, having the smallest RMSE values in training and testing. In addition, the predictions of critical data points are more than acceptable. However, it also has the largest size (1 MB), which is expected. Generally, more accurate models tend to be larger. This observation aligns with findings in other studies that neural networks often achieve higher accuracy at the cost of increased computational complexity [43]. Remark 2 observes that as the overall accuracy of the constructed models increases, the prediction of critical data points improves accordingly.

A prediction model is effective and useful only if it demonstrates good generalization power. Section 5.1 presents a straightforward example involving four specimens with new stratification combinations. This straightforward example makes the presentation easier to understand. It also presents a small optimization problem: finding the best stratification combination that optimizes a specific physical property revealed by a potential mechanical analysis. Such optimization problems have been similarly addressed in recent research focusing on ML-based optimization frameworks for composite materials [47].

Machine learning prediction models can generally replace experimental tests, as they are less expensive and time-consuming. Additionally, these predictions are an excellent tool that can be integrated into the optimization process. As a result, specialists in composite materials can efficiently repeat a cycle of actions involving prediction and mechanical analysis. This approach has been extensively reviewed, especially in the studies by Pathan et al. [45] and Kibrete et al. [42], as a transformative method for reducing experimental costs while maintaining high prediction accuracy.

The generalization power of the ML prediction models is demonstrated in Section 5.2 in a tougher context: specimens with randomly generated layer orientations. Recent studies, such as those conducted by Yi Liang et al., have highlighted similar challenges when applying ML models to composites with random or complex configurations, emphasizing their robustness under such conditions [48].

In addition to noting that the prediction accuracy is quite satisfactory, we emphasize that two conditions must be met to use predictions correctly:

- The specimens must fall within the representation domain of the machine learning model.
- The strain values must be within the range corresponding to the elastic zone.

Typically, there is no established procedure to verify whether the first constraint is strictly satisfied. This fact can be a serious obstacle when using prediction models. Depending on the dataset and practical application, the user can consider a priori a certain ML model representation domain and estimate if this constraint is met. Hybrid approaches combining physics-based simulations with ML have been suggested as potential solutions to address these limitations [42,43].

7. Conclusions

This study was conducted in the context of composite laminates with layers having different orientations. Twelve specimens were tested using the finite element simulator DIGIMAT—VA. The objective of this work was twofold: to perform a comparative mechanical analysis of the composite laminates and to develop ML predictors trained and tested with the data obtained from the tensile tests.

In this specific context, our study presents the following main contributions:

(1) The comparative mechanical analysis of composite laminates with various fiber orientations proved the existence of three classes of laminates.
(2) The development of ML models for composite laminates that are able to predict strain–stress curves.
(3) Particular focus was placed on the implementation aspect; each stage of building the ML models is accompanied by MATLAB scripts and functions provided in the Supplementary Materials.

The first contribution is an example of how materials science specialists must combine different knowledge and disciplines when they understand and create new composite materials with tailored physical parameters. This analysis demonstrated that the considered composite laminates belong to three classes: unidirectional laminates, off-axis oriented laminates, and symmetrically balanced laminates, each having a specific behavior. Beyond the method, this result is relevant to composite materials designers. Moreover, the analysis results enabled us to prepare data for constructing ML models and highlighted the challenges of treating the three classes uniformly.

Regarding the second contribution, we wish to emphasize the two desiderata associated with the development of ML models:

- First, we must demonstrate that our approach is feasible; namely, the strain–stress curves can be determined accurately and quickly by leveraging the ML predictor.

- Secondly, we must construct and analyze a set of ML models for the tensile tests in the context at hand.

Our work contributes to the field by developing and comparing multiple ML approaches, especially in Section 4. Out of the developed ML models, the Regression Neural Network (RNN2) emerged as the superior model, achieving an RMSE of 34.385 in the testing phase—a decrease of 34% compared with the second-best model based on the Support Vector Machine—and $R^2 = 1$. This represents a 95% improvement in prediction accuracy and a 98% reduction in computation time compared to traditional methods while maintaining accuracy within 2% of sophisticated software simulations. The robust generalization capability of our models, particularly for randomly generated layer orientations, was confirmed through extensive testing in Section 5. The excellent results indicate that the first objective has been successfully achieved.

Our third contribution emphasizes the importance of effectively implementing these ML models. Thus, readers can understand and apply the construction procedure to their project. To this end, all algorithms used in this work are fully implemented, and accompanying scripts are provided as Supplementary Materials. Furthermore, all necessary details can be found in the appendices.

The authors' future work will address predicting the entire characteristic of the mechanical load's response, i.e., both elastic and damaged zones. Another direction will be to determine the optimal layer orientations of a composite laminate by utilizing machine learning predictors and selecting a practical optimal criterion. As a general remark, integrating ML models into the design of new composite materials represents a significant advancement, facilitating the discovery of new composite configurations with tailored performance characteristics.

Supplementary Materials: The following supporting information can be downloaded at https://www.mdpi.com/article/10.3390/pr13030602/s1, The archive "ART_Matlb.zip" contains all the files implementing the prediction models and a guide to using the scripts.

Author Contributions: Conceptualization, M.T. and V.M.; methodology, M.T. and V.M.; software, V.M., B.E.B.B. and S.S.; validation, S.S. and B.E.B.B.; formal analysis, M.T. and V.M.; investigation, B.E.B.B. and S.S.; resources, S.S. and V.M.; data curation, S.S. and B.E.B.B.; writing—original draft preparation, V.M., B.E.B.B. and S.S.; writing—review and editing, V.M. and M.T.; visualization, B.E.B.B.; supervision, B.E.B.B.; project administration, M.T. and V.M.; funding acquisition, V.M. All authors have read and agreed to the published version of the manuscript.

Funding: This research was supported by the project COFUND-LEAP-RE-D3T4H2S, Europe Horizon—LEAP-RE program, and the Executive Agency for Higher Education, Research, Development and Innovation Funding (UEFISCDI—Romania, grant 11/2024).

Data Availability Statement: The original contributions presented in this study are included in the article/Supplementary Materials. Further inquiries can be directed to the corresponding author.

Conflicts of Interest: The authors declare no conflicts of interest. The funders had no role in the study's design; in the collection, analyses, or interpretation of data; in the writing of the manuscript; or in the decision to publish the results.

Appendix A

Nomenclature

E_1^t	Longitudinal Young's Modulus in Tension (MPa)
F_1^t	Longitudinal Tensile Strength (MPa)
E_2^t	Transverse Young's Modulus in Tension (MPa)
F_2^t	Transverse Tensile Strength (MPa)

v_{12}^t	Major Poisson's Ratio in Tension
E_1^c	Longitudinal Young's Modulus in Compression (MPa)
F_1^c	Longitudinal Compressive Strength (MPa)
E_2^c	Transverse Young's Modulus in Compression (MPa)
F_2^c	Transverse Compressive Strength (MPa)
v_{12}^c	Major Poisson's Ratio in Compression
G_{12}	Shear Modulus (MPa)
F_{12} (0.2% offset)	Shear Strength at 0.2% Offset (MPa)
F_{12}	Maximum Shear Strength (MPa)
Matrix Properties	
ρ_m	Density of Matrix Material (t/mm^3)
E_m	Young's Modulus of the Matrix (Tension/Compression) (MPa)
v_m	Poisson's Ratio of the Matrix
ρ_f	Density of Fiber Material (t/mm^3)
E_{axial}	Axial Young's Modulus of Fiber (MPa)
$E_{in-plane}$	In-plane Young's Modulus of Fiber (MPa)
$G_{\mathbf{transverse}}$	Transverse Shear Modulus of Fiber (MPa)
$v_{in-plane}$	In-plane Poisson's Ratio of Fiber
$v_{\mathbf{transverse}}$	Transverse Poisson's Ratio of Fiber
F_t^f	Tensile Strength of Fiber (MPa)
F_c^f	Compressive Strength of Fiber (MPa)
G_I	Mode I Fracture Toughness (mJ/mm^2)
G_{II}	Mode II Fracture Toughness (mJ/mm^2)
T_I	Mode I Interlaminar Strength (MPa)
T_{II}	Mode II Interlaminar Strength (MPa)

Appendix B

Details concerning the SW Linear Regression model

The regression model is obtained using the stepwise function.

Model Hyperparameters:

Preset: Step-wise Linear
Initial terms: Linear
Upper bound of terms: Interactions
Maximum number of steps: 1000

Elements of the listing obtained by the call of this function are given below.

Linear regression model:

Table A1. Linear regression coefficients.

Coefficients	Estimate	SE	tStat	p Value
Intercept	66.289	16.348	4.0548	6.4742×10^{-5}
x1	−1.0564	0.90931	−1.1618	0.24629
x2	5.299	1.5056	3.5196	0.00050263
x9	−2.1988	0.58916	−3.7322	0.000229

Table A1. *Cont.*

Coefficients	Estimate	SE	tStat	*p* Value
x12	−1.5438	0.30825	−5.0083	9.6246×10^{-7}
x13	0.42446	0.401	1.0585	0.29071
St	1.2418×10^{5}	1901.7	65.3	7.0906×10^{-174}
x1: 9	0.04764	0.012162	3.9171	0.00011217
x1: t	−1320.7	54.246	−24.346	1.1118×10^{-71}
x2: 13	−0.065347	0.01489	−4.3885	1.607×10^{-5}
x2: t	−176.64	54.038	−3.2687	0.0012122
x9: t	−170.77	29.761	−5.7382	2.4366×10^{-8}
x12: t	666.06	38.01	17.523	3.3209×10^{-47}
x13: t	−307.24	37.886	−8.1096	1.5046×10^{-1}

Details concerning the SVM Regression 1

Hyperparameters Model:

Preset Quadratic SVM
Kernel function: Cubic.
Kernel scale 3.001
Box constraint: Automatic
Epsilon Auto
Standardization data: Yes
Optimizer: Not applicable

Details concerning model RNN 1

Hyperparameters Model:

Preset Narrow NN
Number of fully connected layers: 1
First Layer size 10
Activation ReLU
Iteration limit 1000
Regularization strength (lambda): 0
Standardization data: Yes
Optimizer: Not Applicable

The training of RNN 2 is made using the fitrnet function as below:

```
RegNN = fitrnet(...
    predictors, ...
    response, ...
    'LayerSizes', [166 280 298], ...
    'Activations', 'relu', ...
    'Lambda', 3.6315 × 10^-8, ...
    'IterationLimit', 1000, ...
    'Standardise', true);
```

Appendix C

Table A2. The layers' orientations of the basic and new random specimens and their differences.

	ang1	ang2	ang3	ang4	ang5	ang6	ang7	ang8
PAT2	20.00	−20.00	20.00	−20.00	20.00	−20.00	20.00	−20.00
NEW2	16.80	−20.05	19.67	−22.92	16.85	−22.06	17.68	−16.53
DIFF2	−3.20	−0.05	−0.33	−2.92	−3.15	−2.06	−2.32	3.47
PAT6	70.00	−70.00	70.00	−70.00	70.00	−70.00	70.00	−70.00
NEW6	71.47	−69.27	68.30	−70.32	73.18	−71.33	67.00	−68.95
DIFF6	1.47	0.73	−1.70	−0.32	3.18	−1.33	−3.00	1.05
PAT9	45.00	0.00	−45.00	90.00	45.00	0.00	−45.00	90.00
NEW9	46.59	3.31	−42.30	88.52	48.57	−1.31	−44.54	88.57
DIFF9	1.59	3.31	2.70	−1.48	3.57	−1.31	0.46	−1.43
PAT11	0.00	30.00	0.00	90.00	0.00	−30.00	0.00	30.00
NEW11	−3.49	26.87	−0.14	89.83	2.48	−33.05	3.61	32.53
DIFF11	−3.49	−3.13	−0.14	−0.17	2.48	−3.05	3.61	2.53
	ang9	**ang10**	**ang11**	**ang12**	**ang13**	**ang14**	**ang15**	**ang16**
PAT2	20.00	−20.00	20.00	−20.00	20.00	−20.00	20.00	−20.00
NEW2	23.49	−20.79	16.26	−21.91	16.71	−22.76	17.14	−22.26
DIFF2	3.49	−0.79	−3.74	−1.91	−3.29	−2.76	−2.86	−2.26
PAT6	70.00	−70.00	70.00	−70.00	70.00	−70.00	70.00	−70.00
NEW6	69.55	−73.38	67.92	−67.99	73.58	−72.32	69.08	−70.79
DIFF6	−0.45	−3.38	−2.08	2.01	3.58	−2.32	−0.92	−0.79
PAT9	90.00	−45.00	0.00	45.00	90.00	−45.00	0.00	45.00
NEW9	90.23	−48.59	3.12	45.28	92.95	−41.92	−1.90	42.29
DIFF9	0.23	−3.59	3.12	0.28	2.95	3.08	−1.90	−2.71
PAT11	30.00	0.00	−30.00	0.00	90.00	0.00	30.00	0.00
NEW11	29.95	1.22	−28.86	−0.13	92.30	−0.28	26.92	−0.82
DIFF11	−0.05	1.22	1.14	−0.13	2.30	−0.28	−3.08	−0.82

Table A3. Mechanical properties comparison of unidirectional, off-axis oriented, and symmetric balanced composite laminates.

Unidirectional		Off-Axis Oriented					Symmetric Balanced				
S-1	**S-7**	**S-2**	**S-3**	**S-4**	**S-5**	**S-6**	**S-8**	**S-9**	**S-10**	**S-11**	**S-12**
Laminate strength σ_{max} (MPa)											
2064.24	63.978026	1540.1604	593.258	192.457	101.786	77.31364	1121.76	680.072	499.3297	1327.031	540.079
Laminate stiffness E (MPa)											
130,393.1	9267.9938	87,155.067	47,161.4	17,598.4	10,707.3	9624.426	76,708.3	50,271.8	36,305.75	90,502.45	41,686.8
Laminate strength (First Ply Failure-CLT) (MPa)											
2044.3	63.949701	490.05462	215.782	95.7956	70.0768	65.63686	501.917	340.917	228.9543	628.3132	280.259

Table A3. *Cont.*

Laminate stiffness (CLT) (MPa)											
130,357.1	9266.6908	85,824.211	45,846.3	17,082.9	10,554.5	9578.134	76,823.8	50,489.8	36,274.97	90,767.63	41,706.7
Unidirectional				**Off-Axis Oriented**					**Symmetric Balanced**		
S-1	**S-7**	**S-2**	**S-3**	**S-4**	**S-5**	**S-6**	**S-8**	**S-9**	**S-10**	**S-11**	**S-12**
Young's modulus E11 (MPa)											
130,393.1	9267.9942	87,155.065	47,161.4	17,579.9	10,707.3	9624.426	76,708.4	50,271.8	36,305.75	90,502.45	41,686.8
Maximum strain (ε_{max})											
0.01607	0.0072998	0.0191447	0.01745	0.02818	0.0100	0.00823	0.01558	0.01518	0.01556	0.01539	0.01480
Maximum stress (σ_max) (MPa)											
2064.24	63.9780	1540.1604	593.258	192.452	101.786	77.3134	1121.7	680.0729	499.39	1327.01	540.074

References

1. Nachtane, M.; Tarfaoui, M.; Abichou, A.; Vetcher, A.; Rouway, M.; Aâmir, A.; Mouadili, H.; Laaouidi, H.; Naanani, H. An Overview of the Recent Advances in Composite Materials and Artificial Intelligence for Hydrogen Storage Vessels Design. *J. Compos. Sci.* **2023**, *7*, 119. [CrossRef]
2. Tarfaoui, M.; Nachtane, M.; Goda, I.; Qureshi, Y.; Benyahia, H. 3D Printing to Support the Shortage in Personal Protective Equipment Caused by COVID-19 Pandemic. *Materials* **2020**, *13*, 3339. [CrossRef] [PubMed]
3. El Moumen, A.; Tarfaoui, M.; Lafdi, K. Additive Manufacturing of Polymer Composites: Processing and Modeling Approaches. *Compos. Part B Eng.* **2019**, *171*, 166–182. [CrossRef]
4. Daly, M.; Tarfaoui, M.; Bouali, M.; Bendarma, A. Effects of Infill Density and Pattern on the Tensile Mechanical Behavior of 3D-Printed Glycolyzed Polyethylene Terephthalate Reinforced with Carbon-Fiber Composites by the FDM Process. *J. Compos. Sci.* **2024**, *8*, 115. [CrossRef]
5. Yang, C.; Kim, Y.; Ryu, S.; Gu, G.X. Prediction of composite microstructure stress-strain curves using convolutional neural networks. *Mater. Des.* **2020**, *189*, 108509. [CrossRef]
6. Rajak, D.K.; Pagar, D.D.; Kumar, R.; Pruncu, C.I. Recent Progress of Reinforcement Materials: A Comprehensive Overview of Composite Materials. *J. Mater. Res. Technol.* **2019**, *8*, 6354–6374. [CrossRef]
7. Khammassi, S.; Tarfaoui, M.; Škrlová, K.; Měřínská, D.; Plachá, D.; Erchiqui, F. Poly (Lactic Acid) (PLA)-Based Nanocomposites: Impact of Vermiculite, Silver, and Graphene Oxide on Thermal Stability, Isothermal Crystallization, and Local Mechanical Behavior. *J. Compos. Sci.* **2022**, *6*, 116. [CrossRef]
8. Gan, Y.X. Effect of Interface Structure on Mechanical Properties of Advanced Composite Materials. *Mater. Sci. Eng. A* **2009**, *500*, 79–86. [CrossRef] [PubMed]
9. Fu, S.-Y.; Lauke, B. Effects of Fiber Length and Fiber Orientation Distributions on the Tensile Strength of Short-Fiber-Reinforced Polymers. *Compos. Sci. Technol.* **1996**, *56*, 1179–1190. [CrossRef]
10. Zhang, L.; Ren, C.; Ji, C.; Wang, Z.; Chen, G. Effect of Fiber Orientations on Surface Grinding Process of Unidirectional C/SiC Composites. *Appl. Surf. Sci.* **2016**, *366*, 424–431. [CrossRef]
11. Bert, C.W.; Asme, M. Models for Fibrous Composites with Different Properties in Tension and Compression. *Appl. Mech. Rev.* **1977**, *30*, 1035–1053. [CrossRef]
12. Kugler, S.K.; Kech, A.; Cruz, C.; Osswald, T. Fiber Orientation Predictions—A Review of Existing Models. *J. Compos. Sci.* **2020**, *4*, 69. [CrossRef]
13. El Moumen, A.; Tarfaoui, M.; Lafdi, K. Modelling of the Temperature and Residual Stress Fields During 3D Printing of Polymer Composites. *Int. J. Adv. Manuf. Technol.* **2019**, *104*, 1661–1676. [CrossRef]
14. Tarfaoui, M. Dynamic Composite Materials Characterisation with Hopkinson Bars: Design and Development of New Dynamic Compression Systems. *J. Compos. Sci.* **2023**, *7*, 10. [CrossRef]
15. Valenza, A.; Fiore, V.; Borsellino, C.; Calabrese, L.; Di Bella, G. Failure Map of Composite Laminate Mechanical Joint. *J. Compos. Mater.* **2007**, *41*, 951–964. [CrossRef]
16. Khan, A.; Azad, M.M.; Sohail, M.; Kim, H.S. A Review of Physics-Based Models in Prognostics and Health Management of Laminated Composite Structures. *J. Korean Soc. Precis. Eng.* **2023**, *40*, 1175–1192. [CrossRef]
17. Kim, K.S.; Yoo, J.S.; Yi, Y.M.; Kim, C.G. Failure Mode and Strength of Unidirectional Composite Single Lap Bonded Joints with Different Bonding Methods. *Compos. Struct.* **2006**, *72*, 477–485. [CrossRef]
18. Wang, H.W.; Zhou, H.W.; Gui, L.L.; Ji, H.W.; Zhang, X.C. Analysis of the Effect of Fiber Orientation on Young's Modulus for Unidirectional Fiber Reinforced Composites. *Compos. Part B Eng.* **2014**, *56*, 733–739. [CrossRef]

19. Pinho, S.T.; Darvizeh, R.; Robinson, P.; Schuecker, C.; Camanho, P.P. Material and Structural Response of Polymer-Matrix Fiber-Reinforced Composites. *J. Compos. Mater.* **2012**, *46*, 2313–2341. [CrossRef]
20. Miao, X.Y.; Chen, X. Structural Transverse Cracking Mechanisms of Trailing Edge Regions in Composite Wind Turbine Blades. *Compos. Struct.* **2023**, *308*, 116984. [CrossRef]
21. Garstka, T.; Ersoy, N.; Potter, K.D.; Wisnom, M.R. In Situ Measurements of Through-the-Thickness Strains During the Processing of AS4/8552 Composite. *Compos. Part A Appl. Sci. Manuf.* **2007**, *38*, 2517–2526. [CrossRef]
22. Cirino, M.; Friedrich, K.; Pipes, R.B. The Effect of Fiber Orientation on the Abrasive Wear Behavior of Polymer Composite Materials. *Wear* **1988**, *121*, 127–141. [CrossRef]
23. Giordano, M.; Calabro, A.; Esposito, C.; D'Amore, A.; Nicolais, L. An Acoustic-Emission Characterization of the Failure Modes in Polymer-Composite Materials. *Compos. Sci. Technol.* **1998**, *58*, 1923–1928. [CrossRef]
24. Vahed, R.; Rajani, H.R.Z.; Milani, A.S. Can a Black-Box AI Replace Costly DMA Testing? A Case Study on Prediction and Optimization of Dynamic Mechanical Properties of 3D Printed Acrylonitrile Butadiene Styrene. *Materials* **2022**, *15*, 2855. [CrossRef] [PubMed]
25. Ho, N.X.; Le, T.T.; Le, M.V. Development of an Artificial Intelligence-Based Model for the Prediction of Young's Modulus of Polymer/Carbon-Nanotubes Composites. *Mech. Adv. Mater. Struct.* **2022**, *29*, 5965–5978. [CrossRef]
26. Dotoli, R.; Gerardi, A.; Polydoropoulou, P.; Lampeas, G.; Pantelakis, S.; Rovira, A.C. Virtual Testing Activities for the Development of a Hybrid Thermoplastic Composite Material for the NHYTE Project. *IOP Conf. Ser. Mater. Sci. Eng.* **2021**, *1024*, 012007. [CrossRef]
27. Campbell, F.C. *Structural Composite Materials*; ASM International: Materials Park, OH, USA, 2010; ISBN 978-1615031405.
28. Botelho, E.C.; Figiel, Ł.; Rezende, M.C.; Lauke, B. Mechanical Behavior of Carbon Fiber Reinforced Polyamide Composites. *Compos. Sci. Technol.* **2003**, *63*, 1843–1855. [CrossRef]
29. McGee, S.H.; McCullough, R.L. Characterization of Fiber Orientation in Short-Fiber Composites. *J. Appl. Phys.* **1984**, *55*, 1394–1403. [CrossRef]
30. Blanc, R.; Germain, C.; Da Costa, J.P.; Baylou, P.; Cataldi, M. Fiber Orientation Measurements in Composite Materials. *Compos. Part A Appl. Sci. Manuf.* **2006**, *37*, 197–206. [CrossRef]
31. Rubin, A.M. Common Failure Modes for Composite Aircraft Structures Due to Secondary Loads. *Compos. Struct.* **1992**, *21*, 145–155. [CrossRef]
32. Tezvergil, A.; Lassila, L.V.J.; Vallittu, P.K. The Effect of Fiber Orientation on the Thermal Expansion Coefficients of Fiber-Reinforced Composites. *Dent. Mater.* **2003**, *19*, 471–477. [CrossRef] [PubMed]
33. Yakout, M.; Elbestawi, M.A. Additive Manufacturing of Composite Materials: An Overview. In Proceedings of the 6th International Conference on Virtual Machining Process Technology (VMPT), Montreal, QC, Canada, 29 May–2 June 2017.
34. Tserpes, K.I.; Papanikos, P.; Labeas, G.; Pantelakis, S.G. Multi-Scale Modeling of Tensile Behavior of Carbon Nanotube-Reinforced Composites. *Theor. Appl. Fract. Mech.* **2008**, *49*, 51–60. [CrossRef]
35. Goodfellow, I.; Bengio, Y.; Courville, A. Machine Learning Basics. In *Deep Learning*; The MIT Press: Cambridge, MA, USA, 2016; pp. 95–161. ISBN 978-0262035613.
36. Newbold, P.; Carlson, W.L.; Thorne, B. Multiple Regression. In *Statistics for Business and Economics*, 6th ed.; Pfaltzgraff, M., Bradley, A., Eds.; Pearson Education, Inc.: Upper Saddle River, NJ, USA, 2007; pp. 454–537.
37. Goodfellow, I.; Bengio, Y.; Courville, A. Example: Linear Regression. In *Deep Learning*; The MIT Press: Cambridge, MA, USA, 2016; pp. 104–113. ISBN 978-0262035613.
38. The MathWorks Inc. Stepwise Regression Toolbox Documentation, Natick, Massachusetts: The MathWorks Inc. 2024. Available online: https://www.mathworks.com/help/stats/stepwise-regression.html (accessed on 1 May 2024).
39. The MathWorks Inc. Regression Neural Network Toolbox Documentation, Natick, Massachusetts: The MathWorks Inc. 2024. Available online: https://www.mathworks.com/help/stats/regressionneuralnetwork.html (accessed on 1 May 2024).
40. Mînzu, V.; Arama, I. A Machine Learning Algorithm That Experiences the Evolutionary Algorithm's Predictions—An Application to Optimal Control. *Mathematics* **2024**, *12*, 187. [CrossRef]
41. Mînzu, V.; Arama, I.; Rusu, E. Machine Learning Algorithms That Emulate Controllers Based on Particle Swarm Optimization—An Application to a Photobioreactor for Algal Growth. *Processes* **2024**, *12*, 991. [CrossRef]
42. Kibrete, F.; Trzepieciński, T.; Gebremedhen, H.S.; Woldemichael, D.E. Artificial Intelligence in Predicting Mechanical Properties of Composite Materials. *J. Compos. Sci.* **2023**, *7*, 364. [CrossRef]
43. Sorour, S.S.; Saleh, C.A.; Shazly, M. A review on machine learning implementation for predicting and optimizing the mechanical behaviour of laminated fiber-reinforced polymer composites. *Heliyon* **2024**, *10*, e33681. [CrossRef]
44. Nneji, R.I.; Nneji, G.U.; Monday, H.N.; Chima, W.; Olumba, C. Ensemble Machine Learning Approaches to Predicting Mechanical Properties of Carbon Fiber Reinforced Composites. *World Sci. News* **2024**, *197*, 159–181.

45. Pathan, M.V.; Ponnusami, S.A.; Pathan, J.; Pitsongsawat, R.; Erice, B.; Petričnic, N.; Tagarielli, V.L. Predictions of the Mechanical Properties of Unidirectional Fibre Composites by Supervised Machine Learning. *Scientific Reports* **2019**, *9*, 13964. [CrossRef] [PubMed]
46. Huang, P.; Dong, J.H.; Han, X.C.; Qi, Y.P.; Xiao, Y.M.; Leng, H.Y. Prediction of mechanical properties of composite materials based on convolutional neural network-long and short-term memory neural network. *Metalurgija* **2024**, *63*, 369–372.
47. Parra, D.P.; Ferreira, G.R.B.; Díaz, J.G.; Ribeiro, M.G.d.C.; Braga, A.M.B. Supervised Machine Learning Models for Mechanical Properties Prediction in Additively Manufactured Composites. *Appl. Sci.* **2024**, *14*, 7009. [CrossRef]
48. Liang, Y.; Wei, X.; Peng, Y.; Wang, X.; Niu, X. A review on recent applications of machine learning in mechanical properties of composites. *Polym. Compos.* **2024**, *46*, 1939–1960. [CrossRef]

processes

MDPI

Article

Domain-Specific Manufacturing Analytics Framework: An Integrated Architecture with Retrieval-Augmented Generation and Ollama-Based Models for Manufacturing Execution Systems Environments

Hangseo Choi and Jongpil Jeong *

Department of Smart Factory Convergence, Sungkyunkwan University, 2066 Seobu-ro, Jangan-gu, Suwon 16419, Gyeonggi-do, Republic of Korea; choinara@g.skku.edu

* Correspondence: jpjeong@skku.edu

Abstract: To support data-driven decision-making in a Manufacturing Execution System (MES) environment, a system that can quickly and accurately analyze a wide range of production, quality, asset, and material information must be deployed. However, existing MES data management approaches rely on predefined queries or report templates that lack flexibility and limit real-time decision support. In this paper, we proposes a domain-specific Retrieval-Augmented Generation (RAG) architecture that extends LangChain's capabilities with Manufacturing Execution System (MES)-specific components and the Ollama-based Local Large Language Model (LLM). The proposed architecture addresses unique MES requirements including real-time sensor data processing, complex manufacturing workflows, and domain-specific knowledge integration. It implements a three-layer structure: an application layer using FastAPI for high-performance asynchronous processing, an LLM layer for natural language understanding, and a data storage layer combining MariaDB, Redis, and Weaviate for efficient data management. The system effectively handles MES-specific challenges such as schema relationships, temporal data processing, and security concerns without exposing sensitive factory data. This is an industry-specific, customized approach focusing on problem-solving in manufacturing sites, going beyond simple text-based RAG. The proposed architecture considers the specificity of data sources, real-time and high-availability requirements, the reflection of domain knowledge and workflows, compliance with security and quality control regulations, and direct interoperability with MES systems. The architecture can be further enhanced through integration with various manufacturing systems, an advanced LLM, and distributed processing frameworks while maintaining its core focus on MES domain specialization.

Keywords: manufacturing execution system (MES); retrieval-augmented generation (RAG); MES domain-specific RAG; local large language model (LLM); real-time data processing; FastAPI

1. Introduction

The large volumes of data generated on the manufacturing floor are a key resource for a variety of decision-making processes, including real-time monitoring, quality improvement, and productivity enhancement [1]. Manufacturing data can be broadly classified as facility or MES data. Facility data encompass continuous numerical data—such as temperature, pressure, and vibration at the manufacturing site—and discrete status data, such as equipment status and alarms. These data are collected in real-time by PLCs and sensor networks and transmitted via industrial protocols such as OPC-UA and MQTT [2]. Pre-processing and filtering based on edge computing improve the data quality and enable real-time monitoring with high-frequency updates within milliseconds to seconds.

MES transactional data, which relate to business processes that occur during the execution phase of manufacturing, include various types of data, such as work orders, performance, defects, material receipts and issues, and inventory. As structured relational data generated according to business processes, these data exhibit complex interrelationships. MES data must ensure the consistency of transactions, making traceability and data versioning important elements [3].

Given these characteristics of manufacturing data, traditional query-based systems face significant challenges in providing flexible and real-time access to this information. This paper proposes a novel architecture that addresses these challenges through an integrated approach combining RAG technology with domain-specific optimizations. MES plays an important role in managing information generated from various business areas such as production, quality, facilities, materials, work orders, etc. to support process optimization and faster decision-making. However, traditional MES data management approaches rely on predefined queries or template-based reports, making it difficult to respond flexibly to new queries or uncertain situations and inconvenient for non-technical users to navigate information intuitively.

Recent advances in LLM-based natural language-processing technologies provide the ability to retrieve and analyze information based on natural language queries without directly exposing complex SQL queries or data structures to the user [4]. In particular, RAG technologies do not simply rely on LLM, but dynamically reference external knowledge bases and combine them with LLM to enable more accurate and flexible response generation [5]. In the MES domain, this RAG-based approach allows for easier access to process-, machine-, and time-specific knowledge, as well as the integration of this knowledge into the SQL generation process to create a more user-friendly data utilization experience [6].

While traditional RAG solutions like LangChain provide powerful general-purpose capabilities, there are many reasons why "domain-specific" RAG components are needed for the specialized environment of MES. MES has data with unique characteristics, such as product history, real-time production data, and sensor logs, which require specialized indexing, Extract–Transform–Load (ETL) pipelines, and security policies to handle effectively [7]. There are also unique requirements for MES environments, such as real-time data processing, reflecting domain knowledge, complying with security regulations, integrating with MES systems, and utilizing domain knowledge graphs [8]. LangChain is an open source project released under the Apache License 2.0, which is free for commercial use and modification and provides stable functionality. Based on this, custom components such as MES-specific embedding models, document structure optimization, manufacturing site terminology processing, and time series data processing can be developed.

Recent advances in LLM-based natural language-processing technology have enabled natural language-based information retrieval and analysis without directly exposing complex SQL queries or data structures. In particular, RAG technology dynamically references external knowledge bases and combines with LLM to enable more accurate and flexible response generation. In the MES domain, this RAG-based approach provides a more user-friendly data utilization experience by easily accessing process, equipment, and time-specific knowledge and integrating it into the SQL generation process [9].

In this paper, we propose a practical approach for improving natural language based data accessibility in MES environments through the integration of RAG and local LLM. This study proposes a method to design a domain-specific RAG architecture optimized for MES data processing, present efficient text-to-SQL conversion with error handling, and verify the performance of the system in real manufacturing environments.

This paper is organized into five sections. Section 2 describes the overall structure of the proposed MES-specific RAG-LLM architecture and details the roles of each component and how they interact. In Section 3, we specifically present the implementation of real-time data processing, domain knowledge integration, and error recovery mechanisms, which are the core features of the proposed architecture. In Section 4, we demonstrate the performance superiority of the proposed architecture through experimental results utilizing real-world manufacturing floor data, especially the improved query-processing accuracy, response time, and scalability compared to existing RAG solutions. Finally, Section 5 concludes this work, presents future research directions, and discusses possible further developments of the proposed architecture.

2. Materials and Methods

The core of the proposed framework consists of three main components: the data pipeline, the RAG engine, and the user interface. The data pipeline is responsible for collecting and structuring real-time and historical data from various sources such as MES and equipment sensors, while the RAG engine leverages advanced natural language-processing capabilities to interpret user queries and generate responses through a built-in search mechanism to effectively access stored data. Finally, the user interface provides an intuitive interactive experience for users to enter questions and gain real-time insights.

These components integrate seamlessly to form a data discovery framework that supports data analysis and rapid decision-making in smart manufacturing environments. Each layer of the framework is designed to effectively handle requests and responses between MES data and users throughout the entire process, from data collection to response generation. This enables fast, accurate, and informed decision-making in a smart factory environment. Figure 1 provides a unified view of the hardware infrastructure and data flow processes of the proposed MES-RAG system architecture. At the center of the system is a high-performance application server based on Ubuntu 22.04 LTS with more than 8 CPU cores and 32 GB of RAM. This server works by orchestrating three core layers.

The application layer implements high-performance asynchronous processing through FastAPI and the Uvicorn ASGI server. This layer provides functions such as authentication, API gateway management, and websocket support for real-time communication. The LLM layer integrates the Mistral-7B and CodeLlama models with LangChain to form a RAG pipeline. Here, the query processor analyzes and structures natural language input, the searcher performs context-based information retrieval, and the generator generates responses based on the retrieved knowledge.

There are three data storage tiers. MariaDB manages structured MES data using 500+ GB SSDs, the Weaviate vector database stores vectorized knowledge with 1+ TB SSDs, and the Redis cache utilizes 16+ GB of RAM to support real-time data access. The data flow in the system begins with user queries being entered through the application layer. The input query is processed at the LLM layer, relevant knowledge is retrieved from the storage layer, and finally a response is generated and delivered to the user. This architecture ensures efficient query processing, scalable data management, and rapid user interaction while maintaining the consistency of the data and system reliability.

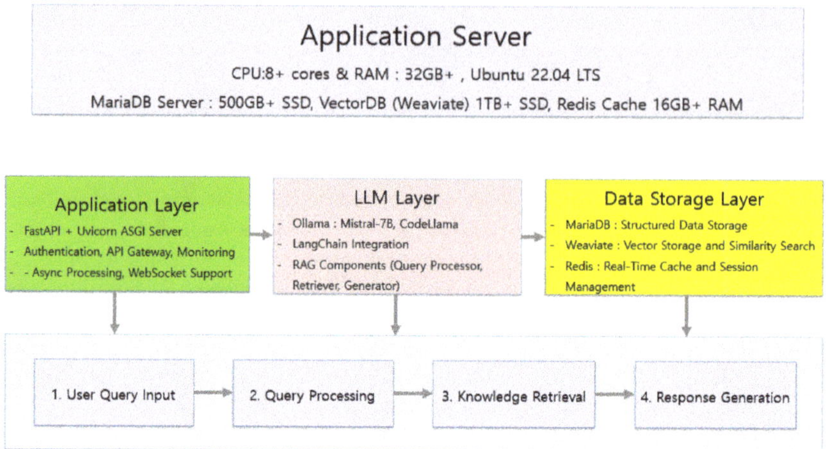

Figure 1. System overview: hardware, software, and data flow.

2.1. Architecture Overview

The proposed framework is implemented using the hardware and software architecture shown in Figure 1, and is designed to support the increasing demand for facility data and MES data integration and statistical retrieval generated by smart factory production processes. It also applies dense path search techniques to improve retrieval accuracy, and introduces the latest Few-Shot learning method to improve model performance [10].

The hardware architecture is centered around an application server with 8+ CPU cores and 32+ GB of RAM, coupled with a MariaDB server with 500+ GB of SSD, a Weaviate vector database with 1TB of SSD, and a Redis cache server with 16+ GB of RAM. This configuration is optimized for processing large amounts of MES data in real-time and building a highly scalable infrastructure while maintaining high performance.

The software architecture consists of four main layers. First, the Application Layer is built on top of FastAPI and the Uvicorn ASGI server and provides functions such as authentication and API gateways, monitoring, asynchronous request processing, and WebSocket support. This allows users to query and analyze data in real-time through a web interface. Second, the LLM Layer uses the Ollama framework, including the Mistral-7B and CodeLlama models. This layer includes RAG components consisting of Query Processor, Retriever, and Generator, which are responsible for processing natural language queries and performing SQL transformations through integration with LangChain. Third, the Data Storage Layer is responsible for relational data storage using MariaDB, vector storage and similarity search using Weaviate, and real-time cache and session management using Redis. This hierarchy enables simultaneous structured and unstructured searches of the data and provides fast query responses [11].

Finally, the Development Environment is based on Anaconda Python 3.10 and includes key packages such as LangChain v03, Transformers 4.25.1, SQLAlchemy 2.0, and FastAPI 0.45.0 to facilitate system-wide development and maintenance. This layered structure optimizes the entire process from data collection to response generation, effectively handling real-time MES data requests and responses for users. This enables fast and accurate decision-making in a smart factory environment.

Figure 2 provides a visual representation of the overall architecture of the proposed MES-RAG system. The architecture consists of three main layers: a data source layer, a data pipeline and processing layer, and an application layer.

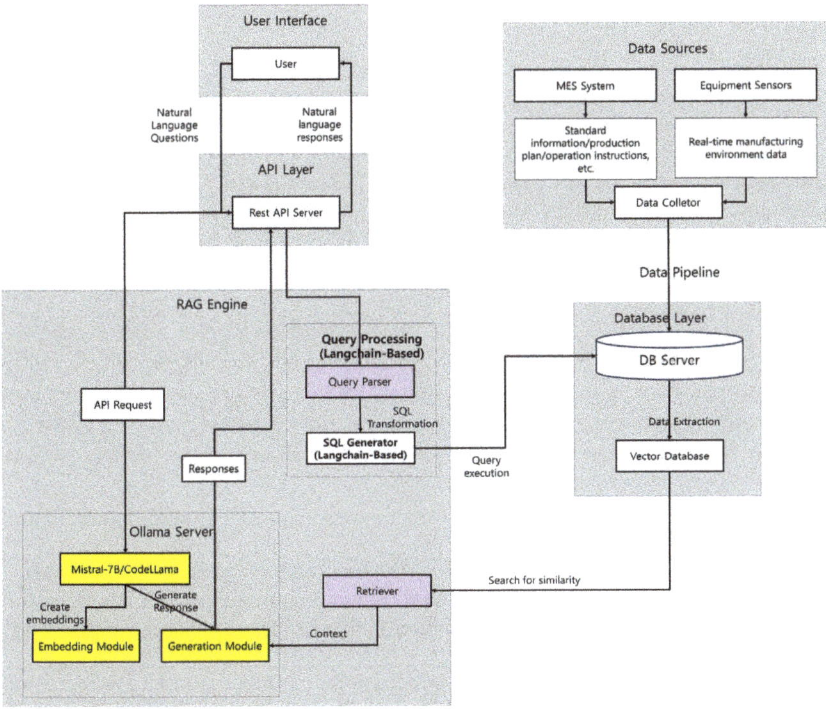

Figure 2. Key components and data flow of the MES-RAG system.

First, the Data Sources Layer collects data generated by the MES system and machine sensors. This layer collects various structured data, such as historical production records, quality information, equipment status, work orders, etc., as well as real-time manufacturing process data to ensure the continuity and consistency of the data. The collected data are processed through the Data Pipeline. During this process, the Data Collector cleanses the data and performs Vector Embedding and relational data storage to transform the data into a searchable form. This provides a structure that can support both structured (SQL-based) and unstructured (vector search) data retrieval.

The RAG Framework handles natural language-based queries. The user enters a question in natural language via the User Interface, which is analyzed by the Query Processor to retrieve the appropriate data, retrieve relevant documents, and generate a final response by the Generator. The final response is returned to the user via the Output Handler, where the LLM Layer performs natural language understanding and generation utilizing a local large language model based on Ollama. Finally, at the Application Layer, a FastAPI-based API server manages natural language query requests and coordinates interactions with the RAG engine and database. This layer is designed to optimize real-time data analytics with asynchronous request processing, API gateways, and websocket support, and to provide fast responses. This layered structure enables real-time data processing and secure data management, enabling fast and flexible data discovery and analysis in manufacturing environments.

2.2. Real-Time Data Processing

2.2.1. Classification of Manufacturing Data

Manufacturing data can be broadly classified as facility or MES data. Facility data encompass continuous numerical data—such as temperature, pressure, and vibration at the manufacturing site—and discrete status data, such as equipment status and alarms. These data are collected in real-time by PLCs and sensor networks and transmitted via industrial protocols such as OPC-UA and MQTT. Pre-processing and filtering based on edge computing improve the data quality and enable real-time monitoring with high-frequency updates within milliseconds to seconds. Furthermore, data pre-processing is important in terms of handling noise and outliers.

MES transactional data, which relate to business processes that occur during the execution phase of manufacturing, include various types of data, such as work orders, performance, defects, material receipts and issues, and inventory. As structured relational data generated according to business processes, these data exhibit complex interrelationships. MES data must ensure the consistency of transactions, making traceability and data versioning important elements.

2.2.2. Data Processing

The data-processing architecture proposed in this study was designed as a multilayered structure to achieve both real-time performance and high-quality data [12]. First, the data collection layer employs edge computing to collect real-time data and apply various protocol conversions and normalization techniques to enhance data quality through initial filtering and pre-processing [13]. This minimizes data-processing delays, reduces the network load, and supports real-time decision making [14].

The collected data are processed in real-time through FastAPI's asynchronous handlers and Apache Kafka in the streaming data-processing layer. Apache Flink is used to validate the data streams and manage ephemeral storage, thereby ensuring the availability and consistency of the data. Redis is also used as an in-memory cache to accelerate the processing of frequently accessed data [15].

Finally, the processed data are stored in a relational database based on the MS SQL Server. This relational database is synchronized with a vector database to support the real-time search and analysis of the data. Advanced analyses—such as similarity search, anomaly detection, and clustering—can be performed using the vector database, and data retention is ensured through the backup and archiving of the data.

2.2.3. End-to-End Data Flow

The proposed architecture extracts data from the MES and production equipment; verifies data formats to ensure the accuracy of the data; ensures the integrity of data through transaction management, concurrency control, and version control to ensure the data remain consistent; and tracks the history of changes in the data. To ensure the transparency and reliability of the data, the architecture supports change-history management, log recording, and the traceability of the data to identify the creation path and clarify the responsibilities. Finally, reliability is enhanced by quantitatively assessing and improving the data's quality using various metrics that measure the completeness, accuracy, consistency, and timeliness of the data.

Figure 3 illustrates the data flow within the system. Data are ingested at the edge, processed through a streaming pipeline, and finally stored in the relational and vector databases. The figure specifically illustrates real-time data filtering at the edge computing layer, streamed processing through Apache Kafka, and performance optimization through

Redis caching. This data flow ensures high performance, real-time processing, and reliable data storage.

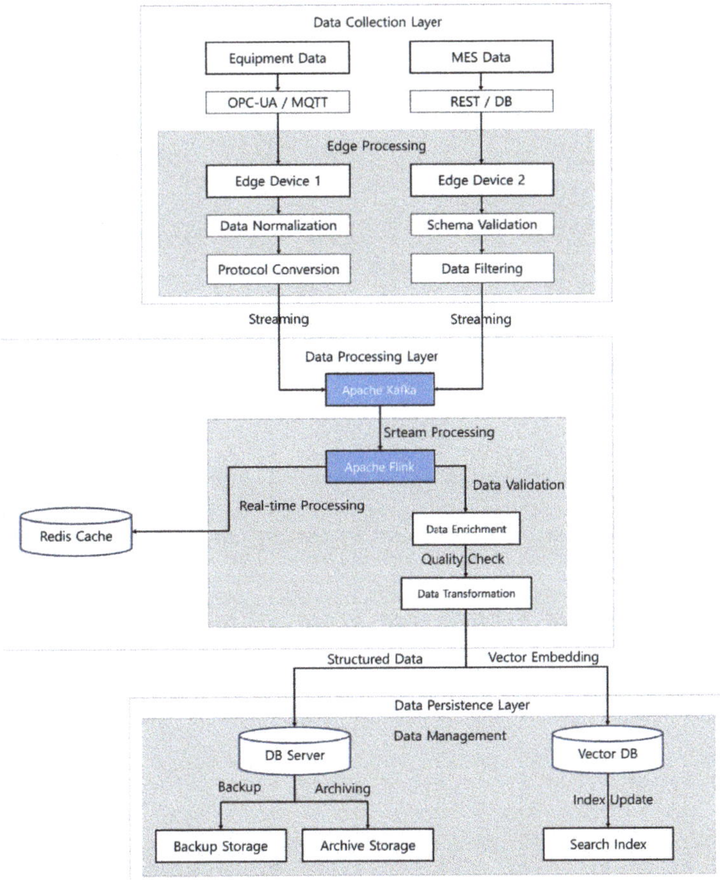

Figure 3. End-to-end data flow diagram.

2.2.4. Data Quality Assurance

The proposed architecture uses several techniques to ensure the quality of the data. First, schema, data type, and constraint validations are performed to ensure the accuracy of the data, thereby maintaining structural consistency and preventing unexpected errors. To ensure data consistency, we implemented transaction management, concurrency control, and versioning. Transactions ensure data integrity by bundling groups of operations into single logical units, concurrency control prevents data conflicts caused by simultaneous access by multiple users, and version control tracks the history of data changes, as well as enabling reversions to previous versions if necessary.

To ensure the transparency and trustworthiness of data, the architecture supports change-history management, audit log recording, and data lineage tracking. Change history management tracks changes in the data, and audit logs similarly record the changes to ensure accountability. The data lineage helps us to understand how data are created and assess their reliability. Finally, we used a variety of metrics to quantitatively assess the data quality. Data are continuously monitored and improved by measuring missing required field rates, missing value rates, data type match rates, value range compliance rates, duplicate data rates, referential integrity compliance rates, data latency, and update cycle compliance rates. This enhances the reliability of the data and improves the accuracy of data-driven decisions. At its core, the architecture ensures data quality as follows:

- Accuracy: schema, data type, and constraint validation;
- Consistency: transaction management, concurrency checking, versioning;
- Transparency: change history management, audit logs, data lineage tracking;
- Reliability: measurement of various quality metrics, continuous improvement.

2.3. RAG-Based Knowledge Retrieval

A key element of the proposed architecture is the RAG-based MES data search engine. The engine is designed to provide users with statistical and analytical data to support on-site decision-making by integrating MES data, as well as various data generated by production equipment, in real-time. It consists of three main components: a query processor, a retrieval, and a generator. Each component works complementarily to provide accurate and meaningful responses to user queries.

The query processor is responsible for converting the natural language query entered by a given user into a form that the system can understand. It maps the user's query into a vector space that the computer can process, helping it retrieve relevant information from the database [16]. The retriever searches the database for relevant information based on vectors generated by the query processor. It uses vector similarity and hybrid search to find the documents most similar to the query, and employs dynamic context windowing to provide the most contextual results. Finally, the generator takes the information retrieved by the retriever and uses the LLM to generate answers in natural, coherent sentences [17]. The generated answers represent accurate and unambiguous responses to the user's query, providing additional information if required.

The main advantage of this system is real-time responsiveness. The system retrieves relevant information and generates an answer as soon as the user enters a query. It also uses an LLM to ensure high-quality answers, and combines vector similarity and hybrid searches to maintain high relevance. In addition, it can answer different types of queries, making it flexible to user needs. The query processor converts natural-language queries entered by users into a form that the system can interpret. In other words, it maps each user's query into a vector space that the computer can process, thereby helping retrieve relevant information from the database.

- Vectorization: This converts the user's query into a high-dimensional vector that numerically represents the meaning of the query, enabling the system to measure the similarity between the query and data;
- Semantic analysis: Determines the core meaning of the query and provides accurate search results;
- Context expansion: Accounts for the query's context to add relevant information, enhancing the accuracy of the search results;
- SQL transformation: Converts natural-language queries into SQL queries for direct access to structured data;
- Hybrid query optimization: Combines structured (SQL) and unstructured (vector) data searches to improve search efficiency.

The retriever is responsible for retrieving relevant information from the database based on the vectors generated by the query processor.

- Vector similarity search: Calculates the similarities between the query vector and document vectors stored in the database to find the most relevant documents;
- Hybrid search: Combines a keyword-based search with a semantic search to increase the search's accuracy and scope;
- Dynamic context window customization: Adjusts the scope of the search based on the context of the query to deliver the most contextual results;
- Caching mechanism: Caches frequently searched query results to improve response time;
- Progressive search: Uses a phased search strategy to balance search result quality and response time.

The generator is responsible for generating responses for the user based on the information retrieved by the retriever.

- Prompt generation: Generates prompts that can be interpreted by the LLM based on the retrieved documents;
- Text generation: Deploys the LLM to generate responses consisting of natural, coherent sentences;
- Answer evaluation: Ensures quality by evaluating whether the generated answers are appropriate for the user's query, grammatically correct, etc.;
- Fact check: Verifies that the generated answers match te information in the retrieved documents;
- Source tracking: Provides the source of the original data upon which the generated answer is based.

Figure 4 shows the interactions between the main components of the RAG system, illustrating how the query processor, retriever, and generator components work together to process a user's natural language query. The query processor vectorizes the user's input, the retriever retrieves relevant information, and the generator generates the final response. Notably, the data flow and processing between each component are represented sequentially, providing a clear understanding of how the RAG system operates as a whole to enable accurate and reliable response generation.

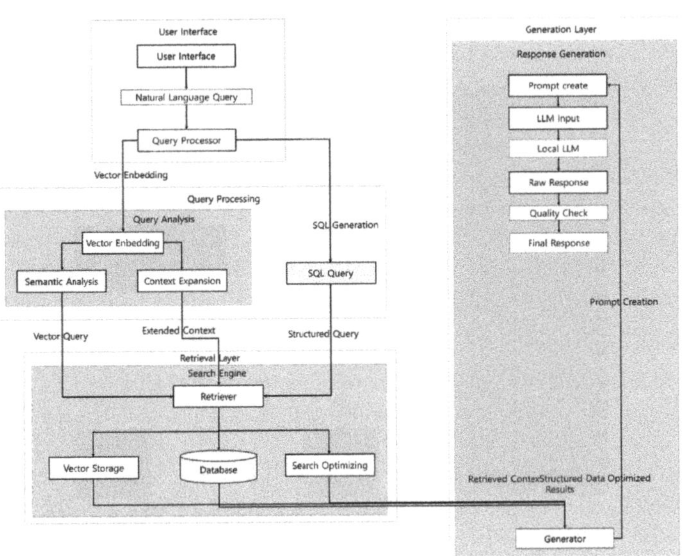

Figure 4. Interacting with RAG Components.

2.4. SQL Generation Pipeline

The SQL generation pipeline uses a multistep approach to improve accuracy. First, it determines the intent of a given natural language query and identifies the relevant tables. It then generates the necessary join conditions, constructs the WHERE clause, and finally assembles and optimizes the final query. A library of query templates specific to the MES domain was constructed to encompass frequently used query patterns such as production performance queries, equipment condition monitoring, and quality data analysis. Thus, an appropriate template is selected and used depending on the context.

2.5. User Interaction

The proposed system implements a conversational interface model that is fundamentally different from traditional MES user interfaces. While traditional MES environments typically use rigid web-based interfaces with predefined menus and fixed data query paths, our approach adopts a more flexible, conversational-based interaction pattern that improves data accessibility and user experience. The system leverages a chatbot-like interface to enable natural language communication between the user and the system. This approach allows users to express their data needs in natural language without the constraints of predefined query templates or fixed menu structures. The interface will be able to incorporate a number of key mechanisms to ensure effective communication, such as continuously maintaining conversational context to help users refine queries and explore data relationships more intuitively [18].

A typical interaction flow begins with the user submitting a natural language query through the chat interface. The system processes this query using the RAG component and may ask for clarification if necessary. Once it understands the intent of the query, it generates and presents a response in an appropriate format. The user can then follow up with related queries or request corrections to the information presented, maintaining an ongoing conversation that enables a more in-depth exploration of the data. This conversational approach offers a number of advantages over traditional interfaces. Users do not need to navigate complex menu hierarchies or understand specific query syntax; the system can guide users through query refinement when necessary and suggest relevant information based on the context of the conversation. In addition, the interface adapts to different user expertise levels, making it accessible to both technical and non-technical users while maintaining the sophisticated data-handling capabilities required in a manufacturing environment.

Error-handling and feedback mechanisms are integrated throughout the interaction process. The system provides immediate feedback on unclear or ambiguous queries and engages users in interactive error resolution through natural dialog. This approach not only helps users formulate more effective queries, but also contributes to the system's continuous learning and improvement process.

2.6. LLM Agents and RAG for Question Answering

Recently, hybrid approaches that combine LLM and RAG have gained traction in the programming domain. Lewis et al. demonstrated that such hybrid approaches outperform traditional LLMs by 15–25% on knowledge-intensive tasks [6]. In particular, there is active research on improving the quality and accuracy of programming questions, which is attributed to the effective combination of code generation and document retrieval. The key to hybrid RAG systems is the balance between accurate information retrieval and context-based response generation. Barron et al. proposed an architecture that combines a domain-specific vector store with a knowledge graph [11]. This approach leverages multi-layered information sources such as code snippets, API documentation, and technical specifications to achieve 83.5% recall. Dynamic search scope adjustment based on question intent and domain-specific embeddings for precision search are key factors in this achievement.

The context quality metrics proposed by Wang et al. play an important role in the pre-verification of code quality and executability [12]. The memory efficiency optimization achieved through dynamic context windowing significantly improved the scalability of the system, and the in-context learning and self-correction mechanism developed by Pourreza and Rafiei improved code generation accuracy by 15–20% [19]. In terms of relevance to this research, the PDF document-processing optimization techniques proposed by Yang et al. directly influenced the design of our system's document-processing pipeline [20]. Notably, the multi-column layout recognition and table information extraction accuracy improvement methodologies have become core components of our system.

Given the specificity of the programming domain, domain knowledge integration is crucial to the design of RAG systems. Developing specialized vector embeddings for each programming language and effectively linking API documentation and code repositories are key. Real-time versioning and updating to reflect the latest information is also essential. In terms of context optimization, dynamic composition with token efficiency is key. The process of reconfiguring the context based on the intent of the question and integrating different forms of information determines the performance of the system. The quality metrics proposed by Iftikhar et al. are an important criterion to ensure the reliability and executability of the code generated in this process [19].

We evaluated and optimized the performance of our system based on the domain-specific benchmarks proposed by Wang et al. and further improved the retrieval accuracy by 18% by adopting the tensor decomposition-based search method proposed by Barron et al. An automated code review system, execution result-based validation, and integration of user feedback enable continuous quality improvements.

3. Results

In this paper presents the necessity and efficiency of RAG- and LLM-based text-to-SQL transformation in the MES environment of smart factories. The proposed architecture can be implemented through Docker and Anaconda virtualization environments and reflects the data-processing requirements of real manufacturing sites. Particular emphasis was placed on the accuracy of domain-specific query processing, the reliability of error recovery mechanisms, and the real-time response performance.

3.1. Related Works

In this paper, we analyze the results of existing research to explore the performance improvement potential of introducing an RAG system and text-to-SQL technology in a MES environment.

RAG techniques combine the ability to generate LLM with external knowledge discovery to generate more accurate and contextualized responses. The paper "Optimizing RAG Techniques for Automotive Industry PDF Chatbots: A Case Study with Locally Deployed Ollama Models" found that, compared to simple RAG models, the advanced RAG system with a customized context compression pipeline significantly improved performance [17]. Specifically, the introduction of self-supervised RAG agents improved answer relevance by 7.3%, 6.3%, and 9.4%, and increased fidelity by 4.6%, 6.8%, and 5.5% on QReCC, CoQA, and self-built datasets, respectively. Contextual accuracy increased by 2.2%, 3.3%, and 2.4%, and contextual recall increased by 2.4%, 2.1%, and 3.3% [17].

In addition, the study "RAG VS FINE-TUNING: Knowledge Injection into Pretrained Language Models" analyzed the combined effect of RAG and fine-tuning. The accuracy of the baseline model improved from 47% to 72% with fine-tuning and increased to 74% when RAG was added [21]. This shows that the combination of fine-tuning and RAG can have a synergistic effect.

The study "Optimizing RAG Techniques for Automotive Industry PDF Chatbots: A Case Study with Locally Deployed Ollama Models" found that introducing a custom function call mechanism to optimize the output of Ollama models resulted in an additional 9.0–13.3% improvement in answer relevance, 6.0–7.1% improvement in fidelity, 1.9–5.3% improvement in contextual accuracy, and 4.6–5.6% improvement in contextual recall [17].

To improve the accuracy of text-to-SQL conversion, recent studies have tried various approaches, including Intermediate Representations (IRNet), Structure-Based Dictionary Learning (STRUG), and the application of LLM. The experimental results of these studies support the design direction of the system we propose in this work. First, the experiments with IRNet showed an overall accuracy improvement of 19.5% compared to the existing SyntaxSQLNet, and a significant performance improvement of 23.3% when the difficult level was Hard, which requires complex SQL structures [22]. This demonstrates that intermediate representations are effective in improving SQL transformation accuracy in a variety of domains. A study applying the STRUG technique achieved 5.5% higher accuracy than the default RAT-SQL model when evaluating the Spider development set. In particular, it showed a performance improvement of 8.4% when the difficultly level was Extra Hard, and the largest improvement in WHERE clause processing from 71.7% to 75.6% [23]. It also maintained 5.7% higher execution accuracy than the BERT-based model on the realistic Spider-Realistic dataset [23]. A recent study utilizing a large language model showed even more remarkable results. The CodeLLaMA-34B model achieved an average accuracy of 20.6% higher than the existing Text-to-SQL model, and outperformed LLaMA-2-CHAT-70B with twice the number of parameters by 12.4% [24]. In particular, one-shot evaluations using GPT-4 achieved up to 80.7% accuracy, depending on the prompt design [24]. These prior studies show the effectiveness of combining intermediate representations with large language models to improve the accuracy of text-to-SQL conversions. The system we propose in this study was also designed based on this approach, and further performance improvements are expected to be achieved by combining it with RAG.

3.2. Experiment Setup

To implement the proposed architecture, we planed a virtual experiment environment based on Docker container and Anaconda for reproducibility and scalability. The MariaDB database and Weaviate vector database were run as containers, and a Python library environment was built in a virtual environment. This setup can allowed to run experiments and verify results under the same conditions, regardless of the hardware specifications or host OS.

The Docker container-based environment was set up to simultaneously run a relational DB and establish a Weaviate environment for the vector search. MariaDB created an initial DB by specifying DB_ROOT_PASSWORD and DB_NAME as environment variables, and loaded tables and sample data for testing via init.sql. Weaviate utilized the latest images to provide a vector embedding-based search (or RAG), and we set the search limits by specifying QUERY_DEFAULTS_LIMIT as an environment variable. By organizing it as a container, multiple services can be managed in isolation without depending on specific libraries, allowing for easy extension or updates in the future.

We created a virtual environment with Anaconda to install major libraries such as FastAPI, Uvicorn, SQLAlchemy, and Weaviate-client, which is important for the following reasons. First, it minimizes unpredictable errors or fluctuations in results by maintaining the compatibility of the versions of the libraries used, regardless of various OS environments. Second, it ensures reproducibility so that other researchers or practitioners can perform experiments under the same conditions by configuring the environment with the same version.

Libraries such as Transformers and LangChain are key components of the local LLM and RAG, used to automate text-to-SQL operations and compute contextual information. By disclosing the above configuration, we can clearly show how the core elements of the proposed architecture (LLM, vector DB, relational DB) interwork, and simplify the installation and configuration procedures for further research or industrial applications. Furthermore, by providing precise version information and execution methods, reproducibility is ensured, allowing for other researchers to achieve the same results and providing a practical guide for implementation or extension.

Finally, the proposed structure suggests that the Mistral-7B and CodeLlama-7B models can be utilized in different roles. It is effective to configure Mistral-7B to be responsible for natural language query understanding and context processing, while CodeLlama-7B is responsible for SQL generation. Both models can be run in a local environment through the Ollama framework, and memory usage can be optimized through the application of 8-bit quantization.

Building SQL query–natural language pair datasets from real-world manufacturing sites is important for configuring models specific to MES environments. These datasets are preferably collected by major business areas, such as production management, quality management, and facility management [25]. Database schema information can be provided in JSON format and should include information about the relationships and constraints between tables. When applying the few-shot prompting technique, it is effective to utilize examples that reflect the specificity of the MES domain. For production, quality, and facility-related queries, including representative examples of each in the prompts enables domain-specific responses to be generated. For handling SQL errors, consider including common error patterns and solutions in the prompts.

3.3. Experiment Methodology

To examine the effectiveness and efficiency of the proposed system, this paper focuses on metrics such as response time, query accuracy, system overhead, precision, speed, and fault tolerance to validate the effectiveness of Text-to-SQL conversion and establish a baseline for comparison with existing MES query systems [26]. Metrics are utilized as a key tool to quantitatively evaluate system performance, identify problems, and highlight the advantages of the proposed system over existing systems. In particular, in text-to-SQL conversion systems, metrics play a role in evaluating performance, analyzing efficiency, verifying reliability, and ensuring comparability. These metrics can be expected to provide quantitative verification, suggest improvement directions, enhance user experience, support real-time decision-making, and provide a basis for comparison. To verify the conversion effectiveness of the Text-to-SQL system, a set of natural language queries for each major business area of the actual manufacturing site was organized. Table 1 shows the test scenarios and evaluation metrics for each business area.

Table 1. MES domain-specific test scenarios and evaluation metrics.

Business Area	Test Case Types	Evaluation Points	Example Queries
Production Management	Performance analysis	Multi-table join, aggregation function	"Production volume and target achievement rate of each product on each line yesterday"
	Quality analysis	Time series processing, grouping	Processes with high frequency of defects by type in the last month"
Materials Management	Inventory Analysis	Conditional filtering, threshold processing	"Unordered items of material below safety stock"
	LOT tracking	hierarchy processing, traceability	"Raw material LOT information of defective finished products"
Facilities Management	Utilization analysis	Real-time data processing	"Which facilities are currently under 70% utilization and Causes"
	Predictive maintenance	Time series pattern analysis	"List of facilities scheduled for next scheduled maintenance"

The transformation performance for each query is evaluated from three aspects: schema-matching accuracy, temporal condition handling, and domain rule reflection. Table 2 shows the detailed evaluation items for each area of evaluation.

Table 2. Text-to-SQL conversion evaluation framework.

Evaluation Area	Evaluation Items	Evaluation Method	Key Indicators
Functional Evaluation	Query-processing accuracy	domain-specific standard query set validation	Schema-matching rate, business rule reflection
	Time-processing accuracy	Time series data-processing validation	Time point interpretation accuracy, period-processing accuracy
	Domain knowledge utilization	Manufacturing-specific requirement validation	Manufacturing terminology interpretation rate, rule application accuracy
Non-functional Evaluation	System performance	System resource monitoring	Response time, CPU/memory usage
	Scalability	Load testing	Concurrent user throughput, data-processing volume
	Stability	Error recovery testing	Automatic recovery rate, system availability

For comparison with existing MES query systems, we set the evaluation criteria as shown in Table 3. This provided objective criteria to demonstrate the practical value of the system.

The proposed evaluation methodology provides a basis for comprehensive system validation during the actual implementation phase. Future work will build on this framework to perform an empirical validation in various manufacturing environments.

Table 3. Comparison analysis with existing MES systems.

Evaluation Item	Existing MES System	Proposed System	Key Differences
Response Time	Template-based fast but slows with complex queries	Consistent response time with RAG-based real-time search	Consistent performance for complex requests via Redis caching and dynamic query processing
Query Accuracy	Dependent on fixed templates, weak with new queries	Enhanced accuracy with domain knowledge reflection	Accurate reflection of domain-specific data structures and user intent
Fault Tolerance	Manual correction needed for errors	Automatic recovery mechanism provided	Enhanced stability through automatic schema-matching and time error recovery
User Accessibility	SQL knowledge required	Natural language-based queries possible	System easily usable by non-experts
System Overhead	High CPU and memory usage	Reduced resource consumption with optimized processing	Efficient resource utilization for large-scale data processing

3.4. Scalability and Performance Metrics

The proposed architecture implements both horizontal and vertical scaling mechanisms to ensure system performance under increasing data volumes and concurrent user queries. In terms of horizontal scalability, the system leverages distributed data processing via MariaDB's primary–secondary replication configuration to enable the distributed processing of read operations [27]. The vector database component utilizes Weaviate's sharding capabilities for the distributed storage and retrieval of large vector data, and the Redis cluster provides an elastic scaling of cache capacity and throughput.

Vertical scalability is achieved through optimized resource allocation for each system component. The application server is configured with 8+ CPU cores and 32+ GB of RAM to ensure sufficient concurrent request-processing capacity. The Vector database utilizes SSD storage to ensure fast search speeds, and the in-memory cache is allocated more than 16 GB of RAM to optimize the cache hit rates. The system's concurrent processing is implemented through FastAPI's asynchronous processing capabilities that efficiently manage concurrent requests. WebSocket support supports distributed real-time data-processing loads, and connection pooling optimizes database connection management. Load-balancing is achieved through request routing and load-balancing via API gateways, and the cache tier reduces thee database load by separating read/write operations.

Data-processing optimization is implemented through several mechanisms. Vector-embedding batch processing improves system resource efficiency, while optimized indexing strategies improve query performance. Time series data partitioning improves search efficiency for temporal data queries [28]. These optimization strategies work together to maintain a consistent performance as data volumes increase. Performance monitoring and management is accomplished by comprehensively tracking key metrics, as shown in Table 4. These metrics provide a quantitative measure of system performance and scalability across multiple dimensions, including response performance, system resource utilization, database performance, and concurrent handling capacity.

Table 4. Performance and scalability monitoring metrics.

Monitoring Area	Key Metrics	Target	Monitoring Method
Response Performance	Average Response Time	Simple Query: <1 s Complex Query: <3 s	APM Tools
System Resources	CPU, Memory Usage	CPU: < 70% Memory: < 80%	System Monitoring
Database	Query Throughput Cache Hit Rate	TPS: 1000+ Cache Hit Rate: 80%+	DB Monitoring Tools
Concurrency	Concurrent Users Requests per Second	100+ Users 500+ Requests/s	Load Testing Tools

Future scalability improvements will focus on three main areas: First, a distributed processing framework will be implemented that integrates Apache Kafka for real-time data pipelines and Apache Flink for extended streaming data processing. Second, high-availability configurations will be enhanced through the deployment of multiple availability zones and the improvement of automatic recovery mechanisms. Third, improvements in performance optimization will be achieved through query pattern analysis and caching algorithms.

With these scalability design principles and performance optimization strategies, we aim to maintain reliable service delivery even as the amount of data and the number of users increase. The monitoring framework provides continuous feedback for system optimization, enabling proactive scaling decisions based on actual usage patterns and performance metrics.

3.5. Data Processing for MES-RAG Integration

3.5.1. Hierarchical Processing Structure

The proposed data-processing architecture is tiered, flexible, scalable, and can operate in real-time. It overcomes the limitations of traditional MES data management structures by taking into account the heterogeneous nature of plant and MES data while implementing optimized processing and storage strategies.

At a foundational level, edge computing-based ingestion processes multi-protocol data via OPC-UA, MQTT, and similar methods at the point closest to the production floor. It performs preprocessing tasks including data normalization, protocol conversion, and noise filtering to improve data quality and reduce transmission load. This enables the efficient handling of high-frequency updates and low-latency data pipelines to enable real-time decision making [29].

In the next step, we implement streaming data processing through a high-performance pipeline using Apache Kafka and Apache Flink to perform real-time data validation, fine-grained quality verification, and enrichment. The Redis in-memory cache improves performance and responsiveness across the data-processing layer by providing fast access to frequently requested information. This tier ensures the availability, consistency, and timeliness of the data for real-time analytics and advanced inferences such as LLM-based data searches and RAG.

The final layer integrates the data persistence and vector database. After preprocessing and streaming analytics, structured data are stored in a relational database such as MS SQL Server to support structured data analysis, transaction history management, and the reflection of business logic. These data are concurrently associated with a vector database to enable unstructured data exploration and analysis, including similarity searches, anomaly detection, and vector embedding-based clustering. A long-term retention strategy with backup, archiving, and active storage ensures reliable management and scalability throughout the data lifecycle.

This tiered processing structure organically combines real-time streaming processing, structured data management, and unstructured searches and analytics to provide a holistic response to the different types of data generated in an MES environment. Ultimately, this is the foundation for enabling natural language-based queries and high-level insights using LLM and RAG.

3.5.2. Data Pipeline

The data pipeline workflow consists of several integration steps. The process starts with extracting real-time and cyclic data from MES and equipment, followed by denoising, outlier detection, protocol conversion, and edge layer normalization. For real-time validation and quality assurance, the system uses Kafka and Flink to perform data validation, quality checks, and transformations through streaming pipelines to maximize the completeness, consistency, and accuracy of the data.

Redis provides caching and fast in-memory processing to improve responsiveness. The transaction management phase stores the cleansed data in a relational database with transaction consistency, concurrency control, versioning, and change history tracking mechanisms to ensure reliable data access for business decisions. The system also supports vector embedding and unstructured analytics by embedding data in a vector database in parallel with SQL Server, allowing users to explore data relationships through natural language queries and perform semantic analysis through LLM and RAG-based approaches.

The pipeline maintains continuous quality assessments and improvements through real-time feedback loops based on data quality metrics, facilitating continuous performance enhancements, quality control, and model retraining to improve the overall reliability and maintainability of the data management and analytics system.

These workflows fulfill the purpose of the architecture: to build high-quality, real-time data pipelines and provide intuitive data access through LLM and RAG, which transcends the limitations of traditional SQL-based queries. The rationale for this architectural design is to establish an intelligent data management and analytics ecosystem that provides rich insights and rapid response with minimal expertise, even in MES environments.

3.6. Text-to-SQL Conversion

To implement text-to-SQL conversion, we first constructed a set of MES domain-specific queries. Our objective was to improve the accuracy of SQL conversion for natural language queries by selecting queries in key business areas (production management, materials management, and task management) that are frequently raised in manufacturing sites and performing instruction tuning based on these queries.

3.6.1. MES Query Set

The query set for each category within the MES domain was designed to include simple queries at the aggregation level as well as complex queries, such as those that identify correlations between quality factors. This enables the local LLM to perform SQL transformations that are domain-specific and not simply based on keyword-matching. Statement tuning is performed by expanding the original query to be more specific, or by inserting additional conditions that leverage domain-specific knowledge. For example, a simple query such as, "What was the production volume yesterday?" can be rewritten as a richer contextual query—such as, "How many units of each product did each production line produce yesterday and how well did they meet their targets?"—and an appropriate SQL query can be presented to guide the model in clearly understanding the target concept and generating sophisticated SQL. This instruction tuning process improves query accuracy by encouraging the local LLM to reflect conditions such as the production line, product, time period, and target versus goal in the SQL.

3.6.2. Error Analysis and Recovery

Even if a set of queries specific to the MES environment is constructed, various errors can occur in the actual operating environment because of incomplete queries or unfamiliarity with the environment. To prevent and quickly respond to these errors, we selected frequently occurring error types and applied the same treatment method as that used for the instruction tuning. The errors were categorized as schema linkage or time-processing errors, and appropriate correction patterns were presented for each error case.

This error handling pattern can also be applied to new queries in a manner similar to statement tuning. For example, given a new query such as 'average uptime per asset for the last week', existing time-processing patterns and asset-specific schema structures can be used to impose appropriate time-range filtering and join conditions. RAG and the LLM enable the automation or semi-automation of these error-handling patterns, providing a flexible basis for responding to new domain queries and error situations. Consequently, given the complex query requirements and error scenarios encountered in MES environments, the instruction-tuning and error-handling strategies proposed in this study support continuous performance improvements and enhance practical applicability.

The following are the main types of SQL errors that can occur in an MES environment and how to avoid them. Table 5 presents specific examples of SQL errors commonly encountered in a Manufacturing Execution System (MES) environment. It demonstrates how incorrect queries can lead to issues like table join errors, time-processing errors, and aggregation errors. For each error type, the table provides an "Incorrect Query", an "Improved Query" that addresses the error, and the "Key Improvements" made to correct the query. Table 6 broadens the scope by outlining common error types in MES SQL queries, the typical issues associated with each type, and the countermeasures that can be implemented to prevent these errors. The table also details the expected positive effects of these countermeasures, such as enhanced data integrity, consistent time management, and improved query reliability.

3.6.3. Query Processing Pipeline Validation

In this section, we illustrate our query-processing pipeline using a combination of RAG and LLM methods. The proposed architecture dynamically references MES domain knowledge through the RAG during the process of converting natural language queries to SQL, and allows the LLM to use this context to form the correct SQL patterns. The process encompasses two phases: context retrieval and SQL pattern-matching.

We validate the effectiveness of the proposed architecture in response to novel, non-predefined types of queries. Our objective was to verify that novel queries can be handled by reusing existing instruction tuning patterns and error handling strategies, and that the proposed architecture adapts to complex requirements with different conditions and time scales.

Table 5. SQL Error examples and improvements in MES environment.

Error Type	Incorrect Query	Improved Query	Key Improvements
Table Join Error	SELECT * FROM PRODUCTION p, QUALITY q WHERE p.date = '2024-01-01'	SELECT * FROM PRODUCTION p INNER JOIN QUALITY q ON p.lot_id = q.lot_id WHERE p.date = '2024-01-01'	Explicit join condition using metadata-driven relationship mapping
Time Processing Error	WHERE create_time = GETDATE()	WHERE CAST(create_time AS DATE) = CAST(GETDATE() AS DATE)	Standardized date format handling with proper type-casting
Aggregation Error	SELECT line_id, SUM(quantity), product_name FROM PRODUCTION	SELECT line_id, SUM(quantity), product_name FROM PRODUCTION GROUP BY line_id, product_name	Proper GROUP BY clause for aggregate functions with non-aggregated columns

In SELECT *, * means "all columns".

Table 6. Error types and countermeasures in MES environment.

Error Type	Common Issues	Countermeasures	Expected Effects
Table Join Errors	Missing join conditions Incorrect relationship mapping	Metadata-based relationship management Automated join condition generation	Enhanced data integrity Reduced join-related failures
Time Processing Errors	Timezone inconsistency Date format mismatches	Standardized time handling library Temporal pattern formalization	Consistent time management Improved temporal queries
Aggregation Function Errors	Invalid GROUP BY clauses Missing aggregation fields	Automated GROUP BY validation Required field verification	Accurate statistical analysis Reliable aggregation results
Syntax Errors	SQL syntax violations Query structure issues	Automated syntax validation Error pattern database	Enhanced query reliability Systematic error prevention

These validation cases demonstrate that the proposed RAG-based LLM can adaptively handle various requirements beyond simple queries, such as complex conditions, temporal patterns, and ratio calculations. Instruction-tuning strategies and error-handling patterns augmented with domain-specific knowledge can be consistently applied to new queries, achieving a high level of query interpretation and transformation ability that effectively reflects the complex data structures and business logic of MES environments.

4. Discussion

In this study, we developed a text-to-SQL transformation using RAG and a local LLM in an MES environment, and identified important implications based on the experimental results obtained from the proposed architecture. First, we clearly demonstrated that the RAG system can effectively reflect the specificity of the MES domain in the SQL generation process. When processing queries specific to each business domain—such as production management, quality control, and facility management—the use of domain-specific knowledge contributes significantly to the improvement of SQL conversion accuracy. This suggests that it is possible to reflect the data structures and business logic of specific industries beyond the domain-adaptation limitations of existing generic text-to-SQL conversions.

We also found that the use of an Ollama-based local LLM practically addresses concerns related to data privacy and system responsiveness [30]. The ability to generate complex SQL queries without exposing sensitive manufacturing site data is an important factor that enhances the proposed architecture's applicability. Furthermore, the error recovery mechanism proposed in this study was shown to be effective in systematically identifying and resolving common errors in MES environments, including schema linkage and time-processing errors. This is significant because it provides a practical method to deal with various types of errors that may occur during SQL generation. However, this study had several limitations. The experimental data did not sufficiently cover all the business areas of real-world MES, and there were limitations in reflecting the specificities of different industries. In addition, a validation in large-scale concurrent user environments or long-term operational situations was not sufficiently performed, and the inability to use large-scale models, stemming from computational resource constraints, limits the potential to validate model performance changes.

Despite these limitations, this study is significant in that it provides a concrete and useful solution to the practical challenge of improving natural language data accessibility in MES environments. Future research should explore various optimization directions, including the standardization of performance evaluation methods, validation for different industrial domains, the application of larger-scale models and distributed processing architectures, and the improvement of caching mechanisms. The practicality and scalability of the system may be further improved by extending the system to different manufacturing industries, supporting multilingualism, and integrating multiple databases.

To further illustrate the advances made by our proposed framework, Table 7 presents a comprehensive comparison between traditional approaches and our solution in MES environments. This comparison reveals several key advantages of our framework. First, the implementation of domain-specific RAG-based natural language processing significantly enhances query flexibility while maintaining high accuracy. Second, our architecture's distributed processing capabilities and containerization ensure both real-time performance and scalability. Third, the automatic error recovery mechanism and local model deployment address critical industrial requirements for reliability and data security.

The comparison demonstrates that while AI-enhanced systems have made progress in user accessibility, they often lack domain-specific optimizations and may compromise on data security due to external AI dependencies. Our framework addresses these limitations while maintaining the benefits of advanced natural language processing. Particularly noteworthy is the framework's ability to combine high security standards typical of traditional MES with the user-friendly interface of modern AI systems, all while adding domain-specific optimizations for manufacturing environments.

Table 7. Comparison of Data Analysis Approaches in MES Environment.

Characteristics	Traditional MES	Template-Based Systems	AI-Enhanced Systems	Proposed Framework
Query Method	Fixed SQL queries	Predefined templates	Basic natural language processing	Domain-specific RAG-based NLP
Real-time Processing	Limited	Partial support	Supported	Fully supported with distributed processing
Scalability	Low	Medium	Medium	High (containerization, distributed processing)
Error Handling	Manual processing	Template-based processing	Basic automation	Automated recovery mechanism
Data Security	High	High	Medium (external AI dependency)	High (local model usage)
Domain Knowledge	Limited	Medium	Limited	High (MES-specific vector search)
User Accessibility	Low (expert required)	Medium	High	High (natural language interface)
Maintainability	Low	Medium	Medium	High (modular design)

5. Conclusions

We developed a text-to-SQL conversion architecture for MES environments using RAG and a local LLM to enable intuitive data access based on natural language while reflecting the domain specificity of manufacturing sites. We found that complex manufacturing data structures can be accurately transformed into the SQL by implementing a query-processing specific to each MES business domain and leveraging the RAG to effectively utilize the domain knowledge. Furthermore, we systematically classified the SQL generation error types commonly encountered in MES environments, such as schema-matching and timing errors, and implemented a recovery mechanism that can actively respond to various error situations. We confirmed that the Ollama-based local LLM can be utilized to prevent the leakage of sensitive production-site data, addressing a critical issue in real-world industrial applications.

The effectiveness of the proposed architecture was verified in terms of domain-specific query-processing accuracy, the reliability of the error recovery mechanism, real-time response performance, and scalability. However, challenges such as long-term validation in real industrial environments, performance verification based on large datasets, and scalability in different manufacturing domains, remain. In future studies, it will be necessary to establish a standardized benchmark dataset to establish a performance evaluation methodology, develop objective performance evaluation metrics, and conduct verifications in different industries. In addition, further approaches to system optimization are possible, such as applying large-scale language models, implementing distributed processing architectures, and improving caching mechanisms.

Future work will extend this research in three main directions. First, in terms of integrating advanced AI techniques, we will enhance the decision support system on the manufacturing floor by leveraging the inference capabilities of large-scale language models. We will optimize the local LLM based on Ollama for the manufacturing domain to improve its performance and extend it to allow for it to handle non-textual data through multimodal analysis. We will also enhance the analytical capabilities of the system by adding advanced analytical functions such as time series prediction and anomaly detection.

Second, we will validate and improve the industry-specific scalability of the proposed architecture. We will evolve the system to reflect the characteristics of different manufacturing industries, such as food, electronics, and automotive industries. This includes effectively integrating the unique production processes, quality control requirements, regulatory compliance, etc., of each industry into the system. In particular, we will study how to systematically structure and utilize industry-specific domain knowledge in a vector database.

Third, we plan to improve the error recovery mechanism to increase the stability and reliability of the system. Beyond the current schema-matching and time-processing error recovery, we will develop an intelligent error-handling system that automatically detects and responds to various anomalies on the production floor. To this end, we will introduce a deep learning-based anomaly detection model and improve the system by learning possible error patterns to which it could proactively respond.

Author Contributions: Conceptualization, H.C. and J.J.; methodology, H.C.; architecture, H.C.; validation, H.C. and J.J.; formal analysis, H.C.; research, H.C.; acquisition of data, H.C.; data curation, H.C.; drafting of the manuscript, H.C.; revision and editing of the manuscript, H.C. and J.J.; visualization, H.C.; supervision, J.J.; project management, J.J. All authors have read and agreed to the published version of the manuscript.

Funding: This research received no external funding. The article publication charge (APC) was waived as part of the publisher's special discount programme.

Data Availability Statement: This research proposes a theoretical architectural framework and no data has been separately generated or analysed. The sample queries and Structured Query Language (SQL) statements presented in this paper are descriptive examples to illustrate the functionality of the proposed architecture. The referenced database schema and table structure are based on a typical MES implementation, but are not related to any specific real-world dataset. The architecture design and implementation details are described in the main body of the thesis.

Acknowledgments: This work was supported by ICT Creative Consilience Program through the Institute of Information & Communications Technology Planning & Evaluation (IITP) grant funded by the Korea government (MSIT) (RS-2020-II201821).

Conflicts of Interest: The authors declare no conflicts of interest.

References

1. Bokrantz, J.; Skoogh, A.; Lämkull, D.; Atieh, H.; Perera, T. Data Quality Problems in Discrete Event Simulation of Manufacturing Operations. *Simul. Trans. Soc. Model. Simul. Int.* **2018**, *94*, 1009–1025. [CrossRef]
2. Chen, B.; Wan, J.; Shu, L.; Li, P.; Mukherjee, M.; Yin, B. Smart Factory of Industry 4.0: Key Technologies, Application Case, and Challenges. *IEEE Access* **2017**, *6*, 6505–6519. [CrossRef]
3. Shojaeinasab, A.; Charter, T.; Jalayer, M.; Khadivi, M.; Ogunfowora, O.; Raiyani, N.; Yaghoubi, M.; Najjaran, H. Intelligent Manufacturing Execution Systems: A Systematic Review. *J. Manuf. Syst.* **2022**, *62*, 503–522. [CrossRef]
4. Ning, Z.; Tian, Y.; Zhang, Z.; Zhang, T.; Li, T.J.-J. Insights into Natural Language Database Query Errors. *ACM Trans. Interact. Intell. Syst.* **2024**, *14*, 25. [CrossRef]
5. Kanburoğlu, A.B.; Tek, F.B. Text-to-SQL: A Methodical Review of Challenges and Models. *Turk. J. Electr. Eng. Comput. Sci.* **2024**, *32*, 403–419. [CrossRef]
6. Lewis, P.; Perez, E.; Piktus, A.; Petroni, F.; Karpukhin, V.; Goyal, N.; Küttler, H.; Lewis, M.; Yih, W.-T.; Rocktäschel, T.; et al. Retrieval-Augmented Generation for Knowledge-Intensive NLP Tasks. In Proceedings of the 34th Conference on Neural Information Processing Systems (NeurIPS 2020), Vancouver, BC, Canada, 6–12 December 2020; pp. 945–964.
7. Chan, A. GPT-3 and InstructGPT: Technological Dystopianism, Utopianism, and "Contextual" Perspectives in AI Ethics and Industry. *AI Ethics* **2023**, *3*, 53–64. [CrossRef]
8. Abro, A.A.; Talpur, M.S.H.; Jumani, A.K. Natural Language Processing Challenges and Issues: A Literature Review. *Gazi Univ. J. Sci.* **2023**, *36*, 1522–1536. [CrossRef]

9. Fürst, J.; Kosten, C.; Nooralahzadeh, F.; Zhang, Y.; Stockinger, K. Evaluating the Data Model Robustness of Text-to-SQL Systems Based on Real User Queries. *ACM Trans. Database Syst.* **2024**, *49*, 1–24.
10. Brown, T.B.; Mann, B.; Ryder, N.; Subbiah, M.; Kaplan, J. Language Models are Few-Shot Learners. *arXiv* **2020**, arXiv: 2005.14165.
11. Barron, R.C.; Grantcharov, V.; Wanna, S.; Eren, M.E.; Bhattarai, M.; Solovyev, N.; Tompkins, G.; Nicholas, C.; Rasmussen, K.Ø.; Matuszek, C.; et al. Domain-Specific Retrieval-Augmented Generation Using Vector Stores, Knowledge Graphs, and Tensor Factorization. *arXiv* **2024**, arXiv:2410.02721.
12. Wang, S.; Liu, J.; Song, S.; Cheng, J.; Fu, Y.; Guo, P.; Fang, K.; Zhu, Y.; Dou, Z. DomainRAG: A Chinese Benchmark for Evaluating Domain-Specific Retrieval-Augmented Generation. *Chin. Nat. Lang. Process.* **2023** , 2024, 89–98.
13. Ali, M.I.; Patel, P.; Breslin, J.G. Middleware for Real-Time Event Detection and Predictive Analytics in Smart Manufacturing. In Proceedings of the 2019 15th International Conference on Distributed Computing in Sensor Systems (DCOSS), Santorini, Greece, 29–31 May 2019; pp. 370–376.
14. Al-Gumaei, K.; Müller, A.; Weskamp, J.N.; Longo, C.S.; Pethig, F.; Windmann, S. Scalable Analytics Platform for Machine Learning in Smart Production Systems. In Proceedings of the 2019 24th IEEE International Conference on Emerging Technologies and Factory Automation (ETFA), Zaragoza, Spain, 10–13 September 2021; pp. 1155–1162.
15. Li, X.; Wan, J.; Dai, H.-N.; Imran, M.; Xia, M.; Celesti, A. A Hybrid Computing Solution and Resource Scheduling Strategy for Edge Computing in Smart Manufacturing. *IEEE Trans. Ind. Inform.* **2019**, *15*, 4225–4234. [CrossRef]
16. Pan, J.J.; Wang, J.; Li, G. Vector Database Management Techniques and Systems. In Proceedings of the Companion of the 2024 International Conference on Management of Data, Santiago, Chile, 9–15 June 2024; pp. 597–604.
17. Liu, F.; Kang, Z.; Han, X. Optimizing RAG Techniques for Automotive Industry PDF Chatbots. In Proceedings of the 2024 3rd International Conference on Artificial Intelligence and Intelligent Information Processing, Tianjin China, 25–27 October 2024; pp. 152–159.
18. Singh, A.; Shetty, A.; Ehtesham, A.; Kumar, S.; Khoei, T.T. A Survey of Large Language Model-Based Generative AI for Text-to-SQL-Benchmarks, Applications, Use Cases, and Challenges. *arXiv* **2024**, arXiv: 2412.0528.
19. Pourreza, M.; Rafiei, D. DIN-SQL: Decomposed In-Context Learning of Text-to-SQL with Self-Correction. *arXiv* **2023**, arXiv:2304.11015.
20. Zhang, L.; Liu, Y.; Fan, Z.; Dai, H. Empower Large Language Model to Perform Better on Industrial Domain-Specific Question Answering. *IEEE Trans. Ind. Inform.* **2024**, *20*, 1523–1538.
21. Wang, B.; Chen, H.; Xu, T.; Li, M.; Zhou, J. RAG vs. Fine-tuning: Pipelines, Tradeoffs, and a Case Study on Agriculture. *Comput. Electron. Agric.* **2024**, *213*, 108156.
22. Deng, X.; Awadallah, A. H.; Meek, C.; Polozov, O.; Sun, H.; Richardson, M. Structure-grounded pretraining for text-to-SQL. *arXiv* **2020**, arXiv:2010.12773.
23. Gao, D.; Wang, H.; Li, Y.; Sun, X.; Qian, Y.; Ding, B.; Zhou, J. Text-to-SQL Empowered by Large Language Models: A Benchmark Evaluation. *arXiv* **2023**, arXiv: 2308.15363. [CrossRef]
24. Guo, J.; Zhan, Z.; Gao, Y.; Xiao, Y.; Lou, J.G.; Liu, T.; Zhang, D. Towards Complex Text-to-SQL in Cross-Domain Database with Intermediate Representation. In Proceedings of the 57th Annual Meeting of the Association for Computational Linguistics, Florence, Italy, 28 July–2 August 2019; ACL: Stroudsburg, PA, USA, 2019; pp. 4524–4535.
25. Zhao, W.X.; Zhou, K.; Li, J.; Tang, T.; Wang, X.; Hou, Y.; Min, Y.; Zhang, B.; Zhang, J.; Dong, Z.; et al. A Survey of Large Language Models. *arXiv* **2024**, arXiv: 2303.18223.
26. Iftikhar, N.; Baattrup-Andersen, T.; Nordbjerg, F.E.; Bobolea, E.; Radu, P.-B. Data Analytics for Smart Manufacturing: A Case Study. In Proceedings of the 8th International Conference on Data Science, Technology and Applications, Prague, Czech Republic, 26–28 July 2019; pp. 392–399.
27. Henzel, R.; Herzwurm, G. Cloud Manufacturing: A State-of-the-art Survey of Current Issues. *Procedia CIRP* **2018**, *72*, 947–952. [CrossRef]
28. Jing, Z.; Su, Y.; Han, Y. When Large Language Models Meet Vector Databases: A Survey. *arXiv* **2024**, arXiv: 2402.01763.
29. Lee, J.; Bagheri, B.; Kao, H.-A. A Cyber-Physical Systems Architecture for Industry 4.0-Based Manufacturing Systems. *Manuf. Lett.* **2015**, *3*, 18–23. [CrossRef]
30. Chen, L.; Zhang, W.; Smith, K. LLM-Based Edge Intelligence: A Comprehensive Survey on Architectures, Applications, Security and Trustworthiness. *IEEE Commun. Surv. Tutor.* **2024**, *26*, 1–28.

processes

Article

Computer Science Techniques Applied to Temperature Control in Biodiesel Production: Mathematical Modeling, Optimization, and Sensorless Technique

Mario C. Maya-Rodriguez [1], Ignacio Carvajal-Mariscal [1,*], Raúl López-Muñoz [2,*], Mario A. Lopez-Pacheco [1] and René Tolentino-Eslava [1]

[1] Escuela Superior de Ingeniería Mecánica y Eléctrica Unidad Zacatenco, Instituto Politécnico Nacional, Mexico City 07738, Mexico; mmayar@ipn.mx (M.C.M.-R.); mlaopezp@ipn.mx (M.A.L.-P.); rtolentino@ipn.mx (R.T.-E.)

[2] Unidad Profesional Interdisciplinaria en Ingeniería y Tecnologías Avanzadas (UPIITA), Instituto Politécnico Nacional, Mexico City 07340, Mexico

* Correspondence: icarvajal@ipn.mx (I.C.-M.); raulopezm@ipn.mx (R.L.-M.)

Abstract: This paper demonstrates that biodiesel production processes can be optimized through implementing a controller based on fuzzy logic and neural networks. The system dynamics are identified utilizing convolutional neural networks, enabling tests of the reactor temperature response under different control law proposals. In addition, a sensorless technique using a convolutional neural network to replace the sensor/transmitter signal in case of failure is implemented. Two optimization functions are proposed utilizing a metaheuristic algorithm based on differential evolution, where the aim is to minimize the use of cooling for the control of the reactor temperature. Finally, the control system proposals are compared, and the results show that a neuro-fuzzy controller without optimization restrictions generated unviable ITAE (1.9597×10^7) and TVU (22.3993) performance metrics, while the restriction proposed in this work managed to minimize these metrics, improving both the ITAE (3.3928×10^6) and TVU (17.9132). These results show that combining the sensorless technique and our optimization method for the cooling stage enables energy saving in the temperature control processes required for biodiesel production.

Keywords: control process; biodiesel; optimization; sensorless technique

1. Introduction

It is important for industrial control systems to maximize profits and the use of resources in their production processes in order to remain competitive. To achieve this, it is necessary to optimize different areas, such as Electricity, Electronics, Mechanics, Administration, and Automatic Control, among others. In the area of Automatic Control, it is possible to contribute to the fulfillment of the optimization objective in the following ways: (1) tuning classical controllers [1]; (2) modeling and identification of parameters [2]; (3) monitoring and prediction of behaviors [3]; and (4) implementation of non-conventional controllers [4]. This work will focus on the modeling and identification of parameters, allowing a mathematical expression of the linear or nonlinear nature of production processes through the application of classical mechanics methods and computer science techniques [2]. As a possible benefit, it is possible to carry out simulations of the behavior and dynamics of a process under different considerations such as operating conditions, different controllers, and performance evaluation, among others. This has to do with purposes associated with the design of production plants as well as with the optimization of already established production processes. The implementation of non-conventional controllers is increasing worldwide and will be seen with greater force in the industry of undeveloped countries, because most of them use classical controllers. Due to the complex nature of these processes, classical controllers quickly reach their limitations and affect the profits of a factory over time through, for example, using more energy than necessary, product quality problems, and difficulty in complying with environmental regulations, among other issues. Controllers based on expert systems with fuzzy logic and those based on neural networks have gained popularity in recent years [5]. However, this type of controller has a fundamental problem—the inherent dependence on hyper-parameters and their initial conditions—as well as the classic problems of stability and convergence. Given the rapprochement between Automatic Control and Computer Science in the last few decades, it is now possible to combine the techniques and reach the optimization objective from a metaheuristics point of view, minimizing the dependence on the selection of initial conditions. Mexico's energy consumption is primarily distributed to the transportation sector as well as heat and electricity generation. All this energy comes from fossil material burning (approximately 60%) [6], renewable sources (9%), and biodiesel (5%) [7], among other sources. Several political reforms in Mexico have been implemented to reduce the pollutant emissions into the atmosphere. Although 35% of the total energy production came from renewable resources by the end of 2024, several factors limited this progress [8], such as high production costs, inadequate infrastructure for the energy production, and the lack of incentives to develop those products [9]. However, biodiesel production has prospered thanks to various recent research studies on its production and its promotion in different applications [6,10,11].

Biodiesel production requires a specific viscosity similar to that of diesel, which is achieved by mixing different substances and heating them under particular conditions for each type of mixture in transesterification reactors [12]. Adequate control of the inside of the reactor, specifically temperature control, is important. Nowadays, most of the theoretical advances in the area of Automatic Control are not being effectively exploited mainly for two reasons. The first reason is due to a lack of collaboration with areas of industrial production, which need to contribute their perspective in order to guarantee the efficient use of resources, as well as the problems and requirements for national growth. The second reason is due to the lack of interest or efforts by researchers to carry out implementations in real-world problems, which sometimes, depending on the observer, are underestimated. However, the union of both aspects will allow us to continue developing and finding ways to achieve industrialization objectives, the optimization of resources, and the satisfaction of human needs in accordance with the corresponding area in question. In this work, an illustration of the above is proposed by combining a problem of interest, such as the generation of biodiesel, convolutional neural networks, metaheuristic algorithms, and control through a neuro-fuzzy approach. One can find works such as [13], where a comparative test of a boiler is presented, proving the maximum boiler efficiency indicators and also the minimum toxicity of exhaust gases discharged into the atmosphere, all this considering the proposed control system.

An important part of the control area is the identification of the system to be controlled, because most control techniques used in the industry require a model to be applied [4]. In a real-world system, to obtain a model from them requires a lot of variables, such as knowledge of the system, physical and environmental conditions, and the type of model to be used, i.e., linear, nonlinear, parametric, non-parametric, continuous or discrete time, among others [14]. The aim of these models is to represent in the best way possible the real system and to be as simple as possible, too. Different techniques of modeling have been developed through the years, each one trying to satisfy a specific need [15]. For example, [16] represents the issues that nonlinear systems could introduce in the modeling process and the excitation signals type that could be used in these systems. Also, in [17], linear models were not enough to capture all the neural activity and relationships in the neural system, so they propose a nonlinear model that helps in this matter and also produces sensitive biomarkers to improve diagnosis in neurological disorders. In [18], they propose the use of neural networks as a modeling method for a nonlinear system with constraints on states, which is paired with a back-stepping algorithm and an intelligent controller.

Neural networks use has drastically increased in the past twenty years with a variety of applications like object recognition [19], prediction [20,21], parametric estimation [22,23] and system control [24]. One of the most popular algorithms within the neural networks is the convolutional neural network (CNN) [25], inspired by the studies of Hubel and Wiesel in the 1950s of the visual processing system of animals [26], and image classification is its main application [27]. A recent deep study of recent advances in the use of CNNs can be found in [28,29], where many modified architectures to specific applications of CNNs are gathered together. As regards system identification using CNNs, Guodong Fan and Xi Zhang present an architecture of CNNs to estimate battery capacity using voltage data from different degradation levels [30]. In [31], a CNN is employed for the rapid prediction of fluvial flood inundation where hydraulic/hydrodynamic models used for this same propose are too computationally demanding. Another technique that can be applied to the identification of systems is sensorless, which has been gaining popularity in pump, motor, fan, and extractor control systems, among others [32]. Since sometimes it is not possible to carry out the measurement of certain variables due to their complexity or a high implementation cost, the estimation and approximation of the variables have an inherent and strong correlation with the identification of a mathematical model and its parameters. It is possible to obtain benefits for the prediction of the behavior of a system or to replace a primary and secondary element (sensor/transmitter) due to an instrument failure or communication failure. The implementation of this type of techniques can help reduce instrumentation costs and avoid unscheduled shutdowns due to instrument replacement; however, this should not be used under any circumstances for security systems.

In [33–35], a variety of works have addressed the tuning of different controllers through experimentation, using random combinations, leaving aside hard optimization problems. According to [36], a neuro-fuzzy controller (NFC) controller is a hybridization between controllers based on fuzzy networks [37,38] and neural networks based on [39]. A fuzzy logic system uses if–then rules to determine the appropriate course of action based on input data. The programmer, using its expertise, initially proposes these rules and is further refined by means of a machine learning algorithm. The rules are based not only on the experience of the programmer but on how the system learns from the data. This process involves tuning the hyper-parameters of the neural network and the elements associated with the inference laws of fuzzy logic. Metaheuristics algorithms provide solutions for optimization problems when analytical or classical methods are not available. This non-viability mainly occurs if the objective function to be optimized is not derivable or when the computing resources for analytical methods are limited.

The main objectives of this work are presented below:

- To identify the dynamics of a biodiesel production system by means of a CNN.
- To tune a neuro-fuzzy controller by applying metahumanistic techniques using the system obtained by means of the CNN.

- To propose the cooling action as a temperature control problem to optimize the energy applied in this stage.
- To apply a sensorless technique based on the implementation of a CNN for the replacement of the control signal in case of failure.

Biodiesel production systems present challenges and obstacles associated with various areas of engineering. Computer science techniques can be used to enhance topics related to Automatic Control, such as artificial neural networks and optimization methods. This work presents solutions to different problems featuring a mathematical model to perform the tuning of the temperature control based on convolutional neural networks and tuning for an NFC counter based on an optimization problem defined from the control objective to minimize the heating stage by means of metaheuristics. In addition, a sensorless technique is implemented in case of failure of the sensor/transmitter element, which consists of using a convolutional neural network to momentarily replace the original signal of the system and avoid problems in the production process. Finally, the result is a model that allows simulations to be carried out to observe the dynamics of the process and thereby evaluate the production process. A method is also proposed for energy saving by minimizing the use of the cooling stage and, finally, a sensorless technique to guarantee the continuity of the production process in the event of sensor/transmitter failure.

2. Materials and Methods

2.1. Problem Description

One of the great challenges in the application of Automatic Control in the industry is the need to carry out tests of the controllers virtually and to allow the evaluation of the effectiveness of the control law through measurements or performance metrics. However, a method based on trial and error is not economically feasible in most cases due to the time it may take and the inputs or raw materials required to carry out such tests. To work virtually, it is necessary to have a dynamic mathematical model that allows emulation of the behavior of the process in a specialized software environment and, from there, to carry out dynamic tests under a specific control proposal to achieve the control objectives based on the need to optimize the industrial process. Determining a mathematical model of an industrial production process is often highly complex and complicated by means of laws based on classical mechanics to obtain the differential equations that describe its behavior over time. One way that has emerged to try to solve this problem is the use of parametric and non-parametric identification techniques, which allow the approximation of dynamics or parameters by means of the input and output signals of a process. In general, industrial processes have nonlinear behaviors, which implies an important challenge to determine a mathematical structure that describes their behavior. One of the most used techniques to determine their dynamics is artificial neural networks; due to their plasticity and flexibility in the learning process, they become strong candidates to be used since in industrial processes, it is possible to collect a large amount of information on the control action (manipulated variable) and the process variable (controlled variable) through a data acquisition system. The system presented in [40] is taken as a case study, which is described through Figure 1, where TC, T_p, T_1 represents the thermal agent temperature, output product temperature and the temperature of the product, respectively. The objective of control is to regulate the temperature of the product (T_1) as close to the reference temperature T_0 as possible. We control the action on the manipulated variable (valve), determined by the control law programmed in the control system, using the information from the temperature variable transducer coming from the reactor. Also, the flow of cold water and steam to the tank is controlled in order to reach the desired temperature. Here,

q_{ar} is the heat of the cold water, q_p is the output product heat and q_{ab} is the steam heat. The detailed description of the physical–chemical process that takes place in the production of biodiesel and the mathematical model can be found in [40].

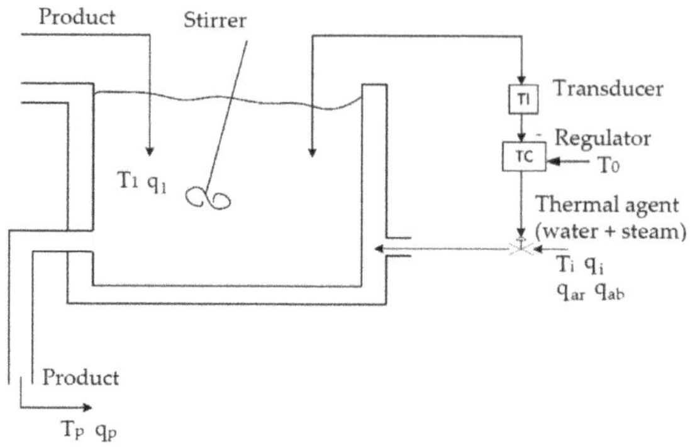

Figure 1. Pilot plant scheme [40].

The transfer function for the tank temperature is described by (1), according to [40]:

$$G_f = \frac{4}{(360s+1)(1143s+1)}. \tag{1}$$

To obtain (1), a cascade of three other transfer functions is needed, i.e., $G_f = G_E G_p G_T$, where these functions represent the dynamic and the static part of the complete system, which is shown next:

$$G_E = \frac{K_E}{T_E s + 1}; \tag{2}$$

$$G_P = \frac{K_p}{T_p s + 1}; \tag{3}$$

$$G_T = 2. \tag{4}$$

The execution element is represented by (2), where $K_E = 2$ and $T_E = 360$; (3) applies to the process with $T_p = 1143\,s$, $K_p = 1$, and for the temperature transducer, (4) is used as a gain. These three transfer functions, working together, ensure the temperature of the tank remains at the same level as long as the vegetable oil and the methanol react with each other and the fatty acids methyl ester and glycerin are produced in the correct way. Although model (1) represents a mathematical model of the system by means of differential equations expressed in the frequency domain through a transfer function, it is worth mentioning that it is a linear approximation that is invariant in time and with initial conditions equal to zero. No model can be an exact copy of the system, so the general considerations that are made to obtain the dynamics of the system generate uncertainty, and therefore, a controller based on that model naturally has problems in the experimental phase. Moreover, the parameters of a mathematical model are time-variant either because of environmental conditions or due to the properties of the system itself. In the case of biodiesel production, this may occur during the process. In the transesterification stage, it is possible to observe adverse phenomena within the equipment and instruments, such as residual material, gas emissions, and physical effects on the tank due to heating even within operating ranges, among other effects. These apparently simple things can cause parameter variation.

For the above reasons, it is preferable to determine the mathematical model of the system in another way. In this work, convolutional neural networks are used, since they have been proven to be an important and novel tool in the area of computer science due to their learning capabilities. To apply the identification process to industrial systems and ensure that the information collected is reliable, it is necessary for the data acquisition system to adequately measure the information of the process variable and the control action according to Theorem 1 and Shannon–Nyquist frequency sampling.

Theorem 1. *The sample rate f_s must be greater than twice the highest frequency component of interest in the signal. This frequency is usually known as the Nyquist frequency f_N [2]:*

$$f_s > 2 * f_N. \tag{5}$$

2.2. Identification Theory Using Convolutional Neural Network

The theory necessary to perform the system identification based on CNN is presented below. As an estimation method for the system, a convolutional neural network has been employed with the following structure: one fully connected layer as the output layer and two convolutional layers one after the other, in which each convolutional layer has 10 hyper-parameter named filters, $F_\ell \in \Re^3$ for $\ell = 1, 2$, and a ReLU activation function is applied to all filters separately.

The output of the CNN $\hat{q_p}$ is the estimated output of the real system q_p, which is calculated below:

$$\hat{q_p} = N_W * \Theta \tag{6}$$

where N_W are the synaptic weights in the output layer and Θ is the input to this layer, while Θ is the concatenation of the second convolutional layer outputs $\theta 2$:

$$\Theta = \left[\theta 2_1^T \;\; \theta 2_2^T \cdots \theta 2_{10}^T\right]^T \tag{7}$$

each $\theta 2_i$ for $i = 1, 2, \cdots, 10$, is calculated as

$$\theta 2_i = \max(F_{2,i} \odot \theta 1_i, 0) \tag{8}$$

where (8) is the ReLU operation over the convolution, $\odot$, of the filter $F_{2,i}$ with the ith-output $\theta 1_i$ of the first convolutional layer.

These outputs $\theta 1_i$ are obtained as follows:

$$\theta 1_i = \max(F_{1,i} \odot u_N, 0) \tag{9}$$

where u_N is the input vector for the CNN. For the training of the CNN, the backpropagation algorithm is employed to update the hyper-parameters of the CNN, in this case, the synaptic weights N_W and the filters $F_{2,i}$ and $F_{1,i}$ [41]. The rules for updating the hyper-parameters for the synaptic weights are, in backward order,

$$N_W(k+1) = N_W(k) - \eta \frac{\partial J}{\partial N_W} \tag{10}$$

with k representing the iteration in which the hyper-parameters are updated, J is the objective function to be minimized, and $J = \frac{1}{2}(\hat{q_p} - q_p)^2$. For the ReLU activation function, the gradient through this operation can be calculated as

$$\frac{\partial \theta h_2}{\partial F_{h,i} \odot \theta h - 1_i} = F_{h,i} \odot \theta h - 1_i \tag{11}$$

with $h = 1, 2$ representing the convolutional layer of the structure, which in case of $h = 1$ $\theta h - 1$ corresponds to the input to the CNN, u_N. For filters $F_{h,i}$, the update rule is

$$\frac{\partial F_{h,i} \odot \theta h - 1_i}{\partial F_{h,i}} = (F_{h,i} \odot \theta h - 1_i) \odot \theta h - 1_i \tag{12}$$

2.3. *System Identification Procedure*

The procedure for the system identification is as follows: 12,000 input–output data from the model are generated, which are used as training information. These data were obtained while the system was in a closed-loop configuration. In Figure 2, the procedure is shown, where the control signal U and the tank temperature output signal q_p are used as input to the CNN to produce a estimate value of the tank temperature $\hat{q_p}$. This value is fed back into the CNN along with q_p as an estimation error to calculate the gradient descendant to update the hyper-parameters of the CNN. This process is completed step by step; we received data from the system in different instants of time, the estimation was calculated, and the CNN was calculated in each one of these iterations. The correct application of Theorem 1 was needed to guarantee that the acquired data from the temperature system were reliable. Sometimes, it is hard to determine the Nyquist frequency in real-world applications. Nevertheless, experimental tests can be carried out through a sinusoidal input signal. For variables such as temperature, the dynamics of these signals are usually slow, so its f_N usually is low, in the order of Hertz, which generates problems in finding or proposing a sampling frequency. In this work, through trial and error, 1 s is enough to appreciate changes in the dynamics of the system.

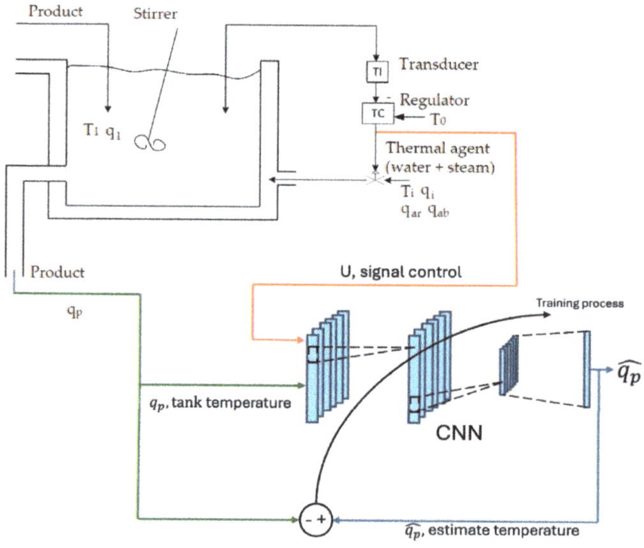

Figure 2. Identification procedure.

For the CNN, the vector u_N is generated in each iteration with the following structure:

$$u_N = \left[U(k)\ \ U(k-1)\ \ \hat{q_p}(k-1)\ \ \hat{q_p}(k-2)\right]^T \tag{13}$$

As a result of the system identification, Figure 3 shows these results, where both signals are practically equal, and the mean square error (MSE) metric was employed and has a measurement of the identification, leading to a value of 5.52×10^{-21}.

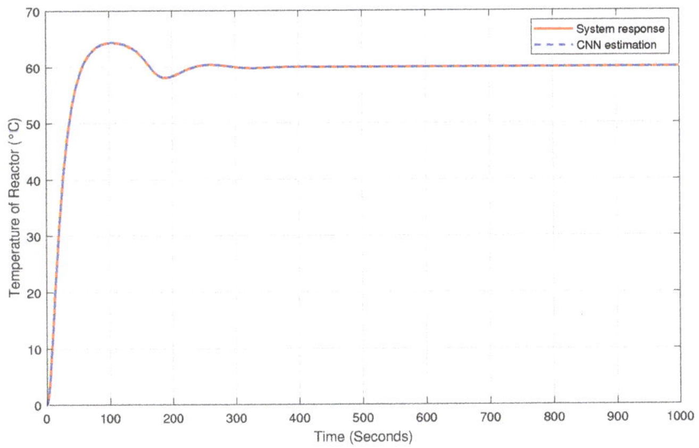

Figure 3. System identification result.

This process can be resumed in Algorithm 1. In it, the first step to propose a CNN structure, i.e., propose the hyper-parameters of the CNN, is performed randomly; there are no specific criteria to chose them. Then, the CNN uses data from the signal U and q_p as input to calculate the estimated value of q_p in that iteration and generate an estimated error, which will be used in the backpropagation algorithm to update the value of the hyper-parameters of CNN. This process is known as the training stage and performed repeatedly until the simulation time is over. Finally, the testing stage is used to verify that the training has been successfully achieved using the MSE metric. If the value of this index is too high, all of the steps are repeated in order to decrease this value.

Algorithm 1 System identification procedure

```
Propose CNN hyper-parameters, F, NN_W.
while i ≤ N do                                          ▷ N -> total of data
    Use control signal U into u_N.
    Run CNN to get q̂_p.
    Use system output g_p into u_N for next iteration.
    Compute estimation error e = q̂_p − q_p.
    Update hyper-parameters F, N_W using gradient descendant method.
end while
Realize a testing stage to verify the training process with aid of MSE index
if MSE ≥ Mm then                 ▷ Mm -> minimum value for acceptance estimation
    Repeat previous steps with a different CNN hyper-parameters selection.
end if
```

2.4. Neuro-Fuzzy Controller

The control used in this work is presented in detail in [42]; below is a brief explanation of its operation and mathematical structure. The error (E) and the error increment over time (ΔE) are the inputs necessary for the neuro-fuzzy controller to work. To determine these signals,

$$E(k) = SP(k) - PV(k), \tag{14}$$

$$\Delta E(k) = \frac{E(k) - E(k-1)}{Ts}, \tag{15}$$

where $SP(k)$ and $PV(k)$ denote the set point and the value of the process variable at time k, $E(k-1)$ represents a time delay of $E(k)$, and Ts is the sampling time.

The architecture of NFC is constituted by five layers. The first one, the input vector $X(k) = [E(k), \Delta E(k)]$, is provided for the fuzzification, mapping these real values to the linguistics applied to make the fuzzification according to the Takagi–Sugeno method [43]. The structure of this layer is defined by (16)

$$\mu_{A_{j,k}} = \Gamma_j(\Lambda_j(k), X_i(k)) \tag{16}$$

where $\mu_{A_{j,k}}$ indicates the membership function quantity selected for the vector $X(k)$. Normally, the selection of the number of membership functions is chosen by trial and error, but in this work, it is part of the tuning carried out by a bilevel optimization approach, including the terms related to $\Gamma_j(\cdot)$ and $\Lambda_j(k)$ that describe the j member functions selected to make the fuzzyfication. For the Gaussian bell described by (17), the set $\Lambda_j(k)$ contains the position and the spread of the Gaussian function in the parameters $\phi_{j,k}$ and $\sigma_{j,k}$, respectively [42].

$$\mu_{A_{j,k}}(X_i(k)) = e^{0.5\left(\frac{X_i(k)\phi_{j,k}}{\sigma_{j,k}}\right)^2} \tag{17}$$

The inferences based on the if–then statements are made in the second layer in order to generate fuzzy sets by means of the vector $X(k)$ for subsequent membership in such a way that the functions are related between each other to be able to determine the possible behavior of the system scenarios. Right away, an output value according to each of the cases established by the set of fuzzy rules denoted by (18) is proposed, where the index $p = 2, 3$ corresponds to the second or third layer, respectively [42].

$$O^p = w(k) = \mu_{A_{j,k}}(X_i(k))^T * \mu_{A_{j,k}}(X_i(k)) \tag{18}$$

The output of the third layer should be normalized according to (19), where the index $l = 1, 2, ..., R$ is defined by the number of fuzzy rules n, since $R = n \times n$ [42].

$$\overline{w}(k) = \frac{w(k)}{\sum_{l=1}^{R} w_l(k)} \tag{19}$$

The fourth layer output (O^4) is generated by the products of the normalized firing strength and the parameter $r_l = \gamma_j(\lambda_j(k))$. This value is computed using (20), with $\beta_n(k) \in \Re^n$, $\gamma_j(\lambda_j(k))$ being the membership function and its parameters that describe the controller actions by considering the output of the third layer, and $\overline{w}_{n,:}(k)$ represents each row of the matrix [42].

$$O^4 = \beta_n(k) = \overline{w}_{n,:}(k) * \gamma_j(\lambda_j(k)) \tag{20}$$

The output of the fifth layer is a scalar value used as a control signal that is obtained by (21),

$$U(k) = \sum \beta(k) \tag{21}$$

A hybridization was made with neural networks to give a learning property to the control system, making it able to learn in a continuous way considering the system is subject to different operating conditions such as disturbances, as mentioned in [44]. The gradient descent method [2] was adapted for tuning the membership functions (Gaussians bells) of the neuro-fuzzy network for the fuzzification and defuzzification stages, as shown, respectively, in [42]. This has led to an improvement in the controller that has resulted in an efficient use of energy. However, it can be used to design a desired dynamic for the controller, which can help avoid unwanted actions by the control signal when trying to bring the behavior to a desired set point.

2.5. Sensorless

In biodiesel production, obstacles or situations can be found that make it difficult to control the reactor temperature. The signal emitted by a sensor/transmitter can be corrupted by noise, it can be miscalibrated, and it can fail due to poor installation or lack of maintenance. When any of these situations occur, the controller will be affected due to the dependence on the measurement of the controlled variable to obtain the error signal. This can lead to a loss of energy, instability in the system, and a loss of raw material, and it can also put operators and process equipment at risk. This work presents a solution that allows to attack this problem through the implementation of the CNN as a virtual sensor, which has the purpose of replacing the sensor/transmitter signal in its absence. Below, in Figure 4, it is shown how the CNN acts in the sensorless system. The CNN for this case, which was previously trained to estimate temperature, uses only information from the controller signal, the set point, and the output of itself as input to estimate this variable, so the controller does not obtain a zero at any time, even if the transducer presents any kind of failure.

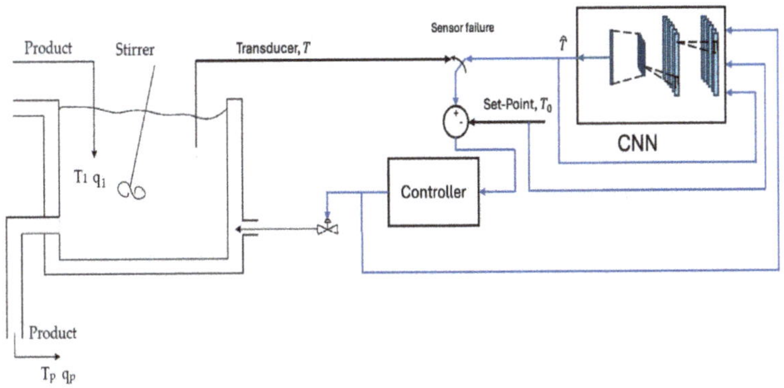

Figure 4. Sensorless system.

In order to show the benefits of the sensorless systems, a simulation was created where the signal from the temperature sensor is missing to compare with real case scenarios where there is a failure in communication with the sensor, such as accidental cable disconnection or breakage. In this case, the controller will send a signal to increase temperature because of the zero signal sent by the sensor, leading to increased energy consumption. For this matter, the plant is controlled with a PID to reach 45 °C with an external disturbance during the simulation. This can be seen in Figure 5; the reactor temperature fluctuations are due to intermittencies in the connection with the sensor. These intermittencies are simulated as occurring repeatedly during certain periods of time in addition to an external disturbance occurring in the system at 5000 s. Both before and after the disturbance, the system will not reach the desired temperature due to the loss of communication with the sensor, potentially compromising the final product properties.

As a solution for this particular problem, a CNN with 2 convolutional layers with 20 filters in each layer and 10 synaptic weights was trained to model the plant and used in parallel with it. In each instance of measurement, the two signals corresponding to the temperature of the reactor can be used as the feedback signal to the controller, preventing the controller from sending an incorrect signal when communication with the sensor is lost, reducing energy consumption and allowing the sensor to be checked without having to stop the process. In both cases, whether the CNN is used or not, 100 s are simulated as the time that the signal sensor is faulty. Figure 6 shows the temperature of the reactor when a CNN is employed; as it can be seen, the controller takes the system to temperature near the defined set point even though the disturbance appears, unlike the previous case.

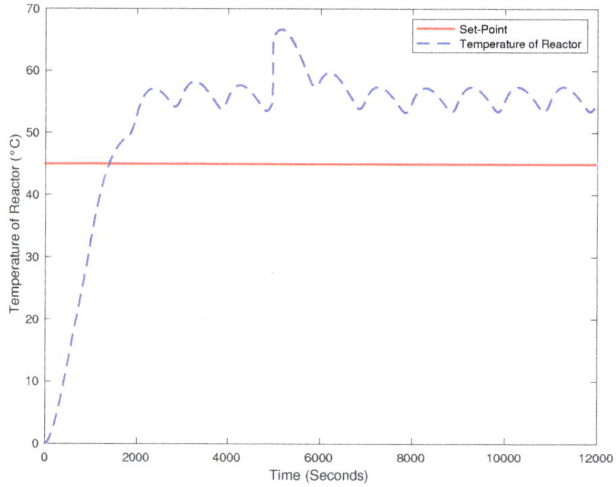

Figure 5. Temperature of reactor with sensor failure communication.

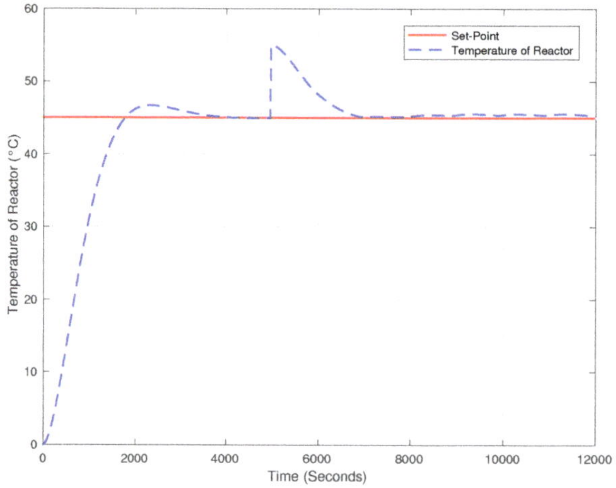

Figure 6. Temperature of reactor with sensor failure communication using CNN.

2.6. Tuning NFC Parameters with a Bilevel Optimization Approach with Differential Evolution (DE)

In [45], it is explained that the tuning of the NFC controller is complex because the configuration space contains vectors with different dimensions that contain both integers and continuous variables, which is a limitation of implementing search algorithms in order to find the best one. In that work, it was proposed to solve the drawbacks following a bilevel optimization approach, which refers to one in which one problem is embedded within another so that the solution of the first (lower problem) restricts the second (upper problem) [46]. As a contribution to this proposal, the modifications to the bilevel optimization approach are presented to tune the NFC with the objective of covering the additional requirement of not exceeding the reference value of the controller.

2.6.1. The Original Problem

The optimization problem presented in [45] that has the objective of tuning the NFC for the tracking task is described below. For both levels of the methodology, the same objective function presented in (22) is considered where e is determined as the difference between a signal control and the result of applying the NFC to the system.

$$f(\vec{x}) = \sum_{i=1}^{k}(|e_i(\vec{x})| + L|\dot{e}_i(\vec{x})|) \tag{22}$$

In (22), the integers decision variables in $\vec{x}$ are the number of membership functions (m), while the weights of the neural network $[w_1, w_2, ..., w_n]$, the parameters of the membership functions ($\phi_{j,k}$) and ($\sigma_{j,k}$), and the learning rate constants η_ϕ, η_σ and η_r are real-value variables. L is a scalar that serves to regulate the influence of the change in the error as a term of the objective function.

The solution spaces differ in the sense that in the first level, part of the solution is completed with random values; on the other hand, in the second level, the random part becomes the solution to be searched using part of the previously found solution as constants. In the first level of optimization, the weights and the initial position of the member functions are not considered as design variables but as a noise vector, resulting in a solution vector and a noise vector structured as in (23) and (24), respectively.

$$\vec{x} = [m, \sigma_1, \sigma_2, \eta_\phi, \eta_\sigma, \eta_r] \tag{23}$$

$$\vec{w} = [w_1, w_2, \ldots, w_{n=m^2}, \phi_1, \phi_2, \ldots, \phi_m] \tag{24}$$

From (23), it is ensured that the size of the solution vector ($\vec{x}$) is fixed. To complete the configuration of the NFC, the vector is combined with randomly generated sets of weights 10 m^2 in size, where the set with the lowest objective function is selected as the fitness for the solution vector. This solution is carried out around the epochs of the DE as an elitist mechanism. Finally, to deal with the integer value (m), the repulsion strategy proposed by Liu et al. [47] is employed.

At the second level, the solution obtained at the first level of optimization allows establishing a restriction at the second level to make the number of functions and weights constant, where the NFC weights are tuned. The structure of the solution vector for the second level is denoted by (25); m has a fixed size and comes from the previous level as a solution. In this second level of optimization, some elements of the solution vector continue as a decision variable; although weights and positions of member functions are greatly important, it is possible that a different configuration would be better for the tracking task.

$$\vec{x} = [\sigma_1, \sigma_2, \eta_\phi, \eta_\sigma, \eta_r, w_1, \ldots, w_{n=m^2}, \phi_1, \ldots, \phi_m] \quad (25)$$

2.6.2. The Proposed Modified Problems

The tuning methodology was modified as the contribution of this work to integrate the additional requirement of not surpassing the point reference in the tracking task. Two new optimization functions were considered. The first one penalizes the value of the objective function every time the controller's action produces a value that exceeds the reference, as shown in (26).

$$f(\vec{x}) = \sum_{i=1}^{k}(|e_i(\vec{x})| + L|\dot{e}_i(\vec{x})| + M\check{e}_i(\vec{x})) \quad (26)$$

where M is a scalar that regulates the influence of the new term on the objective function. $\check{e}_i$ defines the error over the set point and is described by (27) using the description of the NFC.

$$\check{e}_i = \begin{cases} \check{e}_i = PV(k) - SP(K), & 0 \leq PV(k) - SP(K) \\ \check{e}_i = 0, & other. \end{cases} \quad (27)$$

This proposal follows the idea presented in [48] where in order to cover multiple objectives, a common objective function is established, which is a composition of different individual functions whose optimal solution involves covering the objective. As the value of M increases, the overreaction is reduced but does not necessarily reach the optimal value of zero. To force the search algorithm to find solutions with that value, a second approach is proposed.

The second proposal consists of biasing the space of feasible solutions by incorporating the constraint (28) to the original objective function (22). The constraint considers a configuration infeasible when it produces system behavior where the controller action exceeds the set point value when the initial condition begins. Due to the incorporation of this restriction, the feasibility rules [49] as a constraint handler were included in DE.

$$h(\vec{x}) = \sum_{i=1}^{k}\check{e}_i(\vec{x}) = 0 \quad (28)$$

The definition of $\check{e}_i$ implies that it can only have positive or zero values, so the solution of the optimization problem represents the best configuration to perform the monitoring task without exceeding the reference.

2.6.3. Final Methodology

The general process for solving the complete optimization problem of the modified approach is shown in Figure 7, and the operations for the mutation, cross, and selection are described by (29), (30) and (31), respectively. All tuned parameters and its range are shown in Table 1.

$$Mutantx^j = x^{r1} + F(x^{r2} - x^{r3}) \quad (29)$$

$$Newx_i = \begin{cases} Mutantx_i, & rand(0,1) \leq CR \\ Fatherx_i, & rand(0,1) > CR. \end{cases} \quad (30)$$

$$x^{g+1} = \begin{cases} Newx_i, & f(Newx_i) \leq f(x^g) \wedge h(Newx_i) = h(x^g) = 0 \\ x^g, & f(x^g) < f(Newx_i) \wedge h(Newx_i) = h(x^g) = 0 \\ Newx_i, & h(Newx_i) \leq h(x^g) \wedge (h(Newx_i) \neq 0 \vee h(x^g) \neq 0) \\ x^g, & h(x^g) < h(Newx_i) \wedge (h(Newx_i) \neq 0 \vee h(x^g) \neq 0) \end{cases} \tag{31}$$

Table 1. Tuned parameters for the NFC.

Parameter	Range
Parameter of the membership functions ($\phi_{j,k}$ and $\sigma_{j,k}$)	$\phi_{j,k}, \sigma_{j,k} \in [-100, 100]$
Learning rate constants $\eta_\phi, \eta_\sigma, \eta_r$	$\eta_\phi, \eta_\sigma, \eta_r \in [0.0001, 3]$
Member functions number m	$m \in \mathbb{Z} \cap [3, 15]$
Initial weights $\vec{w}$ refereed to r_j	$w_i \in [0.0001, 50]$

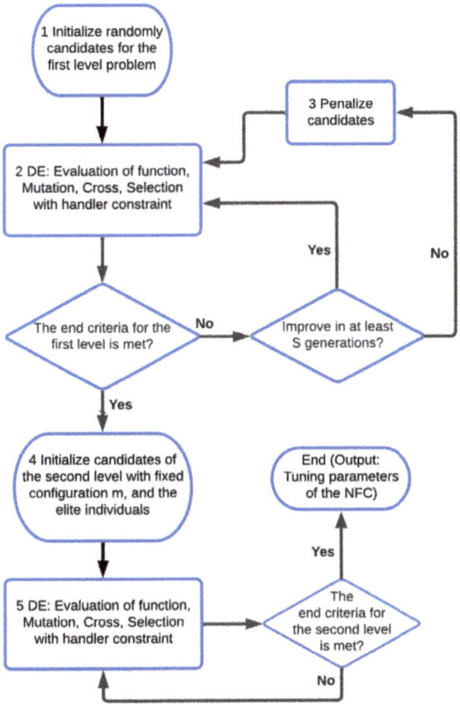

Figure 7. Flowchart of the tuning process.

3. Experimentation and Discussion

3.1. Case of Study

One of the main objectives of biodiesel production is to control the temperature at a desired set point. According to Figure 1, it is possible to perform a heating stage by supplying heat or to carry out a cooling stage, for example, by supplying cold water. When each of the stages occurs, it is due to the nature of the control action determined by the control system based on the temperature variation of the biodiesel production system. One of the main causes that can be easily seen is that if the plant is exposed to an uncontrolled environment, then it is susceptible to environmental factors such as the variation in ambient temperature throughout the day or the season of the year. For example, in the summer, it could be thought that the cooling stage is used more frequently; on the contrary, in the winter, the heating stage could be used more frequently. These environmental variations from the point of view of Automatic Control are considered disturbances. In [45], they are addressed in greater detail. In Section 2.5, it is proposed that the optimization algorithm based on metaheuristics conducts a search and performs a tuning such that the controller in its heating stage mitigates exceeding the desired temperature reference; this leads to energy savings by avoiding the cooling stage being used constantly. Therefore, the above illustrates that considerable efforts can be made to achieve better results from the production processes through not only making improvements in the controller but by making a critical analysis of the needs of a process through understanding it and the industrial objectives.

To illustrate the method proposed in this work, we used the model identified by the CNN. Subsequently, tests were carried out with the proposed functions (26) and (28). The reference of 60 °C was taken; according to the problem posed in [42], at 5000 s, a disturbance due to heat loss of 10 °C is simulated. This may be due to a failure in the control system or some unexpected external agent that causes such a loss, thus affecting the production of biodiesel. Below is the result obtained by the NFC proposed in [45].

In Figure 8, it is observed that in the initial stage, there is an overshoot, causing the cooling stage of the control system to act to bring the reactor temperature to the desired set point. Later, in the disturbance due to heat demand, it is observed that there is no overshoot in the dynamics of the controlled variable and that the temperature recovery time is gradual.

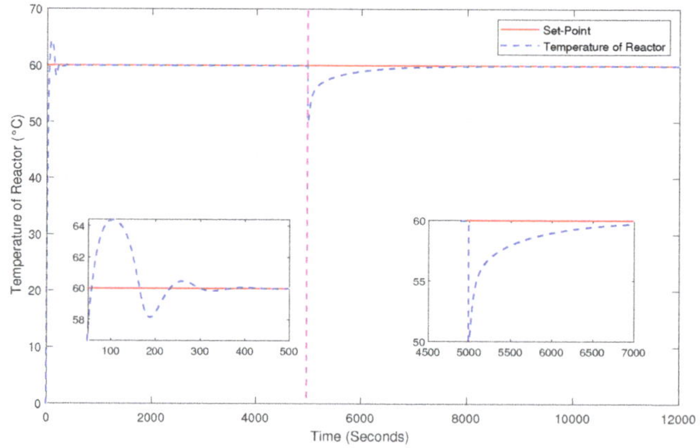

Figure 8. System behavior by NFC without additional considerations.

3.2. First Optimization Proposal

In this experiment, see Figure 9, it is shown how the optimization function proposed in the model (26) penalizes the value of the objective function, as the controller's action produces a value that exceeds the reference. At the beginning of the system response, there is no overshoot, avoiding the cooling stage. However, when the disturbance occurs, it can be observed that in the recovery of the system, there is a time in which the temperature exceeds the proposed reference, so it is necessary to activate the cooling stage. Therefore, speaking in terms of energy savings, it is not desired since this type of system can use cooling towers to supply cold water and lower the reactor temperature.

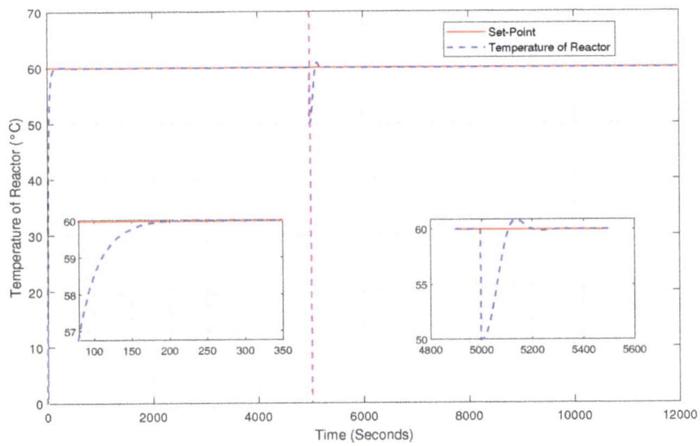

Figure 9. System behavior by NFC with the first optimization proposal.

3.3. Second Optimization Proposal

Figure 10 shows the experimentation considering the second optimization function, which was proposed in model (28), where the constraint considers infeasible a configuration that produces system behavior where the controller action exceeds the set point value when the initial condition begins below. In this case, it is possible to observe that at no time during the experiment, either in the initial stage or at the time of the disturbance, is there a time in which the dynamics of the reactor temperature variable exceed the reference value. Therefore, the activation of the cooling stage was not necessary; this can mean considerable energy savings in biodiesel production systems.

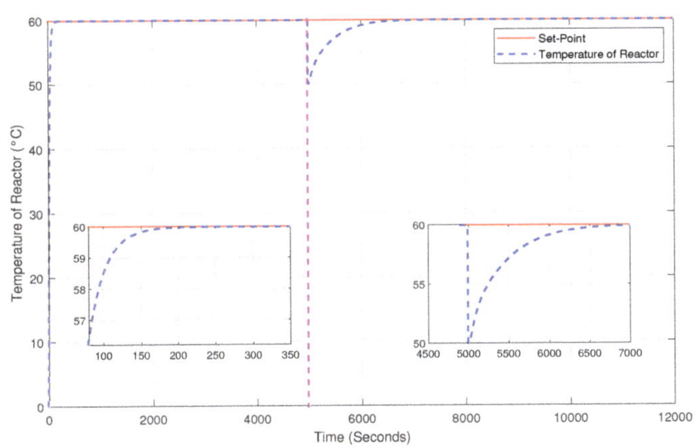

Figure 10. System behavior by NFC with the second optimization proposal.

With this approach, the desired objective was achieved, but due to the stochastic components inherent in metaheuristic techniques, the tuning methodology was repeated 10 times. The results are reported in Table 2. Most of the results are infeasible in the strict sense, but more than half of them have errors less than 1×10^{-6}, so their implementation is viable.

Table 2. Performance of 10 tuned configurations using the second optimization proposal.

Run	Objective Function	Constraint Value	ITAE	TVU
1	8.4832×10^3	0	2.1363×10^7	17.5678
2	5.5300×10^3	10.6989	4.0364×10^6	17.6588
3	4.6494×10^3	3.7802×10^{-5}	4.7453×10^6	30.4145
4	5.3384×10^3	1.9564×10^{-10}	6.2209×10^6	17.5678
5	5.4586×10^3	16.0819	3.8061×10^6	17.7146
6	5.2092×10^3	31.4761	3.3928×10^6	17.9132
7	4.7826×10^3	2.1240×10^{-10}	4.5194×10^6	24.6171
8	5.0259×10^3	31.6624	3.6326×10^6	17.9256
9	4.7149×10^3	8.8660×10^{-5}	4.6459×10^6	29.8925
10	6.9511×10^3	3.1859×10^{-10}	5.6212×10^6	17.5678
Average	5.6143×10^3	8.9919	6.1983×10^6	17.5678
Standard deviation	1.2049×10^3	13.1563	5.4005×10^6	5.3431

3.4. Discussion

The performance metric allows us to interpret the dynamics of the system by quantifying the error signal: the smaller the index, the better the controller performance. There are a variety of metrics that allow the user to obtain many quantitative properties of the performance of the control of dynamic systems [50]. However, when dealing with the temperature variable, it must be taken into consideration that its dynamics are slow, and it is preferable to penalize more strongly the deviations of the controlled variable with respect to the set point in a steady state or in the presence of disturbances with respect to the transient behavior performance. The ITAE metric results are useful since they penalize the error more strongly as time increases; this is preferable for systems with slow dynamics, such as the temperature of the reactor, where the initial error is usually large. Below is the expression of performance metrics [2]:

$$\int_0^{\infty} t|e(t)|\, dx. \tag{32}$$

To obtain the results presented, the integral corresponding to the ITAE index was approximated to the form presented in (33)

$$\frac{1}{2}\sum_{k=1}^{d} ke(k) + (k-1)e(k-1). \tag{33}$$

In many processes, the control signal is an important variable to observe when evaluating the performance of the control system. The TVU index is adequate in this case [2]:

$$\sum_{k=1}^{d} |u(k) - u(k-1)|. \tag{34}$$

Figures 8–10 are helpful in qualitatively judging the behavior of the reactor temperature variable under NFC control with different considerations. However, this is not an analysis that reveals simple proof that one method is better than another. For a numerical comparison point, Table 3 is presented below, which contains the values obtained in each experiment using the ITAE and TVU performance metrics. Identification, system simulation, and controller tuning through optimization were performed using MATLAB R2020a. We used a computing platform with the following specifications: Intel i7 processor at 3.70 GHz, 16 GB of RAM, and Windows 11 operating system.

Table 3. Comparison of controller performance indexes.

Controller	ITAE	TVU
NFC without penalization	1.9597×10^7	22.3993
NFC first optimization proposal	3.5530×10^6	18.3259
NFC second optimization proposal	3.3928×10^6	17.9132

From Table 3, it can be determined that the best control applied is the NFC considering the restriction proposed in the model (28), demonstrating a lower accumulated ITAE error as well as using 20% less energy with respect to the TVU metric. Applied to an industrial control problem, this could mean profits for the company by making better use of the resources involved in the biodiesel production process.

4. Conclusions

This paper presents a way to apply computer science techniques to solve problems in biodiesel production. This includes a conceptual stage featuring the control objective as well as evaluating the results of an experimental phase. The identification phase of the dynamics of a process, by means of a CNN, is crucial to be able to carry out tests through simulations in order to save time and resources. Since this identification is carried out with real data from the industrial system, the possible deviation in the implementation in the real world will be drastically minimized, which is of great importance so that advanced control techniques are more easily accepted by the industrial sector. A sensorless technique is also shown, which allows the use of a virtual sensor executed by a CNN and continues with the control of the process while there is no signal from the sensor/transmitter element.

NFC controllers are flexible and allow for the easier mitigation of changes in the environment where the biodiesel plant is located, as well as disturbances that could arise, such as mechanical or electronic failures. The combination of control objectives with metaheuristic algorithms allows the focus of the controllers to meet optimization challenges, as has been shown in this work, by limiting the use of the cooling stage in the temperature control of the reactor. Finally, the quantitative comparison using the TVU and ITAE performance metrics allows the evaluation of the different proposals and determines that for this case study, the NFC with an optimization function that penalizes any control signal that causes the reactor temperature to exceed the desired set point is the best option, thereby achieving a lower error, using lower energy consumption (20% less). The next work is intended to implement these advances in a biodiesel production plant and carry out an energy study of the experimental results. Future work is intended to carry out experimental tests in an industrial pilot plant for the production of biodiesel from cooking oil and to conduct a study of the energy quality of the controller developed in this research.

Author Contributions: M.C.M.-R. and R.L.-M. worked on all the tasks, I.C.-M. and M.A.L.-P. worked on the literature review, M.C.M.-R. and R.L.-M. conducted experimental studies, I.C.-M. and R.T.-E. performed the supervision; all authors analyzed the results. All authors have read and agreed to the published version of the manuscript.

Funding: This research received no external funding.

Data Availability Statement: The original contributions presented in the study are included in the article; further inquiries can be directed to the corresponding authors.

Acknowledgments: The Consejo Nacional de Humanidades, Ciencias y Tecnologías (CONHACYT) and SECIHTI, for the postdoctoral fellowship awarded to M.C.M-R. (CVU: 706063), that enabled research on biodiesel-based energy alternatives.

Conflicts of Interest: The authors declare no conflicts of interest.

References

1. Borase, R.P.; Maghade, D.K.; Sondkar, S.Y.; Pawar, S.N. A review of PID control, tuning methods and applications. *Int. J. Dynam. Control* **2021**, *9*, 818–827. [CrossRef]
2. Ljung, L. *System Identification Theory for User*; Prentice-Hall: Englewood Cliffs, NJ, USA, 1987.
3. Oluyisola, O.E.; Bhalla, S.; Sgarbossa, F.; Strandhagen, J.O. Designing and developing smart production planning and control systems in the industry 4.0 era: A methodology and case study. *J. Intell. Manuf.* **2022**, *33*, 311–332 . [CrossRef]
4. Schwenzer, M.; Ay, M.; Bergs, T.; Abel, D. Review on model predictive control: An engineering perspective. *Int. J. Adv. Manuf. Technol.* **2021**, *117*, 1327–1349. [CrossRef]
5. Shi, H.; Zhang, L.; Pan, D.; Wang, G. Deep Reinforcement Learning-Based Process Control in Biodiesel Production. *Processes* **2024**, *12*, 2885. [CrossRef]
6. Masera, O.; Rivero, J.C.S. Promoting a Sustainable Energy Transition in Mexico: The Role of Solid Biofuels. *BioEnergy Res.* **2022**, *15*, 1691–1693. [CrossRef]
7. Balance Nacional de Energía: Producción de Energía Primaria. Available online: https://www.gob.mx/sener/articulos/balance-nacional-de-energia-296106 (accessed on 14 June 2022).
8. Orozco-Ramírez, Q.; Cohen-Salgado, D.; Arias-Chalico, T.; García, C.A.; Martínez-Bravo, R.; Masera, O. Production and market barriers of solid forest biofuels in Mexico from the enterprises' perspective. *Madera Bosques* **2022**, *28*, e2812404. [CrossRef]
9. Sosa-Rodríguez, F.S.; Vazquez-Arenas, J. The biodiesel market in Mexico: Challenges and perspectives to overcome in Latin-American countries. *Energy Convers. Manag. X* **2021**, *12*, 100149. [CrossRef]
10. Macías-Alonso, M.; Hernández-Soto, R.; Carrera-Rodríguez, M.; Salazar-Hernández, C.; Mendoza-Miranda, J.M.; Villegas-Alcaraz, J.F.; Marrero, J.G. Obtention of biodiesel through an enzymatic two-step process. Study of its performance and characteristic emissions. *RSC Adv.* **2022**, *12*, 23747–23753. [CrossRef]
11. Boly, M.; Sanou, A. Biofuels and food security: Evidence from Indonesia and Mexico. *Energy Policy* **2022**, *163*, 112834. [CrossRef]
12. Yaqoob, H.; Teoh, Y.H.; Sher, F.; Farooq, M.U.; Jamil, M.A.; Kausar, Z.; Sabah, N.U.; Shah, M.F.; Rehman, H.Z.U.; Rehman, A.U. Potential of waste cooking oil biodiesel as renewable fuel in combustion engines: A review. *Energies* **2021**, *14*, 2565. [CrossRef]
13. Janta-Lipińska, S.; Shkarovskiy, A.; Chrobak, Ł. Improving the Fuel Combustion Quality Control System in Medium Power Boilers. *Energies* **2024**, *17*, 3055. [CrossRef]
14. Ljung, L. Perspectives on system identification. *Annu. Rev. Control* **2010**, *34*, 1–12. [CrossRef]
15. Kozin, F.; Natke, H. System identification techniques. *Struct. Saf.* **1986**, *3*, 269–316. [CrossRef]
16. Nelles, O.; Nelles, O. *Nonlinear Dynamic System Identification*; Springer: Berlin/Heidelberg, Germany, 2020.
17. He, F.; Yang, Y. Nonlinear system identification of neural systems from neurophysiological signals. *Neuroscience* **2021**, *458*, 213–228. [CrossRef]
18. Liu, Y.J.; Zhao, W.; Liu, L.; Li, D.; Tong, S.; Chen, C.P. Adaptive neural network control for a class of nonlinear systems with function constraints on states. *IEEE Trans. Neural Netw. Learn. Syst.* **2021**, *34*, 2732–2741. [CrossRef]
19. Szegedy, C.; Toshev, A.; Erhan, D. Deep neural networks for object detection. In Proceedings of the Advances in Neural Information Processing Systems 26 (NIPS 2013), Lake Tahoe, NE, USA, 5–8 December 2013.
20. Pang, Z.; Niu, F.; O'Neill, Z. Solar radiation prediction using recurrent neural network and artificial neural network: A case study with comparisons. *Renew. Energy* **2020**, *156*, 279–289. [CrossRef]
21. Dolling, O.R.; Varas, E.A. Artificial neural networks for streamflow prediction. *J. Hydraul. Res.* **2002**, *40*, 547–554. [CrossRef]
22. Dong, A.; Starr, A.; Zhao, Y. Neural network-based parametric system identification: A review. *Int. J. Syst. Sci.* **2023**, *54*, 2676–2688. [CrossRef]
23. Lenzi, A.; Bessac, J.; Rudi, J.; Stein, M.L. Neural networks for parameter estimation in intractable models. *Comput. Stat. Data Anal.* **2023**, *185*, 107762. [CrossRef]
24. Liu, Z.; Gao, H.; Yu, X.; Lin, W.; Qiu, J.; Rodríguez-Andina, J.J.; Qu, D. B-spline wavelet neural-network-based adaptive control for linear-motor-driven systems via a novel gradient descent algorithm. *IEEE Trans. Ind. Electron.* **2023**, *71*, 1896–1905. [CrossRef]
25. Derry, A.; Krzywinski, M.; Altman, N. Convolutional neural networks. *Nat. Methods* **2023**, *20*, 1269–1270. [CrossRef] [PubMed]
26. Hubel, D.H.; Wiesel, T.N. Receptive fields of single neurones in the cat's striate cortex. *J. Physiol.* **1959**, *148*, 574–591. [CrossRef]
27. Bharadiya, J. Convolutional neural networks for image classification. *Int. J. Innov. Sci. Res. Technol.* **2023**, *8*, 673–677.
28. Taye, M.M. Theoretical understanding of convolutional neural network: Concepts, architectures, applications, future directions. *Computation* **2023**, *11*, 52. [CrossRef]
29. Krichen, M. Convolutional neural networks: A survey. *Computers* **2023**, *12*, 151. [CrossRef]

30. Fan, G.; Zhang, X. Battery capacity estimation using 10-second relaxation voltage and a convolutional neural network. *Appl. Energy* **2023**, *330*, 120308. [CrossRef]
31. Kabir, S.; Patidar, S.; Xia, X.; Liang, Q.; Neal, J.; Pender, G. A deep convolutional neural network model for rapid prediction of fluvial flood inundation. *J. Hydrol.* **2020**, *590*, 125481. [CrossRef]
32. Rodríguez-Abreo, O.; Velásquez, F.A.C.; de Paz, J.P.Z.; Godoy, J.L.M.; Garcia Guendulain, C. Sensorless Estimation Based on Neural Networks Trained with the Dynamic Response Points. *Sensors* **2021**, *21*, 6719. [CrossRef] [PubMed]
33. Llorente-Vidrio, D.; Ballesteros, M.; Salgado, I.; Chairez, I. Deep Learning Adapted to Differential Neural Networks Used as Pattern Classification of Electrophysiological Signals. *IEEE Trans. Pattern Anal. Mach. Intell.* **2022**, *44*, 4807–4818. [CrossRef]
34. Llorente-Vidrio, D.; Pérez-San Lázaro, R.; Ballesteros, M.; Salgado, I.; Cruz-Ortiz, D.; Chairez, I. Event driven sliding mode control of a lower limb exoskeleton based on a continuous neural network electromyographic signal classifier. *Mechatronics* **2020**, *72*, 102451. [CrossRef]
35. Escobar-Jiménez, R.; Salvide-Hernández, F.; López-Muñoz, R.; Tolentino-Eslava, R.; Maya-Rodriguez, M.C. Monitoring and Prediction of Drinking Water Consumption. In Proceedings of the Telematics and Computing, Cancún, México, 7–11 November 2022; Mata-Rivera, M.F., Zagal-Flores, R., Barria-Huidobro, C., Eds.; pp. 60–75.
36. Pezeshki, Z.; Mazinani, S.M. Comparison of artificial neural networks, fuzzy logic and neuro fuzzy for predicting optimization of building thermal consumption: A survey. *Artif. Intell. Rev.* **2019**, *52*, 495–525. [CrossRef]
37. Pacco, H.C. Simulation of temperature control and irrigation time in the production of tulips using Fuzzy logic. *Procedia Comput. Sci.* **2022**, *200*, 1–12. [CrossRef]
38. Azad, A.S.; Rahaman, M.S.A.; Watada, J.; Vasant, P.; Vintaned, J.A.G. Optimization of the hydropower energy generation using Meta-Heuristic approaches: A review. *Energy Rep.* **2020**, *6*, 2230–2248. [CrossRef]
39. Han, H.; Liu, H.; Li, J.; Qiao, J. Cooperative fuzzy-neural control for wastewater treatment process. *IEEE Trans. Ind. Inform.* **2020**, *17*, 5971–5981. [CrossRef]
40. Stanescu, R.C.; Leahu, C.I.; Soica, A. Aspects Regarding the Modelling and Optimization of the Transesterification Process through Temperature Control of the Chemical Reactor. *Energies* **2023**, *16*, 2883. [CrossRef]
41. Yu, W.; Pacheco, M. Impact of random weights on nonlinear system identification using convolutional neural networks. *Inf. Sci.* **2019**, *477*, 1–14. [CrossRef]
42. Maya-Rodriguez, M.C.; Carvajal-Mariscal, I.; López-Muñoz, R.; Lopez-Pacheco, M.A.; Tolentino-Eslava, R. Temperature Control of a Chemical Reactor Based on Neuro-Fuzzy Tuned with a Metaheuristic Technique to Improve Biodiesel Production. *Energies* **2023**, *16*, 6187. [CrossRef]
43. Bortolet, P.; Palm, R. Identification, modeling and control by means of Takagi-Sugeno fuzzy systems. In Proceedings of the 6th International Fuzzy Systems Conference, Barcelona, Spain, 1–5 July 1997; Volume 1, pp. 515–520. [CrossRef]
44. Huba, M.; Hypiusová, M.; Ťapák, P.; Vrancic, D. Active Disturbance Rejection Control for DC Motor Laboratory Plant Learning Object. *Information* **2020**, *11*, 151. [CrossRef]
45. López-Muñoz, R.; Molina-Pérez, D.; Vega-Alvarado, E.; Duran-Medina, P.; Maya-Rodriguez, M.C. A Bilevel Optimization Approach for Tuning a Neuro-Fuzzy Controller. *Appl. Sci.* **2024**, *14*, 5078. [CrossRef]
46. Dempe, S.; Kue, F.M. Solving discrete linear bilevel optimization problems using the optimal value reformulation. *J. Glob. Optim.* **2017**, *68*, 255–277. [CrossRef]
47. Liu, J.; Wang, Y.; Huang, P.Q.; Jiang, S. Car: A cutting and repulsion-based evolutionary framework for mixed-integer programming problems. *IEEE Trans. Cybern.* **2021**, *52*, 13129–13141. [CrossRef]
48. Starke, S.; Hendrich, N.; Zhang, J. Memetic Evolution for Generic Full-Body Inverse Kinematics in Robotics and Animation. *IEEE Trans. Evol. Comput.* **2019**, *23*, 406–420. [CrossRef]
49. Deb, K. An efficient constraint handling method for genetic algorithms. *Comput. Methods Appl. Mech. Eng.* **2000**, *186*, 311–338. [CrossRef]
50. Dorf, R.C.; Bishop, R.H. *Modern Control Systems*, 14th ed.; Pearson: Boston, MA, USA, 2017.

processes

Article

Sensor Data Imputation for Industry Reactor Based on Temporal Decomposition

Xiaodong Gao [1,2], Zhongliang Liu [1,2], Lei Xu [1,2], Fei Ma [3,*], Changning Wu [4] and Kexin Zhang [4,5,*]

1 China Nuclear Power Engineering Co., Ltd., Beijing 100840, China
2 Engineering Research Center for Fuel Reprocessing, Beijing 100804, China
3 Hangzhou Boomy Intelligent Technology Co., Ltd., Hangzhou 310053, China
4 Institute of Cyber-Systems and Control, Zhejiang University, Hangzhou 310027, China
5 Huzhou Institute of Zhejiang University, Huzhou 313000, China
* Correspondence: maf@boomy.cn (F.M.); zhangkexin@zju.edu.cn (K.Z.)

Abstract: In the processing of industry front-end waste, the reactor plays a critical role as a key piece of equipment, making its operational status monitoring essential. However, in practical applications, issues such as equipment aging, data transmission failures, and storage faults often lead to data loss, which affects monitoring accuracy. Traditional methods for handling missing data, such as ignoring, deleting, or interpolation, have various shortcomings and struggle to meet the demand for accurate data under complex operating conditions. In recent years, although artificial intelligence-based machine learning techniques have made progress in data imputation, existing methods still face limitations in capturing the coupling relationships between the sequential and channel dimensions of time series data. To address this issue, this paper proposes a time series decoupling-based data imputation model, referred to as the Decomposite-based Transformer Model (DTM). This model utilizes a time series decoupling method to decompose time series data for separate sequential modeling and employs the proposed MixTransformer module to capture channel-wise information and sequence-wise information, enabling deep modeling. To validate the performance of the proposed model, we designed data imputation experiments under two fault scenarios: random data loss and single-channel data loss. Experimental results demonstrate that the DTM model consistently performs well across multiple data imputation tasks, achieving leading performance in several tasks.

Keywords: time series imputation; industry system; condition monitoring

1. Introduction

In the process of industry front-end waste treatment, industrial facilities often employ a series of chemical reactions to fully react with hazardous waste and convert it into harmless and environmentally friendly substances. Among these, the industrial reactor is one of the key pieces of equipment at the forefront of industrial fuel reprocessing plants. Its primary function is to receive short segments of waste, dissolve the wasted core within the cladding, produce qualified feed solutions, and discharge the cladding. The reactor equipment consists of the following components: the loading and unloading system, flat trough system, drive system, support system, position confirmation system, and air-lift and slag-discharge system. To ensure the orderly progress of the treatment process, it is essential to obtain accurate and complete sensor data for real-time monitoring of the operational status of the industrial reactor.

 https://doi.org/10.3390/pr13051526

However, in practical applications, issues such as equipment aging, data transmission errors, and storage faults can lead to data loss and abnormal sensor readings in the collected data under actual working conditions. Addressing these missing data are therefore a critical task in the front-end treatment of industry waste. In traditional approaches to handling missing data, common methods include ignoring, deletion, and interpolation. The ignoring method completely disregards missing values without performing any operations on them and directly uses the data containing missing features. The deletion method involves removing the missing values from the dataset, which can result in the loss of a significant portion of the original data's information. As such, this approach is only suitable when the amount of missing data is minimal. With the continuous advancement of science and technology, the demand for accurate data processing outcomes has grown. Consequently, interpolation methods have garnered increasing attention from researchers. This approach involves analyzing the existing data to fill in the missing values, allowing for the use of complete data in subsequent analyses. This reduces the impact of missing data on research and enables more reliable study results.

In the early stages of research, scholars often adopt traditional imputation methods based on statistical analysis. The most basic approaches include mean imputation and median imputation, while these methods are straightforward and easy to implement, they tend to introduce bias, leading to distortions in the data distribution. Additionally, some researchers employ regression-based imputation methods, such as linear regression and logistic regression. However, these methods are highly sensitive to the quality of the dataset. When the dataset lacks completeness, these models struggle to accurately capture the internal relationships within the data, resulting in poor model performance. In summary, these methods, which are based on linear assumptions, are insufficient to fully adapt to the complexities of real-world scenarios and fail to predict the intrinsic relationships among variables effectively.

With the latest advancements and applications of artificial intelligence technology, many researchers have adopted machine learning techniques to construct deep learning models to solve missing data imputation problems. These models can primarily be divided into the following categories, representing algorithms with different processing focuses. CNN-based network models are based on the CNN architecture and integrate various methods for data modeling. However, such methods are limited by the convolutional network itself, which has a restricted receptive field and poor long-sequence perception capabilities. Linear layer-based network models, the most notable of which are Transformer series models, improve on the shortcomings of CNN architecture, such as limited receptive fields and poor sequential modeling capabilities, but they are less effective than CNN-based models in capturing tight coupling relationships between channels. The state-of-the-art machine learning algorithm, DLinear, uses a time series modal decomposition approach, dividing time series data into trend and local information and then conducting deep learning modeling separately. However, this method does not consider the correlations between channels and only employs a channel-independent approach for decoupling computations, ignoring the interference of inter-channel information on local variations. Transformer models, by contrast, focus more on sequence modeling both within and across channels in time series data but lack a long-term view of sequence changes. Although the above methods demonstrate strong performance and potential in sequence modeling and data imputation, they lack a comprehensive and effective approach for capturing and analyzing the coupling relationships between sequences and channels.

To address the challenge of capturing coupling relationships between the sequential and channel dimensions in time series data, this paper proposes a time series decoupling-based data imputation model, referred to as the Decomposite-based Transformer Model

(DTM). By employing a time series decomposite approach, DTM decomposes time series data for separate sequence modeling. Simultaneously, the model utilizes our proposed MixTransformer module to capture inter-channel information and long-term sequence dependencies, enabling deep modeling. To validate the model, this paper designs data imputation experiments under two fault scenarios: random data loss and single-channel data loss. The experimental results demonstrate that the proposed model consistently performs well across multiple data imputation tasks. The contributions of this paper are as follows:

- A channel-level data imputation task is proposed. This task leverages the coupling relationships between channels and the available data to impute missing channel data, thereby enhancing the operational stability of sensor detection and condition monitoring in the industry processing.
- For the proposed imputation task, the DTM model is developed. By integrating time series decomposition with the proposed MixTransformer architecture, the model performs inter-channel and sequence-level modeling of time series data. Experimental results indicate that the proposed model achieves leading performance across multiple imputation tasks.

2. Related Works

2.1. Data-Driven Fault Diagnosis in Industrial Equipment

Condition monitoring and fault diagnosis of industrial equipment are essential for ensuring safe operation and enhancing reliability. This process includes fault detection, identification, localization, and recovery. Currently, fault diagnosis relies on prior analysis of fault modes, allowing for diagnosis based on these results when a fault occurs. However, the diversity of equipment types, complex operating environments (e.g., high temperature, pressure, radiation), and inaccessibility of some equipment present challenges for traditional methods [1,2]. With the rapid development of data-driven technologies, online monitoring, and intelligent inspection systems now gather large-scale operational data via sensors, providing strong support for condition assessment and fault diagnosis. In recent years, data-driven approaches have advanced in reliability analysis, anomaly detection, and intelligent diagnostics, offering new solutions for improving safety and optimizing maintenance in industry processing.

Traditional machine learning methods rely on manually extracted data features. When faced with increasingly complex nonlinear dynamic systems, vast state parameters, and fault information of industrial equipment, they often encounter performance bottlenecks. Therefore, current data-driven monitoring and diagnostic technologies for industrial equipment typically employ neural networks as the mainstream technique. Specifically, Kozma et al. constructed a relatively simple three-layer feedforward neural network to address issues such as coolant boiling monitoring [3], anomaly detection during startup, shutdown, and steady-state operations in power plants [4], as well as anomaly cause localization [5]. These studies indicated that artificial neural networks are faster and more reliable than variance-based statistical methods for anomaly detection. Lee et al. [6] also proposed a fault diagnosis method for Control and Instrumentation (C&I) cable systems based on a simple multilayer perceptron (MLP) and time–frequency domain reflection techniques. This method can detect the presence and location of faults and further distinguish faulty lines in multi-core C&I cables. Mandal et al. [7] addressed online fault detection of thermocouples by proposing a classification method based on Deep Belief Networks (DBNs), using the generalized likelihood ratio test to calculate fault patterns in sensor signals based on fault amplitude. Similarly, Peng et al. [8] developed a fault diagnosis model based on DBN and correlation analysis, which first reduces the dimensionality of features using correlation analysis for feature selection, and then applies DBN for fault recognition. This method demonstrates significant advantages over fault diagnosis models based on backpropagation neural networks and support vector machines.

With the widespread application of Convolutional Neural Networks (CNNs) in image processing, researchers have started introducing them into the fault detection field for industrial equipment to develop higher-performance diagnostic models. Bang et al. [9] proposed a multi-core cable diagnostic method based on reflection measurements. This method converts reflection signals obtained from measurements into images using image processing algorithms and classifies the images using CNNs, thereby enhancing the stability and reliability of multi-core cable system fault detection. Saeed et al. [10] developed an online fault monitoring system that uses CNNs combined with sliding window techniques to identify and evaluate faults such as main feedwater pipe rupture, main pump failure, and pressurizer safety valve failure under different industrial plant conditions. Abdelghafar et al. [11] developed industrial reactors fault detection system based on CNNs, using real-time sensor data to analyze anomalies or faults in reactor operations. The system can prevent catastrophic accidents by detecting faults early, significantly enhancing the safety and reliability of industrial reactors.

Furthermore, since fault diagnosis in industrial equipment often involves large volumes of time series data, many studies have proposed fault detection methods based on Recurrent Neural Networks (RNNs), which have unique advantages in handling such data. However, traditional RNNs have limitations when modeling long-term dependencies. Long Short-Term Memory networks (LSTMs), an improved version of RNNs, overcome this issue, especially in capturing long-term dependencies during the backpropagation process of time series data. Yang et al. [12] proposed an LSTM-based fault diagnosis method, generating fault rankings with probabilities through preprocessing, LSTM networks, and post-processing. Choi et al. [13] proposed a sensor fault detection system framework, using LSTM networks to generate consistency indices to assess sensor reliability and quantify their performance during emergency sequences. This study demonstrated the potential application of this system in handling industrial plant emergencies. To handle untrained faults in industrial plants, Yang et al. [14] first classify major changes that might affect plant status and apply LSTM's autoencoder algorithm for fault diagnosis of typical accidents. Overall, these LSTM and autoencoder-based studies provide new approaches for industrial power plant equipment fault diagnosis, showing significant advantages and potential for dealing with complex time series data and emergencies.

In recent years, Transformer models, which have excelled in natural language processing, have begun to attract attention in the industrial equipment fault diagnosis field. Through the self-attention mechanism, Transformers can effectively extract important features from time series data and apply them to complex fault detection tasks. Zhou et al. [15] proposed a Transformer-based abnormality detection model for reactor cooling pump status monitoring. The model retains the ability of the original Transformer network to capture time dependencies in time series data and enhances the learning of spatial correlations between variables through the attention mechanism. To detect anomalies in industrial data, Trans-MCC [16] employed an unsupervised Transformer framework and modified the loss function using the Maximum Correlation Entropy Criterion (MCC) to enhance robustness. Compared to methods based on CNNs and RNNs, Transformer demonstrates superior modeling capabilities when handling high-dimensional time series data, providing a more efficient solution for fault diagnosis in industrial equipment. These studies highlight the significant advantages of Transformer-based models in addressing complex time series data and diverse fault patterns, indicating their promising potential for widespread application in industrial power plant fault diagnosis.

2.2. Deep Learning-Based Industrial Data Soft Sensing Technology

Industrial data soft sensing technology is a technique that analyzes, models, and predicts large volumes of data from industrial processes to indirectly estimate and monitor physical quantities that are difficult to measure directly. Soft sensors are the core component of soft sensing technology; they use mathematical modeling to infer variables that cannot be directly obtained by leveraging existing measurable data. Traditional soft sensing methods generally rely on classic statistical and machine learning techniques such as regression analysis, Principal Component Analysis (PCA), and Support Vector Machines (SVM). However, these methods often struggle with complex, high-dimensional data, especially when dealing with time series data and nonlinear problems. The introduction of deep learning has brought new breakthroughs to soft sensing technology, enabling it to effectively handle these complex data and provide more accurate predictions.

Autoencoders (AEs) and their variants are widely used in building soft sensors, particularly in semi-supervised learning and handling missing data in industrial processes. For example, NPLVR [17] is a nonlinear probabilistic latent variable regression model that leverages features extracted by a variational Auto-Encoder (VAE). By incorporating supervisory information from label variables into both the encoding and decoding processes, the model effectively extracts nonlinear features for latent variable regression. VW-SAE [18] is a variable-wise weighted stacked autoencoder that uses the linear Pearson correlation coefficient between hidden layer inputs and output labels during pre-training, enabling semi-supervised feature extraction. By assigning weights to variables based on their correlation with the output, VW-SAE emphasizes important features and stacks weighted autoencoders to form a deep network. Furthermore, Wang et al. [19] proposed a generative model, VA-WGAN, based on VAE and Wasserstein Generative Adversarial Networks (WGAN), which can generate distributions from industrial processes that match real data. Additionally, some studies combine autoencoders with other methods to achieve better results. For instance, Yao et al. [20] first use autoencoders for unsupervised feature extraction and then apply Extreme Learning Machines (ELMs) for regression tasks. The experimental results showed that this hybrid approach outperforms using autoencoders alone.

CNNs can capture local dynamic features of process signals in industrial process data or the frequency domain, making them suitable for building soft sensors. Horn et al. [21] used CNNs to extract features from foam flotation sensors, demonstrating good feature extraction speed and predictive performance. For dynamic problems, Yuan et al. [22] proposed a multi-channel CNN for soft sensing applications in industrial dehydrogenation towers and hydrocracking processes. This model learns dynamic features and local correlations of different variable combinations. In the frequency domain, CNNs can exhibit high invariance to signal translation, scaling, and distortion. Based on this, CNN-ELM [23] incorporates convolutional and max-pooling layers to extract high-level features from the vibration spectra of milling machine bearings. These features are then mapped to material levels using an Extreme Learning Machine (ELM), achieving accurate and efficient measurements.

RNNs and their variants, such as LSTM networks, have also been applied in practical cases. For example, Ke et al. [24] built an LSTM-based soft sensor model that can handle the strong nonlinearity and dynamic characteristics of industrial processes. Similarly, SLSTM [25], based on LSTM, is a supervised network that learns dynamic hidden states using both input and quality variables. This approach has proven effective in the penicillin fermentation process and industrial dehydrogenation towers. Raghavan et al. [26] introduced a variant of RNN, the Time-Delayed Neural Network (TDNN), which outperformed traditional Extended Kalman Filters and feedforward neural networks in state estimation of an ideal reactive distillation column. Moreover, Yin et al. [27] proposed an integrated semi-supervised model combining self-supervised autoencoders (SAEs) with bidirectional LSTM. This method not only extracts and utilizes time behaviors from both labeled and unlabeled data but also considers the time dependencies of the quality indicators themselves.

2.3. Time Series Data Imputation and Prediction Techniques

Missing data poses significant challenges to statistical analysis and machine learning, often leading to biased outcomes and inaccurate results. Traditional imputation methods, such as mean imputation, hot-deck imputation [28], and multiple imputation by chained equations [29], are simple and easy to implement but show limitations when dealing with high-dimensional, complex, or nonlinear data. Advanced imputation methods, including KNN imputation [30,31], decision trees [32], random forests [33], and SVM [34,35], can capture complex relationships between variables but often suffer from high computational costs or sensitivity to parameter tuning. In recent years, deep learning-based imputation methods have demonstrated significant advantages in missing data processing due to their strong feature modeling capabilities and adaptability to complex data.

AEs, a class of neural networks capable of learning compressed representations of data, have been widely applied to missing data imputation. For example, Vincent et al. [36] proposed a denoising autoencoder that reconstructs partially corrupted input data, enabling the model to learn robust representations for missing values. This method effectively captures nonlinear patterns and complex structures in time series data, showing high accuracy and robustness in imputation tasks. Similarly, Li et al. [37] introduced a method combining VAE with shift correction to address specific missing values in multivariate time series. By correcting the probability distribution deviations caused by concentrated missingness, this approach significantly improves the accuracy and robustness of imputation.

Generative Adversarial Networks (GANs) have also gained considerable attention in the field of missing data imputation. In this regard, GAIN [38] leverages the generator to impute missing values based on observed data, producing a complete vector, while the discriminator identifies which components are observed and which are imputed. The model also incorporates a hint mechanism to further enhance imputation accuracy. Similarly, imputeGAN [39] utilizes an iterative optimization strategy to handle long sequences of continuous missing values in multivariate time series. This model ensures both the generalizability of the approach and the reasonableness of the imputation results. Additionally, Khan et al. [40] proposed a method using GANs to generate synthetic samples, which improved imputation performance for mixed datasets. By employing Tabular GAN and Conditional Tabular GAN to generate synthetic data, their experiments demonstrated that incorporating synthetic samples can significantly enhance imputation accuracy in scenarios with high missing rates.

Transformer models have shown great potential in missing data imputation, particularly in handling complex patterns and long-term dependencies in time series data. MTSIT [41] leverages the Transformer architecture to perform unsupervised imputation by jointly reconstructing and imputing stochastically masked inputs. Unlike traditional Transformer models, MTSIT uses only the encoder part to reduce computational costs and is specifically designed for multivariate time series data. Building on this, ImputeFormer [42] introduces a low-rankness-induced Transformer model that combines the advantages of low-rank models with deep learning. By capturing spatiotemporal structures, ImputeFormer strikes a balance between strong inductive bias and model expressivity, demonstrating superior imputation accuracy, efficiency, and versatility across diverse datasets such as traffic flow, solar energy, smart meters, and air quality. These studies highlight how Transformer models offer innovative and effective solutions for time series data imputation, excelling in both accuracy and efficiency.

Multivariate Coupled Time Series Representation and Prediction Techniques

Traditional univariate time series forecasting methods typically process multivariate time series (MTS) independently and learn temporal dependencies for each TS separately using classical approaches such as Autoregressive Integrated Moving Average (ARIMA) [43], as well as deep learning models like Recurrent Neural Networks (RNN [44], TCN, Transformers, and others. However, these are unsuitable for complex industrial scenarios where multifaceted temporal couplings exist, as the independence assumption may result in critical information loss. To address this, recent studies have increasingly focused on multivariate time series forecasting to capture interdependencies among variables. For instance, Qin et al. [45] proposed a dual-stage attention-based RNN to automatically learn nonlinear relationships in multivariate TS. Bai et al. [46] employed a gated spatio-temporal graph convolutional network to capture spatial and temporal correlations in passenger demand sequences for multi-step forecasting. Wu et al. [47] introduced a graph learning method that automatically extracts unidirectional relationships between variables and combines temporal convolution with graph convolutional networks for multi-series forecasting. Zhang et al. [48] proposed integrating a graph structure with the Transformer model to effectively identify and model complex relationships among sequences. He et al. [49] introduced adversarial learning to enable fairness modeling for MTS prediction, achieving intrinsic feature extraction of MTS through a recurrent graph convolutional network. Wang et al. [50] incorporated multi-channel distribution information into feature vectors to achieve time series forecasting in an industrial context. These studies explore methods to incorporate the unique characteristics of multivariate time series into models, offering innovative and effective solutions for time series data representation and forecasting, while demonstrating outstanding performance in terms of accuracy and efficiency.

3. Descriptions and Analysis of the Continuous Reactor

In large-scale waste reprocessing plants, the industrial reactor is one of the critical process equipment. Its primary function is to process fuel from waste assemblies, enabling the recovery and reuse of materials. This process is of significant importance for improving energy utilization efficiency, reducing waste generation, and ensuring the sustainable development of energy.

The operation of the industrial reactor requires precise control of various process parameters, such as dissolution temperature, acidity, and stirring speed, to ensure efficiency and safety during the dissolution process. Additionally, the gases and liquid products generated during dissolution need to be effectively separated and treated to meet the requirements of subsequent process stages.

Moreover, the design and operation of the industrial reactor must take into account safety and protection requirements to ensure the safety of operators and the environment. In the context of the back-end waste processing scenarios studied in this research, the monitoring and control of the industrial reactor are essential components for realizing the intelligent operation of the entire reprocessing plant. By applying digital technologies, real-time monitoring, fault diagnosis, and optimized control of the dissolution process can be achieved, thereby enhancing production efficiency and product quality while reducing operational risks.

Since this study focuses on addressing data missing issues in the monitoring of industrial reactor operating conditions, we have limited the scope of our research to the transmission components of the industrial reactor. This approach ensures the generalizability of the findings while maintaining safety and privacy. The stable and uniform transport of waste to subsequent processing stages is also a critical aspect of operating condition monitoring. Therefore, we selected the data monitoring of waste transport components as the research focus of this study.

A schematic diagram of the relevant components of the industrial reactor targeted in this study is shown in Figure 1. The primary devices include a motor for driving, a shaft for connection and transmission, a coupling, and a worm reduction box. The motor drives a large wheel through these components to facilitate the transportation of industrial waste materials.

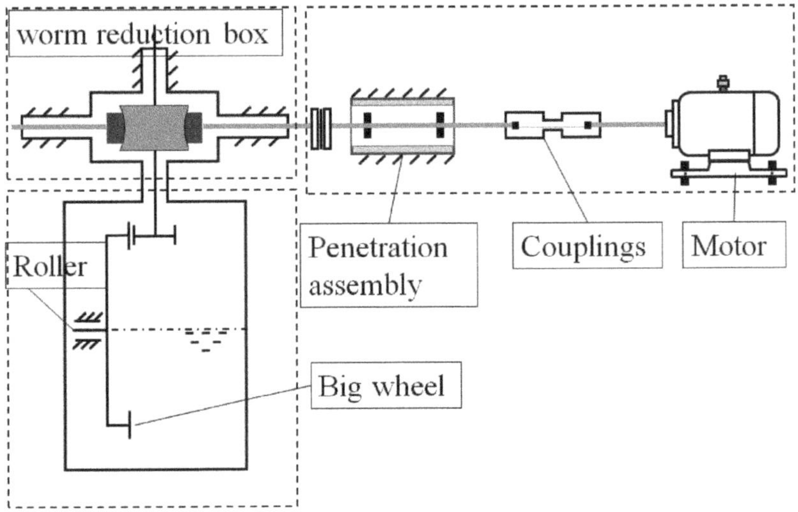

Figure 1. The relevant components of the industrial reactor in this paper.

4. Materials and Methods

4.1. Problem Definition

Assume that there are C channels of sensor data, represented as follows:

$$X = (X^1, X^2, ..., X^C)$$

where the data length of each channel is T, simulating a Ts-second data monitoring process. Thus, the i-th channel of X is expressed as $X^i = [X^i_1, X^i_2, ...X^i_T]^T$. To simulate scenarios of data missing and channel missing, we designed a masking matrix $M = (M^1, M^2, ..., M^C)$, where M has the same shape as X. This study uses M to index whether the corresponding elements in X have missing data. The elements of M are defined as follows:

$$M^c_t = \begin{cases} 0 & if\ X^c_t\ is\ missing. \\ 1 & if\ X^c_t\ is\ not\ missing. \end{cases} \quad (1)$$

Here, based on Equation (1), we use M to compute the data in cases where data missing occurs.

$$\tilde{X}^c_t = (M \times X)^c_t = \begin{cases} X^c_t & if\ M^c_t\ = 1. \\ 0 & if\ M^c_t\ = 0. \end{cases} \quad (2)$$

Figure 2 illustrates the process of simulating data missing in the data preprocessing stage of the imputation problem in this paper, where the original data are processed based on the masking matrix.

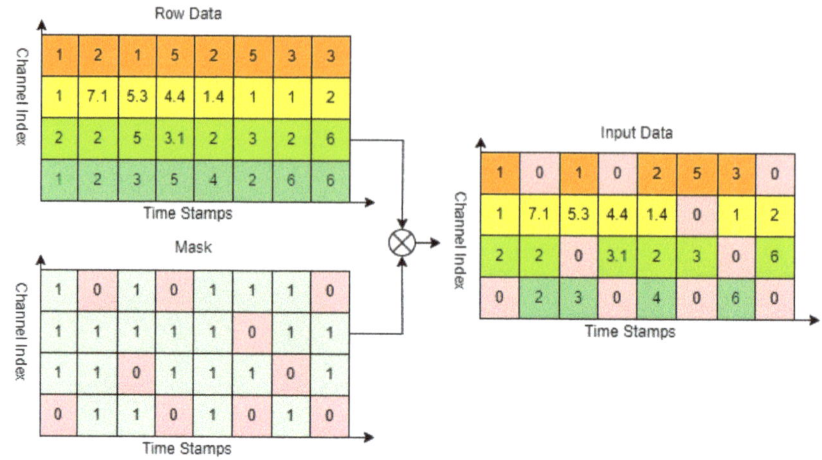

Figure 2. The schematic diagram for data simulation in the imputation problem.

In the subsequent experimental setup, we processed the data based on random data point missing and random channel missing scenarios. Specifically, in the random data point missing imputation task, the elements of M were randomly set to 0 with a certain proportion. In the random channel missing imputation task, one of the all channels in M was randomly set to 0. The objective of the experiment is to train and obtain an optimal mapping $f : \tilde{X} \to X$, such that the imputation error is minimized. The optimization objective is as follows:

$$\min_{f}[M \times (||f(\tilde{x}) - x||_2^2)] \tag{3}$$

4.2. Proposed Method

The overall architecture of the DTM model proposed in this paper is shown in Figure 3. In current deep learning algorithms, Batch Normalization (BN) has been proven to be an effective preprocessing method that helps the model understand and extract features from time series data. In the data point imputation task, the masked data are also normalized using BN, and de-normalization is applied at the output layer to reconstruct the statistical features of the data. However, in the channel imputation task, data that are completely zero can introduce incorrect statistical information, so BN is not applied in this case.

Additionally, this paper follows the time series decomposition approach, splitting the data into trend and seasonal components. On the one hand, the trend component, which occupies the dominant part of the time series, contains a significant amount of low-frequency data and is highly influenced by the coupling relationships between channels. On the other hand, the seasonal component contains less information and has a higher noise content, making it sensitive to channel variations and difficult to model effectively. Therefore, in capturing the coupling relationships between channels, this paper mainly focuses on the coupling information within the trend component.

For time series reconstruction, Zeng et al. (2023) [51] indicated that good data prediction performance can be achieved using only linear layers. Based on this, we input the features of the two components into independent linear layers for reconstruction, aiming to achieve high data modeling accuracy with relatively few operations. As a result, the reconstructed sequences contain the sequence information of their respective dimensions and, through a shared encoder, also capture deep information from the other dimension. In modeling the trend component, our proposed MixTransformer module not only performs data modeling at both the sequence and channel levels for the trend component but also captures and utilizes the coupling relationships between the two dimensions for data modeling.

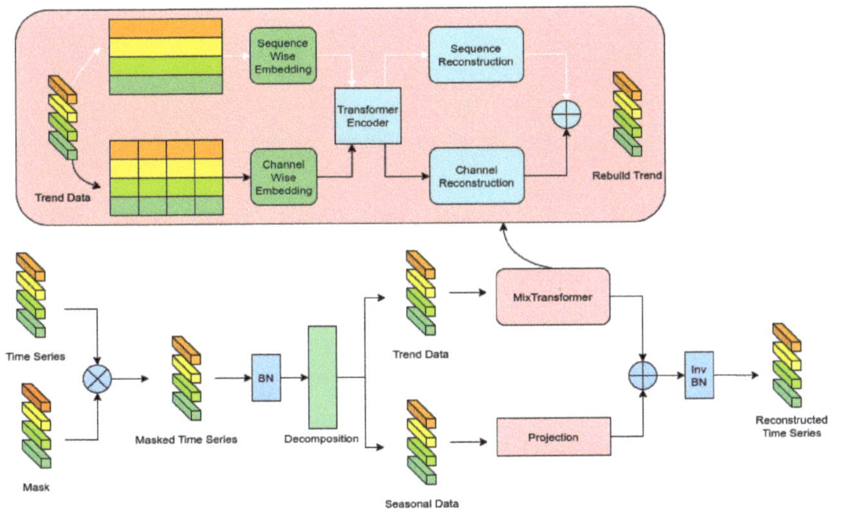

Figure 3. The architecture of the proposed method.

4.2.1. Time Series Decomposite

For data preprocessing, Wu et al., 2021 [52] were the first to propose the use of time series decomposition in time series forecasting, and this has now become a common method in time series analysis. This approach enhances the predictability of the original data. This method applies a sliding average kernel to the input sequence to extract the trend component of the time series. The difference between the original sequence and the trend component is regarded as the seasonal component.

The formula for time series decomposition is as follows:

$$X^c_{trend,t} = \frac{1}{w}\sum_{j=0}^{w}\tilde{X}^c_{t+j} \quad (4)$$

$$X^c_{seasonal,t} = \tilde{X}^c_t - X^c_{trend,t} \quad (5)$$

where t represents the timestamp of X, and c represents the channel index of the sequence.

Building on the decomposition scheme, Zhou et al., 2022 [53] further proposed using a mixture of experts strategy. Its core idea is to combine the trend components extracted by moving average kernels with varying kernel sizes. The method adopted in this paper is based on the decomposition approach used in Autoformer to reduce computational overhead. Figure 4 illustrates the time series decomposition process, while Figure 5 shows the corresponding spectrum. The original data represent a sine function with added Gaussian white noise following $\mathcal{N}(0, 0.2)$. The "Trend data" denotes the trend sequence obtained after decomposition, and the "Seasonal data" represents the local sequence.

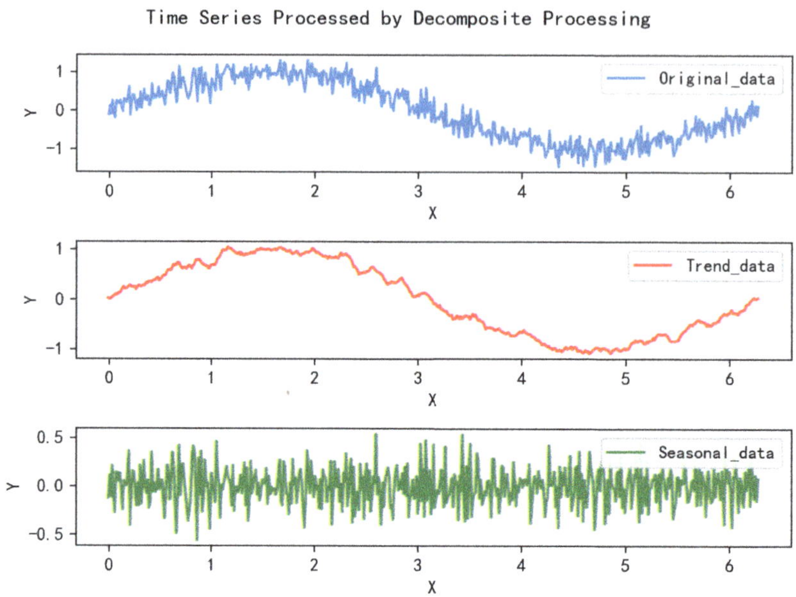

Figure 4. The diagram of the time series decomposition process.

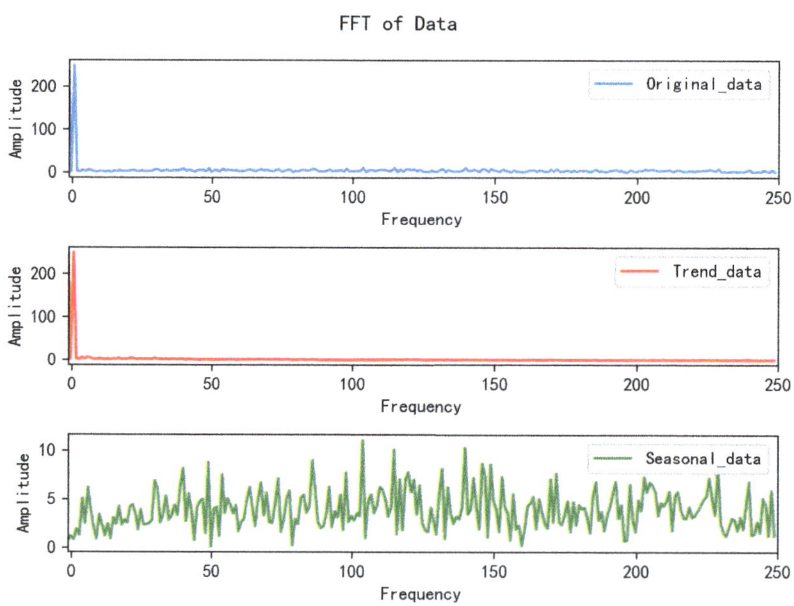

Figure 5. The spectrum diagram of the decomposition process.

Figure 6 illustrates the denoising results of the temporal decomposition module on a sinusoidal function under varying noise levels. To visually demonstrate the denoising performance, we assume all noise to be additive white Gaussian noise with zero mean, where different variances correspond to distinct noise intensities. By comparing the mean square error (MSE) between the trend component output by the temporal decomposition module and the original noise-free sequence, it is evident that the time series decomposition method effectively suppresses additive white Gaussian noise. Specifically, the MSE values remain significantly lower across all tested noise levels, demonstrating the method's robustness in preserving the intrinsic signal structure while isolating stochastic noise components.

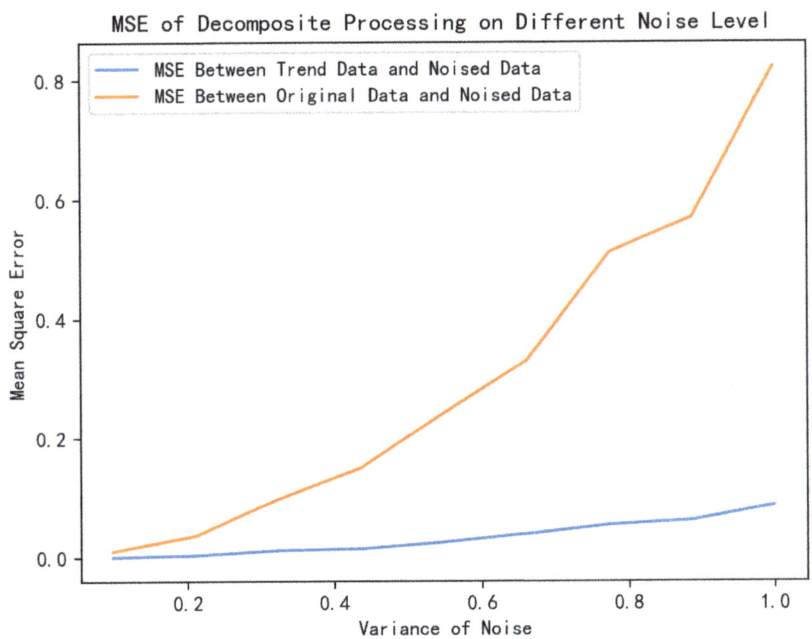

Figure 6. The influence of AWGN on time series decomposition.

From Figures 4 and 5, it can be observed that the "Trend data" preserves the overall trend of the original sequence, which includes the majority of the low-frequency components. On the other hand, the "Seasonal data" reflects the short-term variations of the sequence, primarily capturing the high-frequency components. This part of the data has a relatively low amplitude and contains less critical information from the original sequence. Therefore, it indicates that the focus of our time series imputation task should be on the "Trend data".

Figure 7 depicts the Pearson correlation coefficients between Seasonal Data, Trend Data, and Original Data under the same conditions as in Figure 6. As the noise variance increases, the correlation between the Trend data and the Noised data gradually decreases, indicating that the proportion of the data information captured by the Trend component decreases. Conversely, the correlation coefficient for Seasonal Data increases, suggesting that the Seasonal component retains a growing share of the information. Furthermore, the correlation coefficients between Seasonal and Trend remain consistently low, demonstrating that the two components are approximately orthogonal. These observations illustrate that the time series decomposition method effectively separates the data into two uncorrelated components while minimizing information loss.

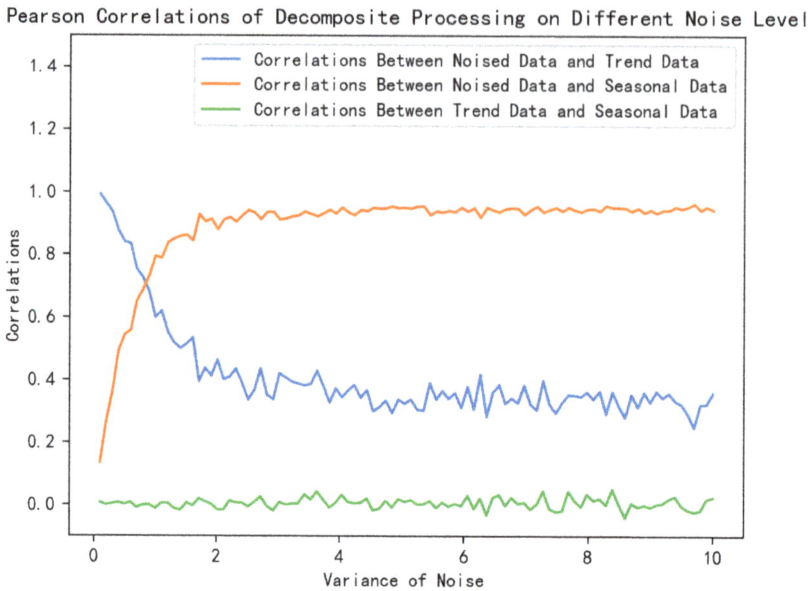

Figure 7. The influence of AWGN on the correlation between different portions.

4.2.2. MixTransformer

To enable the model to fully capture the inter-channel correlations and sequential variations of the trend components in time series data, we propose the MixTransformer module, whose main architecture is shown in Figure 3. Trend data are embedded into the feature space separately along the channel dimension and the sequence dimension through word embedding. These features are further extracted using a shared encoder. Additionally, the shared encoder leverages internal vectors to achieve indirect coupling and interaction between inter-channel and sequential information, thereby capturing the complex features of time series data. The extracted features are subsequently processed by a projection layer for sequence reconstruction, completing the reconstruction of the original trend sequence information.

For positional encoding along the sequence dimension, we adopt the encoding method used in Transformer models. For the input sequence $X \in R^{T\times C}$, after undergoing time series decomposition processing, we obtain $X_{trend} \in R^{T\times C}$ and $X_{seasonal} \in R^{T\times C}$ with unchanged shapes. Here, X_{trend} is calculated through a 1D convolutional layer and Formula (7), respectively, to embed the channel dimension, resulting in $Emb_{trend} \in R^{T\times D_{model}}$ as Formula (6).

$$Emb_{trend} = Conv_{1D}(X_{trend}) + Pos(X_{trend}) \tag{6}$$

Here, $Conv_{1D}()$ represents a 1D convolution applied to the last dimension of X_{trend}, and $Pos()$ denotes Position Embedding, whose encoding value for time step t is defined by Equation (7).

$$P_t^i = f(t)^i := \begin{cases} sin(w_k \cdot t) & if\ i = 2k \\ cos(w_k \cdot t) & if\ i = 2k+1 \end{cases} \tag{7}$$

where

$$w_k = \frac{1}{10,000^{2k/d}}$$

We transposed the X_{trend} and performed the same operations to obtain positional encoding along the channel dimension. The encoded sequence, after convolutional processing to extract the corresponding dimensional features, incorporates sequence information through positional encoding. Here, we employed two different 1D convolutional layers to encode both the temporal and channel dimensions to transform the time series into tokens for input to the Transformer, while ensuring that the data can be encoded independently across channels in the temporal dimension and independently across time in the channel dimension. Therefore, we obtain the data embedding $Emb'_{trend} \in R^{C \times D_{model}}$ corresponding to the transpose of X_{trend}.

In traditional Transformer-based models, single-dimensional sequences are directly used for downstream tasks or reconstructed in an encoder–decoder architecture after encoding. These methods lack the computation of coupled information across multiple dimensions, leading to suboptimal performance.

To address this issue, we utilized a shared encoder for data encoding. According to the attention computation Formula (8), during the calculation of self-attention or multi-head attention, W_Q , W_K , and W_V are trained to learn feature representations corresponding to the input vectors. Therefore, for sequence-wise embedding vectors and channel-wise embedding vectors, the shared encoder weights W_Q , W_K , and W_V enable indirect interactions between the two dimensions. This ensures that deep-level coupling information can be extracted without losing the original embedding information, thereby enhancing the sequence construction process.

$$Attention(Q, K, V) = softmax(\frac{QK^T}{\sqrt{d_k}})V \tag{8}$$

where $Q = W_Q X$, $V = W_V X$, $K = W_K X$. W_Q, W_K, W_V are learnable parameters. We employed the Transformer module as the shared encoder. Based on the aforementioned analysis, the embedding data from the temporal dimension and the channel dimension can achieve indirect interaction during the training process. Through the shared encoder, we obtain the vectors $Enc_{Channel} \in R^{T \times D_{model}}$ and $Enc_{Temporal} \in R^{C \times D_{model}}$, which are mapped into the encoding space for both the channel and temporal dimensions.

Finally, the vectors encoded by the shared encoder are processed through separate linear layers to reconstruct the original trend sequence along the temporal and channel dimensions. The methods for sequence reconstruction and channel reconstruction are described in Formulas (9) and (10), respectively. These reconstructed components are then summed to obtain the final reconstructed trend sequence. Thus, the reconstructed trend data are calculated by Formula (11). Following the principles outlined in the DLinear paper, linear layers are sufficient for sequence construction tasks while significantly reducing computational overhead. Algorithm 1 demonstrates the detailed training process of the proposed DTM model.

$$X_{seq} = (Enc_{Temporal} W_{seq} + b_{seq})' \tag{9}$$

$$X_c = Enc_{Channel} W_c + b_c \tag{10}$$

$$Rec\_X_{trend} = X_{seq} + X_c \tag{11}$$

where $X_{seq} \in R^{T \times C}$, $X_c \in R^{T \times C}$, $W_{seq} \in R^{D_{model} \times T}$, $W_c \in R^{D_{model} \times C}$, $b_{seq} \in R^{C \times T}$, $b_c \in R^{T \times C}$, symbol $'$ means Transpose process.

Algorithm 1 Training Process of DTM

Require: Time-series data X, mask matrix M, batch size B, learning rate η, number of epochs E
Ensure: Trained DTM model parameters θ
Initialize the model parameters θ of DTM
for epoch $e = 1, 2, \ldots, E$ **do**
  Shuffle the training dataset (X, M)
  for each batch (X_b, M_b) in (X, M) **do**
    Apply the mask M_b to the data X_b to simulate missing data, producing $\tilde{X}_b$
    Perform data preprocessing:
      Normalize $\tilde{X}_b$ if required
      Decompose $\tilde{X}_b$ into trend and seasonal components using temporal decomposition
    Reconstruct the decomposed trend components using the **MixTransformer module**:
      Perform embedding in both sequence and channel dimensions
      Pass embeddings through the shared encoder to extract features
      Decode the features using linear layers to reconstruct the sequence and channel dimensions
    Reconstruct the seasonal components using the projection layer
    Combine reconstructed components to obtain the interpolated output $\hat{Y}_b$
    Compute the loss $\mathcal{L}$ using the MSE loss function.
    Backpropagate the loss $\mathcal{L}$ to update model parameters θ using gradient descent with learning rate η
  end for
end for
return Trained DTM model parameters θ

4.3. Comparative Discussion with Existing Models

Compared to models such as Transformer and Autoformer, DTM addresses the limitation of handling only channel-independent time series by performing separate feature extraction along the temporal and channel dimensions, while Transformer-based architectures primarily rely on self-attention mechanisms for global temporal dependencies, they inherently treat multi-channel data as isolated sequences, neglecting critical inter-channel correlations. In contrast, DTM explicitly decouples temporal dynamics and channel-wise interactions through dual-path encoding, enabling effective modeling of strongly coupled sensor data prevalent in industrial scenarios. When compared to CNN-based models like MICN and TCN, DTM overcomes the suboptimal performance caused by limited receptive fields through decomposition modules and Transformer modules.

To the best of our knowledge, the model most similar related to DTM is Crossformer. Both models utilize data from both temporal and channel dimensions to capture latent information in MTS. However, Crossformer employs a patch-based DSW embedding for encoding, whereas DTM adopts a CNN-based embedding approach. This distinction arises because DTM targets interpolation tasks, requiring point-to-point or point-to-segment feature extraction, while Crossformer focuses on prediction tasks, emphasizing segment-to-segment data interaction.

Additionally, Crossformer uses Two-Stage Attention (TSA) layer to process temporal and channel dimensions sequentially in a serial manner. In contrast, DTM's MixTransformer computes temporal and channel dimensions in parallel. Consequently, the method proposed in this work exhibits significant distinctions from existing approaches in both architecture design and task-specific optimization.

4.4. Time Complexity Discussion

The analysis of our approach on time complexity is described below. We assumed that the input sequence length is L, the embedding dimension is d_{model}, and the number of encoder layers is n_l. Then: For the time series decomposition component, the time complexity is $O(L)$. For the temporal embedding component, which employs a Convolutional Neural Network (CNN)-based approach, the time complexity across both the sequence and channel dimensions is $O(L \cdot k \cdot C \cdot d_{model})$. For the shared encoder, following Transformer's method, the time complexity is $O(L \cdot L \cdot n_l)$. For the projection layer, the time complexity is $O(L \cdot d_{model})$. Thus, the overall time complexity of DTM is: $O(L \cdot (n_l \cdot L + 1 + (k \cdot C + 1) \cdot d_{model}))$.

5. Results

5.1. Overview of Data

The data used in the experiment were all collected from the constructed engineering prototype of the industrial reactor. These datasets include operational signals such as eddy current displacement, vibration, motor torque, and motor position obtained during 480 h of simulated operation of the prototype, enabling monitoring of the operating status of key components of the device. For the purposes of this study, it is assumed that only the eddy current displacement data, wheel motor torque, and wheel torque meter torque among the monitored data may encounter issues of data loss or channel loss. All the data used in the experiment were resampled to a frequency of 20 Hz.

Table 1 lists the sensor variables used, where the subscript numbers in the variables indicate sensors of the same type installed at different locations. Reference [54] indicates that heatmaps can visualize the correlations among multiple datasets; therefore, this study also employs heatmaps to represent the interrelationships among multi-channel data. Figure 8 is presented as a heat map to visualize the Pearson correlation coefficients between channels, demonstrating their coupling characteristics. The colors in the figure represent Pearson's correlations between different variables within the range of −1 to +1: lighter shades near zero indicate no significant relationship between variables, while darker shades approaching 1 signify strong correlations. Darker red hues indicate stronger positive correlations between variables, whereas darker blue hues denote stronger negative correlations. The symbols on the X and Y axes correspond to different channel names. This heat map visually demonstrates the interaction relationships among the channel data in the dataset used.

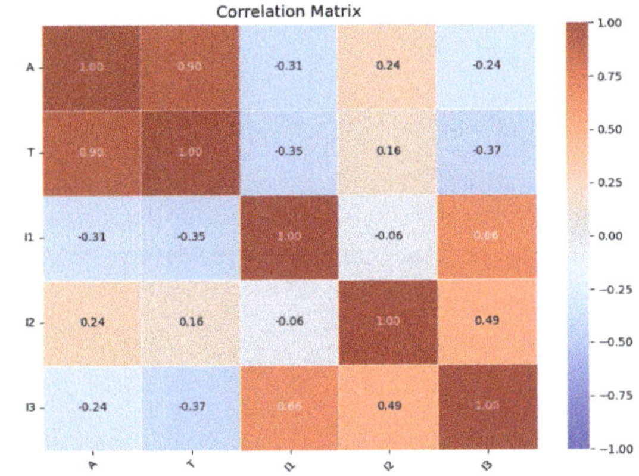

Figure 8. The correlation coefficient between channels.

Table 1. The sensor variable used in this article.

Physical Meanings of Variables	Variables
Torque of Rotary motor	A
Torque of wheel torque meter	T
Eddy current of large wheel surface gear	I_1, I_2, I_3

5.2. Experimental Setup

To simulate the data missing scenarios encountered in real-world situations, we applied random masking to data points and channels from the experimental prototype data to represent two types of anomalies: missing data records and sensor failures. Specifically, data point masking was conducted with proportions of 10% and 25%, while sensor failures were represented by random single-channel masking. Each batch of input data had a batch size of 128, a sequence length of 100, and 5 data channels. The proposed model used the Mean Square Error (MSE) as the loss function, and the final evaluation metrics included both MSE, R^2 and Mean Absolute Error (MAE). The experiments were conducted on an RTX 4060 Ti GPU. To fully reflect the performance of time series imputation, the evaluation metrics selected for this study were MSE, MAE, and R^2. Lower MSE and MAE values indicate better model performance, while an R^2 value closer to 1 reflects a stronger model fit. The formulas for those evaluation metrics are shown below:

$$MAE = \mathbb{E}(\sum_{c=1}^{C}\sum_{t=1}^{T} M \times |X_t^c - f(\tilde{x})_t^c|) \tag{12}$$

$$MSE = \mathbb{E}(\sum_{c=1}^{C}\sum_{t=1}^{T} M \times ||X_t^c - f(\tilde{x})_t^c||_2^2) \tag{13}$$

$$R^2 = 1 - \frac{\sum_{c=1}^{C}\sum_{t=1}^{T} M \times ||X_t^c - f(\tilde{x})_t^c||_2^2}{\sum_{c=1}^{C}\sum_{t=1}^{T} M \times ||X_t^c - \bar{X}_t^c||_2^2} \tag{14}$$

Since we aim to demonstrate that our proposed model exhibits sufficient stability and superiority compared to various types of models in the proposed numerical imputation tasks, we conducted extensive comparisons with a wide range of advanced models including CNN-based Model: MICN (2023) [55], TCN (2018) [56]; MLP-based Model: DLinear (2022) [51] and LightTS (2023) [57]; Transformer-based Model: Reformer (2020) [58], Informer (2021) [59], Pyraformer (2022) [60], Autoformer (2021) [61], FEDformer (2022) [53], Transformer (2017) [62], Crossformer (2023) [63], iTransformer(2023) [64] and ETSformer (2022) [65]; Other advanced Model TimesNet (2023) [66] and FiLM (2022) [67]. Overall, a total of 15 models are included for a comprehensive comparison.

5.3. Experimental Result

As described in Section 4.1, we designed three different imputation experiments. To enable a horizontal comparison, we also employed several state-of-the-art time series models for the same experimental tasks. The hyperparameters of each model were carefully adjusted to ensure optimal results. After multiple rounds of experiments, the results of DTM and the other models are presented in Tables 1–4.

Table 2. The experimental results of masking rate is 10%.

Model	MSE	MAE	R^2
Dlinear	0.1853	0.1769	0.8151
TCN *	0.4381	0.3505	0.5622
TimesNet	0.1487	0.1573	0.8516
Transformer † ◇	0.2647	0.2268	0.7367
Autoformer †	0.4684	0.4709	0.5314
Crossformer †	0.2588	0.2280	0.7412
ETSformer †	0.2411	0.2211	0.7593
FEDformer †	0.1907	0.2053	0.8095
FiLM	0.1852	0.1777	0.8134
Informer †	0.2722	0.2339	0.7254
iTransformer †	0.1871	0.1790	0.8131
LightTS	0.2030	0.1920	0.7974
MICN *	0.1532	0.1678	0.8470
Pyraformer †	0.2175	0.2146	0.7927
Reformer †	0.2404	0.2150	0.7572
DTM	0.1765	0.1712	0.8237

* CNN based model; † Transformer based model; ◇ Baseline model.

Table 3. The experimental results of masking rate is 25%.

Model	MSE	MAE	R^2
Dlinear	0.2045	0.1907	0.7956
TCN *	0.4677	0.3523	0.5325
TimesNet	0.1605	0.1678	0.8397
Transformer † ◇	0.2816	0.2367	0.7198
Autoformer †	0.4404	0.4288	0.5597
Crossformer †	0.2700	0.2335	0.7303
ETSformer †	0.2824	0.2407	0.7177
FEDformer †	0.2139	0.2148	0.7863
FiLM	0.2054	0.1917	0.7941
Informer †	0.2991	0.2473	0.6985
iTransformer †	0.2035	0.1909	0.7966
LightTS	0.2085	0.1940	0.7915
MICN *	0.1867	0.1995	0.8136
Pyraformer †	0.2263	0.2344	0.7739
Reformer †	0.2592	0.2307	0.7380
DTM	0.1905	0.1815	0.8097

* CNN based model; † Transformer based model; ◇ Baseline model.

To demonstrate the model's performance on the interpolation task under varying missing data ratios, we conducted additional experiments with missing ratios of 40%, 50%, and 60%, and compared the results with those of the Pyraformer and Reformer models. The experimental results are presented in Figure 9.

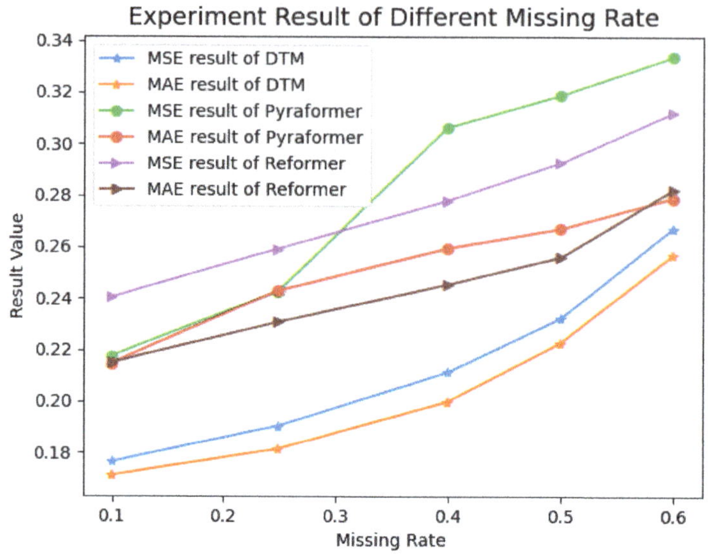

Figure 9. The experimental results under different missing rates.

Table 4. The experimental results of randomly masking one channel.

Model	MSE	MAE	R^2
Dlinear	0.9897	0.5398	−0.0002
TCN *	0.5790	0.4510	0.4051
TimesNet	0.9483	0.5193	0.0395
Transformer † ◇	0.6151	0.4457	0.3939
Autoformer †	1.0856	0.5948	−0.0522
Crossformer †	0.6170	0.4870	0.4042
ETSformer †	0.4800	0.4317	0.5134
FEDformer †	0.8616	0.5130	0.1232
FiLM	0.9603	0.5302	−0.0001
Informer †	0.5019	0.4324	0.4914
iTransformer †	0.8834	0.4992	0.0657
LightTS	0.6264	0.4583	0.3701
MICN *	0.8026	0.5352	0.2164
Pyraformer †	0.5071	0.3840	0.5339
Reformer †	0.4773	0.3886	0.5079
DTM	0.4748	0.4031	0.5504

* CNN based model; † Transformer based model; ◇ Baseline model.

5.4. Impact of Inter-Channel Correlations on DTM

In this section, we investigate how inter-channel correlation levels affect DTM. We designed the following comparative experiments: First, we selected five channels with low Pearson correlation coefficients from the Weather (https://www.bgc-jena.mpg.de/wetter/, accessed on 10 May 2025) dataset, as illustrated in Figure 10. Additionally, we chose six extra channels that exhibited high correlation coefficients with the aforementioned five channels, as shown in Figure 11. Subsequently, we employed DTM to conduct two sets of interpolation tasks: (1) Low-correlation scenario: Experiments using only the five low-correlation channels. (2) Coupled multi-channel scenario: Experiments using the combined data of all 11 channels, but focusing solely on imputation the original 5 low-correlation channels.

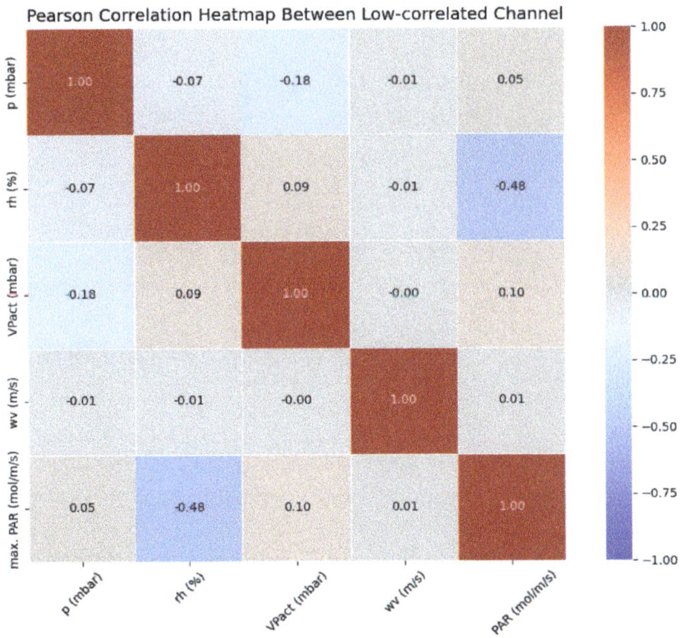

Figure 10. The heat map of low-correlated channel.

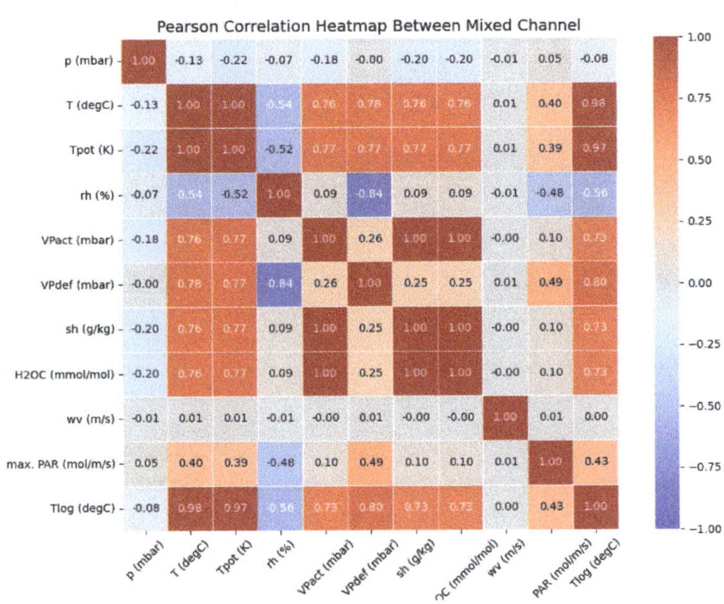

Figure 11. The heat map of the mixed channel.

The final experimental results are presented in Table 5, validating the advantages of incorporating multi-channel coupling information.

Table 5. The experimental result of different correlated data on DTM.

Metrics	10%, 5-5	25%, 5-5	Single, 5-5	10%, 11-5	25%, 11-5	Single, 11-5
MSE	0.2963	0.3161	1.4684	0.0535 ↓	0.0807 ↓	0.1910 ↓
MAE	0.3236	0.3427	0.9430	0.1463 ↓	0.1691 ↓	0.2983 ↓

X%, A-B means experiment with masking rate = X%, using A Channels data to imputation B Channels, single means randomly masking one channel, symbol ↓ denotes the a lower evaluation function value under the same masking strategy.

6. Discussion

We evaluated the proposed model and several state-of-the-art models on the same imputation tasks. The training performance comparison metrics used in the experiments were Mean Square Error (MSE), Mean Absolute Error (MAE), and R^2. The final experimental results are shown in Tables 2–4. Based on the experimental data, our proposed model consistently achieved the leading performance under various conditions, successfully meeting the target objectives of the tasks.

Compared to Transformer-based models, our proposed model outperformed the best-performing Reformer model in the channel-level imputation task, reducing the MSE by 0.0028. In the random data imputation task, our model also surpassed the best-performing iTransformer, achieving at least a 0.01 reduction in MSE. As shown in Figure 9, we observe that across data missing rates ranging from 0.1 to 0.6, both the MSE and MAE of DTM remain consistently lower than those of Reformer and Pyraformer, demonstrating the stable performance superiority of our proposed model.

Compared to the non-Transformer-based and CNN-based methods, such as TimesNet, its performance in scenarios with 10% and 25% masking rates surpasses DTM, demonstrating TimesNet's robust sequence reconstruction capabilities. However, its performance significantly deteriorates in the random channel masking task. We also identified several models with performance deterioration patterns similar to TimesNet. Their R^2 values approach or even fall below 0, indicating that the imputation capability of these models is statistically comparable to naive mean-based imputation within the corresponding channel. This is likely because random value masking causes minimal disruption to the statistical information of individual channels, whereas the random channel masking task completely eliminates the statistical information of entire channels, presenting a substantial challenge for models equipped with sequence modeling capabilities. In contrast, DTM consistently ranks among the top-performing models across all three tasks in terms of R^2 values, demonstrating its robustness in effectively overcoming these challenges.

When compared to non-Transformer models such as DLinear and TCN, our proposed model exceeded their performance in at least one or more tasks. Even in tasks where it did not outperform these models, it maintained a comparable level of performance, demonstrating that our model is better suited to adapt to complex fault scenarios.

According to Figures 10 and 11, we can observe that after adding six new channels, nearly all channels now possess corresponding highly correlated channels that reflect their coupling relationships. The experimental results in Table 5 demonstrate that the model performance shows significant improvement after channel augmentation. This indicates that our model performs poorly when handling data with low correlation relationships, while the introduction of coupling-enhanced channels substantially enhances its capability. These findings verify our model's ability to achieve data interpolation through latent coupling relationships.

7. Conclusions

We summarize as follows: In the random data missing and random channel missing imputation tasks for sensor data in the industrial reactor, the DTM model proved to be a robust solution for completing the tasks. Among the three imputation tasks, DTM achieved leading performance in one task and ranked third in the other two. Our proposed model, DTM, integrates channel decoupling with sequence modeling, enhancing the model's ability to capture multidimensional coupling relationships in multichannel data. Compared to the baseline model, DTM demonstrated improvements across multiple experimental metrics in various tasks. Specifically, MSE and MAE were reduced by up to 33.3% and 24.5%, respectively, while the R^2 value increased by a maximum of 39.73%, highlighting its statistical superiority. Experimental results demonstrate that DTM outperforms most Transformer-based and CNN-based algorithms in the imputation scenarios proposed in this paper.

Author Contributions: Conceptualization, X.G. and Z.L.; methodology, X.G.; software, F.M.; validation, Z.L.; formal analysis, Z.L.; investigation, L.X.; resources, C.W.; data curation, F.M.; writing—original draft preparation, X.G.; writing—review and editing, X.G. and C.W.; visualization, K.Z.; supervision, C.W.; project administration, F.M.; funding acquisition, F.M. All authors have read and agreed to the published version of the manuscript.

Funding: This research received no external funding

Data Availability Statement: The raw data supporting the conclusions of this article will be made available by the authors on request.

Conflicts of Interest: Authors Xiaodong Gao, Zhongliang Liu and Lei Xu were employed by the company China Nuclear Power Engineering Co., Ltd. and Engineering Research Center for Fuel Reprocessing. Author Fei Ma was employed by the company Hangzhou Boomy Intelligent Technology Co., Ltd. The remaining authors declare that the research was conducted in the absence of any commercial or financial relationships that could be construed as a potential conflicts of interest.

References

1. Qi, B.; Liang, J.; Tong, J. Fault diagnosis techniques for nuclear power plants: A review from the artificial intelligence perspective. *Energies* **2023**, *16*, 1850. [CrossRef]
2. Tonday Rodriguez, J.C.; Perry, D.; Rahman, M.A.; Alam, S.B. An Intelligent Hierarchical Framework for Efficient Fault Detection and Diagnosis in Nuclear Power Plants. In Proceedings of the Sixth Workshop on CPS&IoT Security and Privacy, Salt Lake City, UT, USA, 14–18 October 2024; pp. 80–92.
3. Kozma, R.; Nabeshima, K. Studies on the detection of incipient coolant boiling in nuclear reactors using artificial neural networks. *Ann. Nucl. Energy* **1995**, *22*, 483–496. [CrossRef]
4. Nabeshima, K.; Suzudo, T.; Suzuki, K.; Türkcan, E. Real-time nuclear power plant monitoring with neural network. *J. Nucl. Sci. Technol.* **1998**, *35*, 93–100. [CrossRef]
5. Nabeshima, K.; Suzudo, T.; Seker, S.; Ayaz, E.; Barutcu, B.; Türkcan, E.; Ohno, T.; Kudo, K. On-line neuro-expert monitoring system for borssele nuclear power plant. *Prog. Nucl. Energy* **2003**, *43*, 397–404. [CrossRef]
6. Lee, C.K.; Chang, S.J. Fault detection in multi-core C&I cable via machine learning based time-frequency domain reflectometry. *Appl. Sci.* **2019**, *10*, 158.
7. Mandal, S.; Santhi, B.; Sridhar, S.; Vinolia, K.; Swaminathan, P. Nuclear power plant thermocouple sensor-fault detection and classification using deep learning and generalized likelihood ratio test. *IEEE Trans. Nucl. Sci.* **2017**, *64*, 1526–1534. [CrossRef]
8. Peng, B.S.; Xia, H.; Liu, Y.K.; Yang, B.; Guo, D.; Zhu, S.M. Research on intelligent fault diagnosis method for nuclear power plant based on correlation analysis and deep belief network. *Prog. Nucl. Energy* **2018**, *108*, 419–427. [CrossRef]
9. Bang, S.S.; Shin, Y.J. Classification of faults in multicore cable via time–frequency domain reflectometry. *IEEE Trans. Ind. Electron.* **2019**, *67*, 4163–4171. [CrossRef]
10. Saeed, H.A.; Wang, H.; Peng, M.; Hussain, A.; Nawaz, A. Online fault monitoring based on deep neural network & sliding window technique. *Prog. Nucl. Energy* **2020**, *121*, 103236.

11. Abdelghafar, S.; El-shafeiy, E.; Mohammed, K.K.; Drawish, A.; Hassanien, A.E. CNN-Based Fault Detection in Nuclear Power Reactors Using Real-Time Sensor Data. In *Business Intelligence and Information Technology*; Proceedings of BIIT 2023; Springer: Berlin/Heidelberg, Germany, 2023; pp. 639–649.
12. Yang, J.; Kim, J. An accident diagnosis algorithm using long short-term memory. *Nucl. Eng. Technol.* **2018**, *50*, 582–588. [CrossRef]
13. Choi, J.; Lee, S.J. Consistency index-based sensor fault detection system for nuclear power plant emergency situations using an LSTM network. *Sensors* **2020**, *20*, 1651. [CrossRef]
14. Yang, J.; Kim, J. Accident diagnosis algorithm with untrained accident identification during power-increasing operation. *Reliab. Eng. Syst. Saf.* **2020**, *202*, 107032. [CrossRef]
15. Zhou, G.; Zheng, S.; Yang, S.; Yi, S. A Novel Transformer-Based Anomaly Detection Model for the Reactor Coolant Pump in Nuclear Power Plants. *Sci. Technol. Nucl. Install.* **2024**, *2024*, 9455897. [CrossRef]
16. Yi, S.; Zheng, S.; Yang, S.; Zhou, G.; He, J. Robust transformer-based anomaly detection for nuclear power data using maximum correntropy criterion. *Nucl. Eng. Technol.* **2024**, *56*, 1284–1295. [CrossRef]
17. Shen, B.; Yao, L.; Ge, Z. Nonlinear probabilistic latent variable regression models for soft sensor application: From shallow to deep structure. *Control Eng. Pract.* **2020**, *94*, 104198. [CrossRef]
18. Yuan, X.; Huang, B.; Wang, Y.; Yang, C.; Gui, W. Deep learning-based feature representation and its application for soft sensor modeling with variable-wise weighted SAE. *IEEE Trans. Ind. Inform.* **2018**, *14*, 3235–3243. [CrossRef]
19. Wang, X.; Liu, H. Data supplement for a soft sensor using a new generative model based on a variational autoencoder and Wasserstein GAN. *J. Process Control* **2020**, *85*, 91–99. [CrossRef]
20. Yao, L.; Ge, Z. Deep learning of semisupervised process data with hierarchical extreme learning machine and soft sensor application. *IEEE Trans. Ind. Electron.* **2017**, *65*, 1490–1498. [CrossRef]
21. Horn, Z.; Auret, L.; McCoy, J.; Aldrich, C.; Herbst, B. Performance of convolutional neural networks for feature extraction in froth flotation sensing. *IFAC-PapersOnLine* **2017**, *50*, 13–18. [CrossRef]
22. Yuan, X.; Qi, S.; Shardt, Y.A.; Wang, Y.; Yang, C.; Gui, W. Soft sensor model for dynamic processes based on multichannel convolutional neural network. *Chemom. Intell. Lab. Syst.* **2020**, *203*, 104050. [CrossRef]
23. Wei, J.; Guo, L.; Xu, X.; Yan, G. Soft sensor modeling of mill level based on convolutional neural network. In Proceedings of the The 27th Chinese Control and Decision Conference (2015 CCDC), Qingdao, China, 23–25 May 2015; pp. 4738–4743.
24. Ke, W.; Huang, D.; Yang, F.; Jiang, Y. Soft sensor development and applications based on LSTM in deep neural networks. In Proceedings of the 2017 IEEE Symposium Series on Computational Intelligence (SSCI), Honolulu, HI, USA, 27 November–1 December 2017; pp. 1–6.
25. Yuan, X.; Li, L.; Wang, Y. Nonlinear dynamic soft sensor modeling with supervised long short-term memory network. *IEEE Trans. Ind. Informatics* **2019**, *16*, 3168–3176. [CrossRef]
26. Raghavan, S.V.; Radhakrishnan, T.; Srinivasan, K. Soft sensor based composition estimation and controller design for an ideal reactive distillation column. *ISA Trans.* **2011**, *50*, 61–70. [CrossRef]
27. Yin, X.; Niu, Z.; He, Z.; Li, Z.S.; Lee, D.h. Ensemble deep learning based semi-supervised soft sensor modeling method and its application on quality prediction for coal preparation process. *Adv. Eng. Inform.* **2020**, *46*, 101136. [CrossRef]
28. Andridge, R.R.; Little, R.J. A review of hot deck imputation for survey non-response. *Int. Stat. Rev.* **2010**, *78*, 40–64. [CrossRef]
29. Van Buuren, S. Multiple imputation of discrete and continuous data by fully conditional specification. *Stat. Methods Med. Res.* **2007**, *16*, 219–242. [CrossRef]
30. Pujianto, U.; Wibawa, A.P.; Akbar, M.I. K-nearest neighbor (k-NN) based missing data imputation. In Proceedings of the 2019 5th International Conference on Science in Information Technology (ICSITech), Yogyakarta, Indonesia, 23–24 October 2019; pp. 83–88.
31. Troyanskaya, O.; Cantor, M.; Sherlock, G.; Brown, P.; Hastie, T.; Tibshirani, R.; Botstein, D.; Altman, R.B. Missing value estimation methods for DNA microarrays. *Bioinformatics* **2001**, *17*, 520–525. [CrossRef]
32. Stekhoven, D.J.; Bühlmann, P. MissForest—non-parametric missing value imputation for mixed-type data. *Bioinformatics* **2012**, *28*, 112–118. [CrossRef]
33. Doreswamy, I.G.; Manjunatha, B. Performance evaluation of predictive models for missing data imputation in weather data. In Proceedings of the 2017 International Conference on Advances in Computing, Communications and Informatics (ICACCI), Udupi, India, 13–16 September 2017; pp. 1327–1334.
34. Mallinson, H.; Gammerman, A. *Imputation Using Support Vector Machines*; Department of Computer Science Royal Holloway, University of London Egham: London, UK, 2003.
35. Noor, T.H.; Almars, A.; Gad, I.; Atlam, E.S.; Elmezain, M. Spatial impressions monitoring during COVID-19 pandemic using machine learning techniques. *Computers* **2022**, *11*, 52. [CrossRef]
36. Vincent, P.; Larochelle, H.; Bengio, Y.; Manzagol, P.A. Extracting and composing robust features with denoising autoencoders. In Proceedings of the 25th International Conference on Machine Learning, Helsinki, Finland, 5–9 July 2008; pp. 1096–1103.
37. Li, J.; Ren, W.; Han, M. Variational auto-encoders based on the shift correction for imputation of specific missing in multivariate time series. *Measurement* **2021**, *186*, 110055. [CrossRef]

38. Yoon, J.; Jordon, J.; Schaar, M. Gain: Missing data imputation using generative adversarial nets. In Proceedings of the International Conference on Machine Learning, Stockholm, Sweden, 10–15 July 2018; pp. 5689–5698.
39. Qin, R.; Wang, Y. ImputeGAN: Generative adversarial network for multivariate time series imputation. *Entropy* **2023**, *25*, 137. [CrossRef]
40. Khan, W.; Zaki, N.; Ahmad, A.; Masud, M.M.; Ali, L.; Ali, N.; Ahmed, L.A. Mixed data imputation using generative adversarial networks. *IEEE Access* **2022**, *10*, 124475–124490. [CrossRef]
41. Yıldız, A.Y.; Koç, E.; Koç, A. Multivariate time series imputation with transformers. *IEEE Signal Process. Lett.* **2022**, *29*, 2517–2521. [CrossRef]
42. Nie, T.; Qin, G.; Ma, W.; Mei, Y.; Sun, J. ImputeFormer: Low rankness-induced transformers for generalizable spatiotemporal imputation. In Proceedings of the 30th ACM SIGKDD Conference on Knowledge Discovery and Data Mining, Barcelona, Spain, 25–29 August 2024; pp. 2260–2271.
43. Stellwagen, E.; Tashman, L. ARIMA: The Models of Box and Jenkins. *Foresight Int. J. Appl. Forecast.* **2013**, *30*, 28–33.
44. Feng, J.; Li, Y.; Zhang, C.; Sun, F.; Meng, F.; Guo, A.; Jin, D. Deepmove: Predicting human mobility with attentional recurrent networks. In Proceedings of the 2018 World Wide Web Conference, Lyons, France, 23–27 April 2018; pp. 1459–1468.
45. Qin, Y.; Song, D.; Cheng, H.; Cheng, W.; Jiang, G.; Cottrell, G.W. A dual-stage attention-based recurrent neural network for time series prediction. In Proceedings of the 26th International Joint Conference on Artificial Intelligence, Melbourne, Australia, 19–25 August 2017; pp. 2627–2633.
46. Bai, L.; Yao, L.; Kanhere, S.S.; Wang, X.; Sheng, Q.Z. STG2Seq: Spatial-Temporal Graph to Sequence Model for Multi-step Passenger Demand Forecasting. In Proceedings of the Twenty-Eighth International Joint Conference on Artificial Intelligence, Macao, China, 10–16 August 2019; pp. 1981–1987. [CrossRef]
47. Wu, Z.; Pan, S.; Long, G.; Jiang, J.; Chang, X.; Zhang, C. Connecting the dots: Multivariate time series forecasting with graph neural networks. In Proceedings of the 26th ACM SIGKDD International Conference on Knowledge Discovery & Data Mining, Virtual Event, CA, USA, 6–10 July 2020; pp. 753–763.
48. Zhang, Z.; Meng, L.; Gu, Y. SageFormer: Series-Aware Framework for Long-Term Multivariate Time-Series Forecasting. *IEEE Internet Things J.* **2024**, *11*, 18435–18448. [CrossRef]
49. He, H.; Zhang, Q.; Wang, S.; Yi, K.; Niu, Z.; Cao, L. Learning Informative Representation for Fairness-Aware Multivariate Time-Series Forecasting: A Group-Based Perspective. *IEEE Trans. Knowl. Data Eng.* **2024**, *36*, 2504–2516. [CrossRef]
50. Wang, C.; Wang, H.; Zhang, X.; Liu, Q.; Liu, M.; Xu, G. A Transformer-Based Industrial Time Series Prediction Model With Multivariate Dynamic Embedding. *IEEE Trans. Ind. Inform.* **2025**, *21*, 1813–1822. [CrossRef]
51. Zeng, A.; Chen, M.; Zhang, L.; Xu, Q. Are transformers effective for time series forecasting? In Proceedings of the AAAI Conference on Artificial Intelligence, Washington, DC, USA, 7–14 February 2023; Volume 37, pp. 11121–11128.
52. Wu, H.; Xu, J.; Wang, J.; Long, M. Autoformer: Decomposition Transformers with Auto-Correlation for Long-Term Series Forecasting. In *Advances in Neural Information Processing Systems 34 (NeurIPS 2021)*; Ranzato, M., Beygelzimer, A., Dauphin, Y., Liang, P., Vaughan, J.W., Eds.; Curran Associates, Inc.: Red Hook, NY, USA, 2021; Volume 34, pp. 22419–22430.
53. Zhou, T.; Ma, Z.; Wen, Q.; Wang, X.; Sun, L.; Jin, R. FEDformer: Frequency enhanced decomposed transformer for long-term series forecasting. In Proceedings of the 39th International Conference on Machine Learning (ICML 2022), Baltimore, MD, USA, 17–23 July 2022.
54. Binali, R. Experimental and machine learning comparison for measurement the machinability of nickel based alloy in pursuit of sustainability. *Measurement* **2024**, *236*, 115142. [CrossRef]
55. Wang, H.; Peng, J.; Huang, F.; Wang, J.; Chen, J.; Xiao, Y. MICN: Multi-scale Local and Global Context Modeling for Long-term Series Forecasting. In Proceedings of the 11th International Conference on Learning Representations, Kigali, Rwanda, 1–5 May 2023.
56. Bai, S.; Kolter, J.Z.; Koltun, V. An Empirical Evaluation of Generic Convolutional and Recurrent Networks for Sequence Modeling. *arXiv* **2018**, arXiv:1803.01271.
57. Campos, D.; Zhang, M.; Yang, B.; Kieu, T.; Guo, C.; Jensen, C.S. LightTS: Lightweight Time Series Classification with Adaptive Ensemble Distillation. *Proc. ACM Manag. Data* **2023**, *1*, 171:1–171:27. [CrossRef]
58. Kitaev, N.; Kaiser, L.; Levskaya, A. Reformer: The Efficient Transformer. In Proceedings of the International Conference on Learning Representations, Addis Ababa, Ethiopia, 26–30 April 2020.
59. Zhou, H.; Zhang, S.; Peng, J.; Zhang, S.; Li, J.; Xiong, H.; Zhang, W. Informer: Beyond Efficient Transformer for Long Sequence Time-Series Forecasting. In Proceedings of the The Thirty-Fifth AAAI Conference on Artificial Intelligence, AAAI 2021, Virtual Conference, 2–9 February 2021; AAAI Press: Vancouver, BC, Canada, 2021; Volume 35, pp. 11106–11115.
60. Liu, S.; Yu, H.; Liao, C.; Li, J.; Lin, W.; Liu, A.X.; Dustdar, S. Pyraformer: Low-Complexity Pyramidal Attention for Long-Range Time Series Modeling and Forecasting. In Proceedings of the International Conference on Learning Representations, Virtual Event, 25–29 April 2022.

61. Wu, H.; Xu, J.; Wang, J.; Long, M. Autoformer: Decomposition Transformers with Auto-Correlation for Long-Term Series Forecasting. In Proceedings of the Advances in Neural Information Processing Systems, Virtual Event, 6–14 December 2021.
62. Vaswani, A. Attention is all you need. In *Advances in Neural Information Processing Systems 30*; Curran Associates Inc.: Red Hook, NY, USA, 2017; ISBN 9781510860964.
63. Zhang, Y.; Yan, J. Crossformer: Transformer Utilizing Cross-Dimension Dependency for Multivariate Time Series Forecasting. In Proceedings of the International Conference on Learning Representations, Kigali, Rwanda, 1–5 May 2023.
64. Liu, Y.; Hu, T.; Zhang, H.; Wu, H.; Wang, S.; Ma, L.; Long, M. iTransformer: Inverted Transformers Are Effective for Time Series Forecasting. *arXiv* **2023**, arXiv:2310.06625.
65. Woo, G.; Liu, C.; Sahoo, D.; Kumar, A.; Hoi, S. Etsformer: Exponential smoothing transformers for time-series forecasting. *arXiv* **2022**, arXiv:2202.01381.
66. Wu, H.; Hu, T.; Liu, Y.; Zhou, H.; Wang, J.; Long, M. TimesNet: Temporal 2D-Variation Modeling for General Time Series Analysis. In Proceedings of the International Conference on Learning Representations, Kigali, Rwanda, 1–5 May 2023.
67. Zhou, T.; Ma, Z.; Wen, Q.; Sun, L.; Yao, T.; Yin, W.; Jin, R. Film: Frequency improved legendre memory model for long-term time series forecasting. *Adv. Neural Inf. Process. Syst.* **2022**, *35*, 12677–12690.

processes

MDPI

Article

AI-Driven Maintenance Optimisation for Natural Gas Liquid Pumps in the Oil and Gas Industry: A Digital Tool Approach

Abdulmajeed Almuraia, Feiyang He * and Muhammad Khan *

Centre for Life-Cycle Engineering and Management, Cranfield University, Bedford MK43 0AL, UK
* Correspondence: feiyang.he@cranfield.ac.uk (F.H.); muhammad.a.khan@cranfield.ac.uk (M.K.)

Abstract: Natural Gas Liquid (NGL) pumps are critical assets in oil and gas operations, where unplanned failures can result in substantial production losses. Traditional maintenance approaches, often based on static schedules and expert judgement, are inadequate for optimising both availability and cost. This study proposes a novel Artificial Intelligence (AI)-based methodology and digital tool for optimising NGL pump maintenance using limited historical data and real-time sensor inputs. The approach combines dynamic reliability modelling, component condition assessment, and diagnostic logic within a unified framework. Component-specific maintenance intervals were computed using mean time between failures (MTBFs) estimation and remaining useful life (RUL) prediction based on vibration and leakage data, while fuzzy logic- and rule-based algorithms were employed for condition evaluation and failure diagnoses. The tool was implemented using Microsoft Excel Version 2406 and validated through a case study on pump G221 in a Saudi Aramco facility. The results show that the optimised maintenance routine reduced the total cost by approximately 80% compared to conventional individual scheduling, primarily by consolidating maintenance activities and reducing downtime. Additionally, a structured validation questionnaire completed by 15 industry professionals confirmed the methodology's technical accuracy, practical usability, and relevance to industrial needs. Over 90% of the experts strongly agreed on the tool's value in supporting AI-driven maintenance decision-making. The findings demonstrate that the proposed solution offers a practical, cost-effective, and scalable framework for the predictive maintenance of rotating equipment, especially in environments with limited sensory and operational data. It contributes both methodological innovation and validated industrial applicability to the field of maintenance optimisation.

Keywords: NGL pumps; predictive maintenance; maintenance optimisation; AI-based diagnostics; remaining useful life (RUL); MTBFs estimation; fuzzy logic; rule-based system; oil and gas industry; digital tool validation

1. Introduction

Natural Gas Liquid (NGL) pumps are critical in the oil and gas industry, particularly in driving NGL fractionation processes. The failure of these pumps can lead to significant operational disruptions if uninterrupted production is halted across a full 24 h period [1]. Given the high stakes involved, ensuring the reliability and availability of NGL pumps is significant. Traditional maintenance practices, often reliant on manual methods and expert opinions, lack the structured frameworks to optimise pump uptime effectively. This limitation stresses the urgent need to develop innovative methods to predict and prevent equipment failures more accurately.

A substantial amount of research has been conducted on maintenance optimisation, covering various approaches to enhancing equipment reliability and maintenance efficiency [2–5]. Studies have explored methods including Failure Modes and Effects Analysis (FMEA) [6]; Reliability, Availability, and Maintainability (RAM) analysis [7]; as well as condition assessments combined with data analytics techniques [8]. These contributions have provided structured frameworks that guide maintenance decisions and improve industry asset performance. More recently, data-driven techniques have gained attention for enhancing maintenance decision-making. Methods such as predictive maintenance based on data analytics and digital twin technology have been introduced, enabling real-time condition monitoring and proactive maintenance planning [9–13]. Furthermore, various statistical and machine learning approaches have been applied to identify faults early and improve maintenance scheduling accuracy, resulting in better reliability and reduced operational costs [14–17]. These data-driven strategies offer advantages over traditional maintenance methods, especially for managing complex systems with limited or uncertain data availability [18–22].

A range of studies have been conducted across general industrial applications to evaluate the effectiveness of predictive models in maintenance planning. Using machine learning integrated with building information modelling, Cheng et al. [19] developed a predictive maintenance framework for mechanical, electrical, and plumbing (MEP) systems. Falamarzi et al. [23] applied artificial neural networks (ANNs) and support vector regression (SVR) for predicting tram track gauge deviations, though their model lacked a comprehensive performance analysis. Similarly, Susto et al. [24] and Susto and Beghi [25] proposed predictive systems for epitaxy processes, employing ridge regression and support vector machines (SVMs) but without a comparative evaluation or uncertainty quantification. Mathew et al. [26] implemented a support vector regression kernel to estimate the remaining useful life (RUL) for turbofan engines, whereas Amruthnath and Gupta [17] used unsupervised learning methods for early fault detection in industrial assets. Despite their technical strengths, these studies did not clearly define the operational contexts or validate models across varied working conditions.

Although these general models illustrate the utility of predictive techniques, they are typically not tailored to the specific operational requirements of rotating machinery, such as pumps and compressors, which often operate under different load profiles, environmental conditions, and maintenance constraints.

To address this limitation, a body of research has specifically focused on rotating equipment [27]. Janssens et al. [28] used a CNN with thermal imaging data to detect anomalies in machinery, though the absence of an equipment-specific context limited the study's practical relevance. Sampaio et al. [29] developed an ANN model for motor failure prediction but without a robust model comparison or a sensitivity analysis. Bekar et al. [30] designed an intelligent predictive method for motors, and Praveenkumar et al. [31] used support vector machines to diagnose gearbox faults. Similarly, Prytz et al. [21] applied random forest models to predict compressor failures using historical vehicle data. While these methods showed promise, many lacked depth in their performance validation and scalability.

For more complex rotating systems, Durbhaka and Selvaraj [32] analysed vibration signals in wind turbines using several classifiers, including k-NN, SVMs, and k-means clustering. Yet, the study's focus on a single data source constrained its generalisability. Su and Huang [33] and Butte [34] applied AI-based approaches to detect faults in exhaust fans.

Their findings highlighted the importance of combining physical parameters and extended datasets to improve accuracy. Abu-Samah et al. [25] introduced a hybrid model using Bayesian networks and multi-gene genetic programming to monitor pump conditions. However, they did not compare their approach against other modelling techniques.

Despite these contributions, current AI-based maintenance models show several limitations that are particularly critical in the context of NGL pump operations. Most existing models focus primarily on failure prediction, often without supporting comprehensive maintenance routines that consider equipment availability and maintenance costs [28–30,33]. In addition, many models rely on large, high-quality datasets for training, yet in practical applications such as NGL pumps, both historical and real-time data are often limited in quantity and resolution, which poses challenges for their implementation [23,33,35]. Although some studies cover diagnostics or prediction [36–40], few offer integrated solutions that address both dimensions within a single framework [21,31]. Furthermore, cost and downtime considerations are not typically incorporated into these models despite their importance in industrial maintenance decision-making. Lastly, while many existing approaches target rotating machinery, none are specifically tailored to the unique characteristics and constraints of NGL pumps.

To address these challenges, this study proposes a practical, Artificial Intelligence (AI)-supported digital tool specifically designed for the maintenance optimisation of NGL reflux pumps in the oil and gas sector. Unlike existing predictive models, the tool utilises historical failure records and real-time sensor inputs to evaluate component conditions, estimate reliability, and generate optimised maintenance plans over a 10-year operational horizon. In addition to producing maintenance schedules, the tool also incorporates diagnostic functionality and provides actionable recommendations to improve equipment availability while controlling maintenance costs. A simplified schematic of the proposed framework is shown in Figure 1.

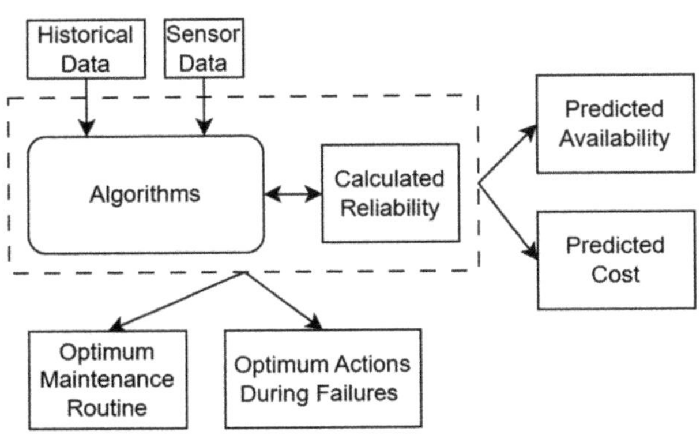

Figure 1. A schematic representation of the proposed AI-supported maintenance optimisation model for NGL pumps.

2. Method

This study adopts a structured methodology for developing and validating a digital maintenance optimisation tool tailored to NGL reflux pumps. The methodological framework, summarised in Figure 2, consists of four main stages: dynamic optimisation model development, risk-based prioritisation using FMEA, integrated maintenance routine optimisation, and tool development and validation.

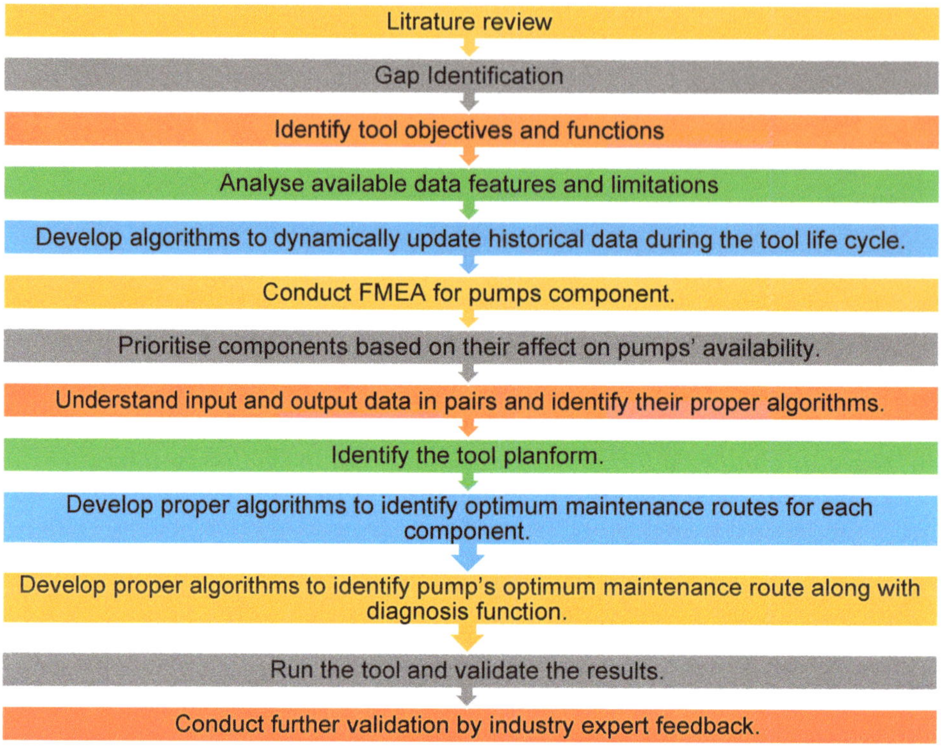

Figure 2. Flow chart of the framework and tool development.

2.1. Dynamic Optimisation of Maintenance Intervals

The first stage involves the development of a dynamic optimisation model that updates the Original Equipment Manufacturer (OEM)-recommended maintenance schedule based on actual operational conditions. This model incorporates both real-time sensor data and historical failure records to estimate the current mean time between failures (MTBFs) of individual pump components. Based on this analysis, the maintenance intervals are dynamically adjusted to better reflect the actual degradation behaviour of the system. This adaptive approach helps minimise unnecessary maintenance actions and avoids unexpected failures. A visual representation of this dynamic optimisation process is shown in Figure 3.

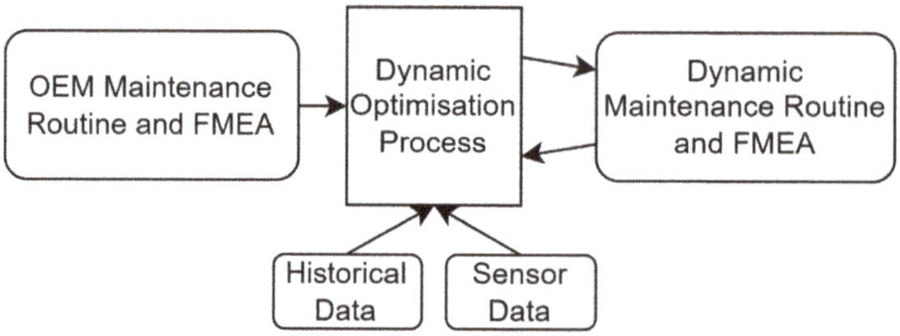

Figure 3. Dynamic optimisation flow diagram.

2.2. Risk-Based Prioritisation Using FMEA

Following the dynamic adjustment of component-specific maintenance intervals, an FMEA was developed to prioritise pump components according to their risk levels. Each component was evaluated based on its probability of failure, impact severity, and detection likelihood. The risk priority number (RPN) was then calculated to rank components by criticality. This ranking informs the maintenance planning by identifying which components require closer monitoring and more frequent interventions.

2.3. Integrated Maintenance Optimisation Logic

An optimum maintenance strategy for NGL pumps must satisfy three primary criteria. It should ensure optimum availability while maintaining reliability that meets operational requirements. In addition, it should minimise the total maintenance cost, accounting for both direct service activities and the associated operational losses during downtime. Furthermore, the strategy should be capable of reducing the occurrence of unplanned failures by supporting accurate diagnostics and appropriate maintenance actions when failures do occur. These criteria form the basis for the integrated optimisation logic presented in this section.

After establishing risk-based maintenance intervals, a system-level optimisation process was applied. The objective was to coordinate and integrate individual component schedules into a unified maintenance strategy that maximises availability and reduces the total cost. Simply applying each component's dynamic schedule independently does not guarantee overall optimisation. Therefore, this stage involved aligning maintenance activities during common downtime windows, where feasible. This integrated scheduling approach reduces the number of shutdowns and avoids redundant tasks. This approach is illustrated conceptually in Figure 4, where maintenance activities across different components are overlapped within shared downtime windows to maximise system availability and minimise operational disruption.

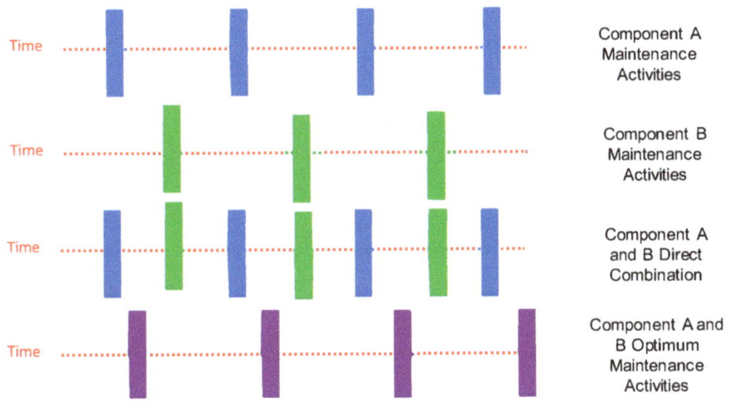

Figure 4. Overlaying activities concept.

The optimisation procedure employs an iterative process to identify combinations of component frequencies that satisfy user-defined reliability thresholds. An optimisation factor is computed for each combination, reflecting both the cost and operational impact. The combination yielding the lowest optimisation factor is selected. The total cost is calculated as the product of maintenance cost and operational loss cost associated with each maintenance plan. This process is summarised in Figure 5.

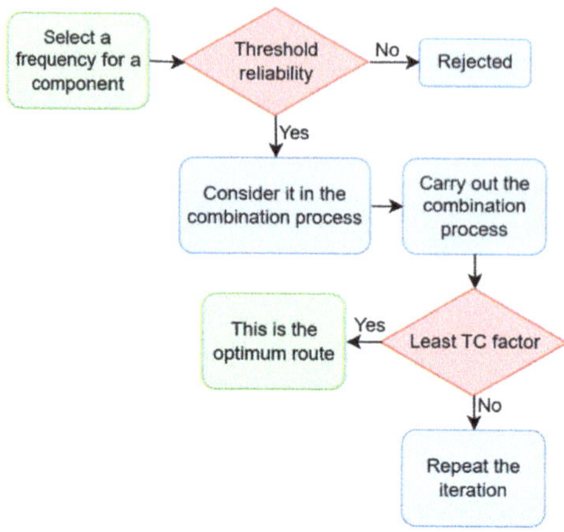

Figure 5. Optimisation logic flow diagram.

2.4. Handling Unplanned Failures

In addition to scheduled maintenance, the tool also supports optimising responses to unplanned failures. The tool interprets real-time sensor signals using embedded diagnostic logic to identify failure symptoms and generate appropriate maintenance actions. A structured database is incorporated into the system, mapping symptoms to possible causes and recommended interventions. This enables rapid decision-making during unexpected events and supports more accurate fault isolation.

2.5. Tool Development and Validation

The final stage of the methodology involved developing and validating the digital tool. The prototype was implemented using Microsoft Excel Version 2406 as the development platform. Data input included historical failure records, sensor readings, failure symptoms, and corresponding maintenance recommendations. The tool executes iterative calculations to determine optimal maintenance schedules and diagnoses.

Validation was performed in two stages. First, the tool was tested using operational data from Saudi Aramco's single pump unit (G221) to verify its accuracy and practicality in a real-world setting. The pump unit is compliant with API 610/ISO 13709 [41]. Its specifications are shown in Table 1. Second, oil and gas maintenance experts were consulted to review the tool's logic, usability, and applicability. Their feedback informed further model refinement and ensured its alignment with industry practices.

Table 1. G221 pump unit components and their specifications.

Pump Components	Specification Data	
Pump Inboard Bearing (PIB)	Manufacturer	Byron Jackson® (Woodland, CA, USA)
Pump Outboard Bearing (POB)	Type–size	DSJH-10X14X20L
Motor Inboard Bearing (MIB)	Pump type	Horizontal
Motor Outboard Bearing (MOB)	Capacity	4715 usgpm
Inboard Mechanical Seal (IMS)	Rated speed	1770 rpm
Outboard Mechanical Seal (OMS)	Maximum allowable speed	1800 rpm
Coupling	Total weight	4873 lbs

Although the tool was validated using operational data from a single NGL pump unit (G221), the underlying algorithmic structure and diagnostic logic were intentionally designed to be adaptable to other rotating equipment types exhibiting similar degradation modes.

3. Digital Tool Development

This section outlines the development of a digital tool designed to support the maintenance optimisation for NGL pumps. The tool functions as a digital twin by simulating pump performance and predicting future maintenance needs using both historical and real-time data. Its primary aim is to operationalise the proposed methodology, validate its feasibility, and deliver actionable insights for improving pump availability and reducing maintenance costs.

3.1. Functional Design of the Digital Tool

The tool was built to address the limitations identified in Section 1 directly. It performs six essential functions: (1) optimising maintenance routines by balancing availability and cost, (2) identifying critical components through FMEA-based risk prioritisation, (3) assessing component condition based on sensor inputs, (4) predicting the remaining useful life (RUL), (5) reducing downtime via diagnostic support and recovery recommendations, and (6) acting as a digital twin to simulate maintenance scenarios and evaluate their impact on pump performance.

3.2. Data Structure: Inputs and Outputs

The tool is structured around two categories of data: input and output. The input data include both historical records and real-time sensor data, which serve as the basis for algorithmic calculations. The historical data consist of failure and replacement records, particularly for bearings, mechanical seals, and couplings. The real-time data include monthly vibration readings and mechanical seal leakage pressures. These inputs are processed to generate outputs such as MTBFs, RUL estimates, component condition assessments, maintenance diagnostics, and ultimately an optimised maintenance routine.

Figure 6 illustrates the hierarchical relationship between output functions, while Table 2 summarises each output and its corresponding inputs.

Table 2. Output summary.

Output	Description	Related Input
MTBFs	Components' mean time between failures	Failures Record (history)
Bearing RUL Function	The function relates the vibration reading with the bearings' RUL	Failures Record (history)
Estimated Lifespan/Shutdown Frequency	This is the estimated replacement time for each component	MTBFs, Reliability Threshold
Bearing RUL	The estimated bearings' RUL based on the current vibration reading	Bearing RUL Function, Sensor Data
Components' Current Condition	The current health condition of components based on the sensor data	Senor Data
Diagnosis and Recommended Actions	Failures diagnosis and recommended actions to reduce pumps' downtime	Current Components' Condition, Sensor Data
Optimised Maintenance Routine	The optimum maintenance routine in terms of pumps' availability and cost	Estimated Lifespan/Shutdown Frequency, Bearings' RUL

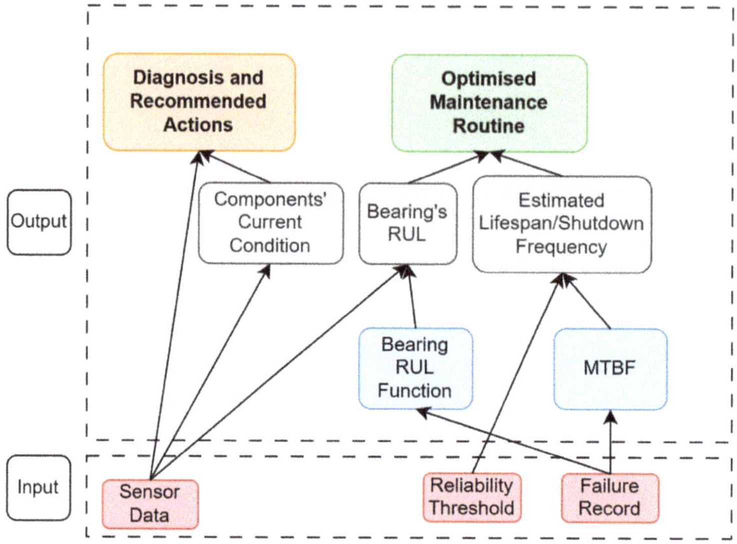

Figure 6. The tool's output map.

3.3. *AI Algorithm Determination for Processing Inputs*

Each tool function is associated with one or more processing algorithms to generate a reliable output. The selection of algorithms is based on the characteristics of available data, the statistical behaviour of pump components, and the relationships between inputs and the required outputs.

3.3.1. Input Data Sources and Limitations

The tool uses historical failure records and real-time sensor data. Table 3 presents an overview of the features and limitations of these datasets. The historical failure records include each component's year of failure and replacement. These records are available starting from 2011. On average, each component has experienced only two to three failure incidents, making the dataset small in size and limited in statistical richness. Historical sensor data are also sparse and consist mostly of bearing vibration measurements taken before and after replacements. Regular condition-monitoring data for healthy operating states are not available.

Table 3. Input data features and limitations.

Input Data	Features	Limitations
Components' Failure Record	Accurate Used easily to calculate MTBFs Single dimensional	Very limited size
Sensors' Historical Data	Shows certain patterns Single dimensional	Very limited size
Real-Time Sensor Data	Accurate Detailed to support diagnosis Single dimensional	Limited databases

The real-time sensor data include monthly checks of bearing vibration and mechanical seal leakage. Bearing conditions are assessed using portable vibration analysers, while leakage is monitored by observing pressure on a dedicated gauge. Due to database storage constraints, only abnormal readings and failure-related data are recorded and stored.

3.3.2. Algorithm Allocation by Function

Each tool function is supported by a specific algorithm selected according to the type and quality of data available. Figure 7 outlines the main processing modules and their associated computational methods.

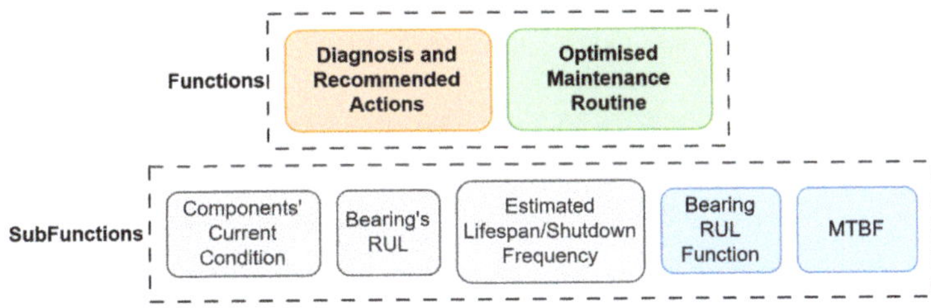

Figure 7. Functional mapping of the tool and algorithm allocation.

- **MTBFs Estimation**

The MTBFs for each component was calculated using historical failure records. The tool continuously updates the MTBFs value as new failure data become available. The calculation follows a deterministic approach, applying the following equation:

MTBFs (years) = Total operational time (years)/Number of recorded failures

- **Bearing RUL Estimation**

Although the historical data do not directly link the bearing condition and the RUL, the degradation behaviour of bearings is well established. Previous studies [34] have demonstrated empirical correlations between the RUL and vibration levels. One such relationship is given by Equation (1).

$$\text{RUL} = \frac{1}{V^n} \quad (1)$$

where V represents the current bearing vibration (in/sec), and n is a constant derived experimentally, which is influenced by the initial condition of the bearing and is suitable for the bearings used in NGL pumps, given their size, load, and operating speed.

When this equation is applied to bearings with known lifespans and documented initial vibration levels, a linear relationship is observed between the n value and the initial vibration reading at the replacement time. This linear pattern is used by the tool to estimate n from the initial vibration reading. The tool then applies Equation (1) along with the current real-time vibration measurement to estimate the RUL of the bearing.

Figure 8 presents this linear relationship as observed in pump G221. The linear regression model is updated continuously as new data become available.

- **Component Lifespan Calculation**

The tool uses historical failure data to estimate the operational lifespan of each component. Based on these estimates, it determines suitable shutdown intervals for scheduled replacements. This estimation process follows the exponential reliability model, where the reliability function is expressed as follows:

$$Reliability = e^{-t/MTBF} \tag{2}$$

From this, the expected lifespan corresponding to a specific reliability threshold is calculated as follows:

$$Lifespan = -MTBF \times Ln(Reliability\ threshold) \tag{3}$$

This method allows users to define an acceptable minimum reliability level, which the tool uses to determine replacement frequencies. The calculation is carried out using a deterministic algorithm.

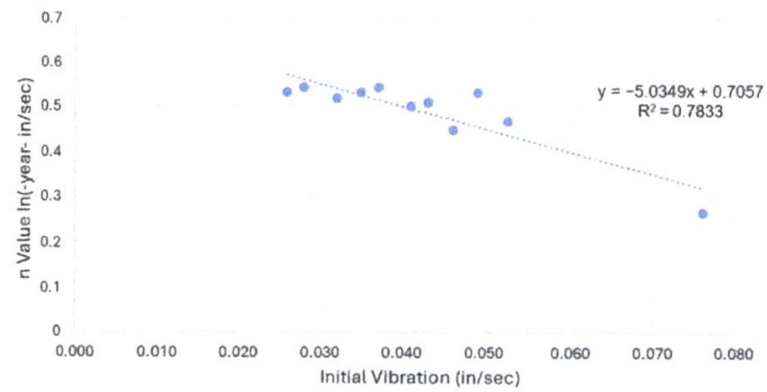

Figure 8. Linear relationship between parameter n and initial vibration levels. Dotted line represents the linear regression fitting.

- **Component Condition Assessment**

The tool assesses the condition of key components using real-time sensor data. For bearing evaluation, the ISO 20816 standard [42] is applied using a rule-based algorithm with defined alarm and fault thresholds set at 0.08 in/sec and 0.17 in/sec, respectively.

For mechanical seals, there are no explicitly defined leakage limits in either API 682 or Saudi Aramco Safety Standards [43]. Instead, these standards define overlapping pressure ranges for normal (0–7.5 psiG), alarming (5–10 psiG), and faulty (10–20 psiG) operating states. To interpret these ranges, the tool uses a fuzzy logic algorithm with triangular and trapezoidal membership functions. Figure 9 illustrates the membership structure.

The fuzzy logic system classifies mechanical seal conditions into three sets: normal, alarming, and faulty. Each set is defined by a specific membership function. The normal condition is represented by a triangular function, which assigns a membership value of one when the leakage pressure is less than or equal to 5 psiG. This value decreases linearly between 5 and 10 psiG, reaching zero beyond 10 psiG. The selection of 10 psiG as the upper bound ensures a smooth transition between the condition states.

The alarming condition is modelled using a trapezoidal membership function. It begins with a membership value of zero for leakage pressures at or below 5 psiG, then increases linearly from 5 to 7.5 psiG. The function maintains a value of one between 7.5 and 12.5 psiG and decreases linearly from 12.5 to 15 psiG, returning to zero beyond 15 psiG.

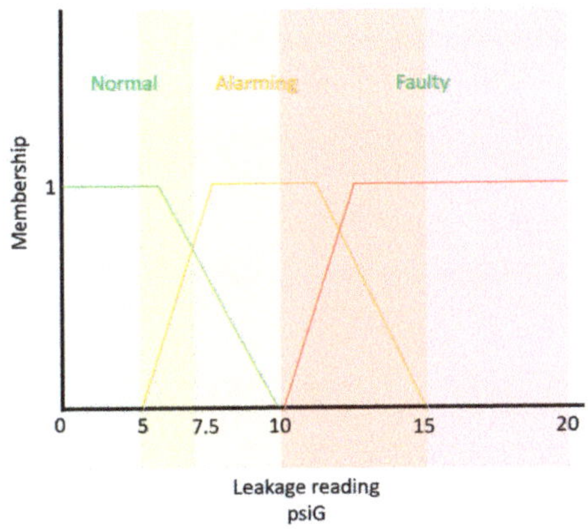

Figure 9. Fuzzy membership functions. Fuzzy sets: normal, alarming, faulty.

The faulty condition is also defined using a trapezoidal function. It starts with a membership value of zero for leakage pressures at or below 12.5 psiG, increases linearly up to 15 psiG, and maintains a membership value of one for any pressure equal to or greater than 15 psiG.

A defuzzification process is used to derive a single output value from the fuzzy sets. This involves calculating a weighted average based on representative midpoints for each fuzzy set—3.75, 10, and 17.5 psiG, respectively. The defuzzification formula is shown in Equation (4):

$$Crisp = \frac{(Normal\ Membership \times 3.75 + Alarming\ Membership \times 10 + Faulty\ Membership \times 17.5)}{Noraml\ Membership + Alarming\ Membership + Faulty\ Membership} \quad (4)$$

Based on the resulting crisp value, the condition is classified into one of three discrete states: values less than or equal to 5 indicate a normal condition; values between 5 and 12.5 indicate an alarming condition; and values above 12.5 are classified as faulty.

- **Diagnostic Functions and Maintenance Recommendations**

This function is designed to reduce downtime by diagnosing faults and recommending actions for bearing and mechanical seal replacements. Coupling replacements are excluded, as their procedures are typically straightforward and do not require AI-based decision-making.

For bearing diagnostics, the tool uses vibration frequency and bearing clearance data to identify possible failure causes and propose corrective actions. A rule-based algorithm based on if–then logic is employed to guide this analysis. The decision structure is presented in Figure 10.

In the case of mechanical seals, the diagnostic process is informed by measurements such as casing concentricity and squareness, along with visual inspections of seal flashing lines. These inputs help identify causes of failure and support the recommendation of appropriate maintenance actions. Like the bearing module, the seal diagnosis function uses a rule-based if–then algorithm to process input data and determine the output. The diagnostic framework is summarised in Figure 11.

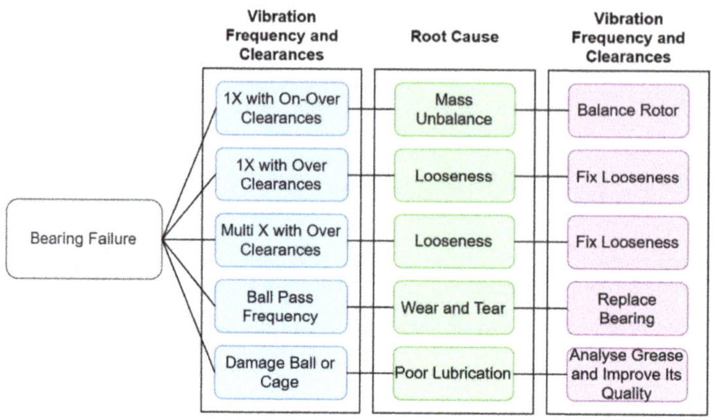

Figure 10. Logic tree for failure causes and recommended actions for bearings.

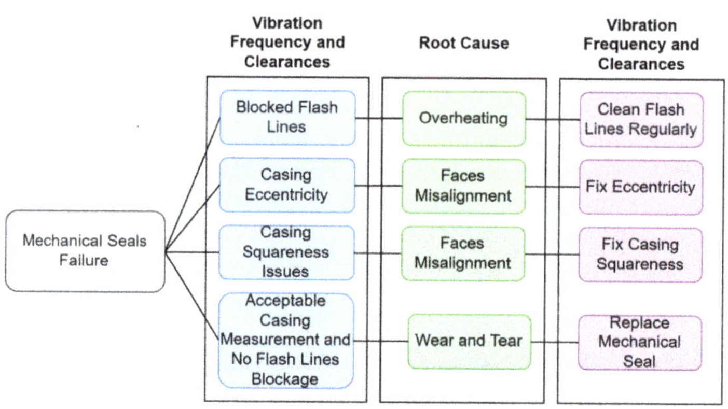

Figure 11. Logic tree for failure causes and recommended actions for mechanical seals [39].

- **Maintenance Schedule Optimisation**

This function represents a core element of the tool and is responsible for optimising maintenance strategies based on the estimated lifespan of each component and predefined reliability thresholds. The tool considers three major components—bearings, mechanical seals, and couplings—each with its own replacement frequency derived from lifespan calculations.

Two types of maintenance schedules are evaluated. The first is the Optimal Replacement Frequency approach, which treats each component independently and determines its most effective replacement interval to maximise its lifespan. The second is the Unified Maintenance Schedule, in which the tool identifies the component with the shortest lifespan and aligns the replacement of other components with this timeframe. This approach seeks to consolidate maintenance actions within shared downtime periods, thereby reducing the total number of shutdowns and minimising cumulative disruption. This concept is illustrated in Figure 6.

The tool performs a cost analysis for both strategies, incorporating both the direct maintenance costs and the indirect operational losses associated with downtime. The total routine cost is calculated using the following equation:

$$\text{Total routine cost} = \text{Total maintenance cost} + (\text{Total shutdown days} \times \text{Daily operational cost}) \quad (5)$$

Users are required to input values such as the shutdown duration for each component, maintenance costs, and daily operational losses due to unavailability. A simple if–then algorithm is then used to compare all options and select the maintenance routine that yields the lowest total cost. This schedule is applied to optimise maintenance planning over a 10-year period extending to the year 2034.

3.4. Development Platform and Validation

The tool was implemented using Microsoft Excel. It uses built-in functions and logical operations to analyse data and produce recommendations. The tool was validated through two procedures: a case application using data from pump G221 and expert reviews by oil and gas industry engineers to assess its usability and practical relevance.

4. Case Study Results and Discussion

The proposed methodology and the developed tool were tested using data from NGL Pump G221. This section presents the case study's outcome and discusses the results' technical validity. The analysis focuses on generating the optimised maintenance plan and the accuracy of the computed reliability metrics.

4.1. MTBFs of Components

The computed MTBFs for the pump's bearings shows close agreement with the OEM's five-year estimated lifetime. Specifically, the inboard bearing recorded an MTBFs of 6.5 years, while the outboard bearings showed an MTBFs of 4.3 years. These values are compared with OEM estimates in Figure 12.

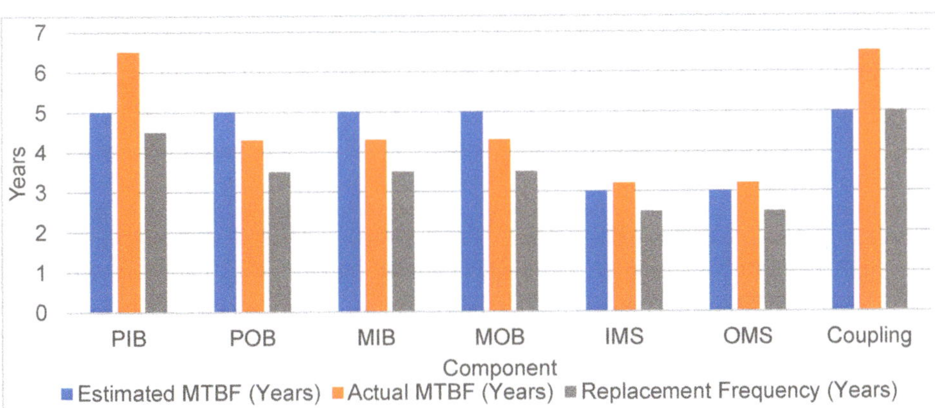

Figure 12. Components' MTBFs comparisons and replacement frequencies—replacement frequencies are shown to be shorter than MTBFs values.

The slight variation in bearing MTBFs values may be attributed to differences in lubrication management and installation accuracy. Each bearing operates with its own oil reservoir, making lubrication levels and prioritisation critical to operational life. Additionally, factors such as alignment precision and clearance tolerances during installation contribute significantly to overall bearing reliability.

For the mechanical seals, the tool estimated an MTBFs of 3.2 years, which closely aligns with both API 682 and OEM expectations of approximately three years. The minor deviation may be explained by the high purity of process conditions maintained at the Saudi Aramco NGL facility.

Although the coupling does not have a specified OEM service life, the estimated MTBFs of 6.5 years is considered acceptable compared to values reported in the literature, which suggest an average lifespan of approximately five years [44]. These findings indicate that the historical input data are consistent and technically adequate for generating reliable MTBFs estimates.

4.2. Optimised Component Replacement Frequencies

This part of the case study evaluates the relationship between reliability thresholds and optimised component replacement intervals. A reliability threshold of 60% was applied to the analysis. The resulting replacement frequencies were determined based on optimising component lifespan while maintaining an acceptable probability of failure.

In the oil and gas industry, unplanned downtime generally incurs higher costs than scheduled component replacement. Therefore, selecting an appropriate reliability threshold is essential to reduce unexpected outages while maintaining cost-effectiveness. Although a 60% reliability implies a 40% probability of failure within the replacement interval, the historical records confirm that no components have failed within the predicted replacement times. This outcome validates the practical suitability of the selected reliability threshold.

Moreover, the historical failure data display limited variation, reinforcing the use of MTBFs as a dependable estimate for lifespan calculation. This provides additional confidence that the computed replacement frequencies, which are intentionally shorter than the MTBFs, are both realistic and conservative. Figure 12 illustrates the comparison between the calculated MTBFs values and the replacement frequencies generated by the tool.

4.3. Component Condition

The tool assessed the condition of the pump's major components—bearings, mechanical seals, and coupling—using current sensor data and corresponding evaluation algorithms.

4.3.1. Bearings

The analysis provides two key outputs: the current condition of each bearing and its estimated RUL. All the bearings are within acceptable condition limits, as confirmed by vibration readings that comply with ISO and Saudi Aramco standards. Due to the limited historical dataset, a rule-based algorithm was applied instead of machine learning models.

RUL estimations were generated using the current vibration values and Equation (1), which incorporates the parameter "n". This parameter has a documented correlation with bearing vibration and can potentially be predicted using regression-based machine learning if sufficient data become available. In this study, a linear relationship was observed between initial vibration readings and "n", yielding an R^2 value of 0.78 (as shown in Figure 8), which supports the use of this approach.

Although the maintenance plan calls for all the bearings to be replaced in 2024, the tool recommends extending the service life of three bearings until 2026. These bearings were replaced in 2023 and have operated for only one year; however, statistical and reliability calculations suggest scheduling their replacement during the planned shutdown in 2024. In contrast, the outboard (OB) bearing, replaced in 2022, shows a shorter RUL and does not qualify for the extension.

4.3.2. Mechanical Seals

The inboard (I) and outboard (O) mechanical seals (MSs) were assessed based on leakage readings. Although the seals are typically replaced on a schedule due to their unpredictable failure behaviour, the tool offers real-time condition insights to prevent unexpected failures. Fuzzy logic was applied to evaluate their condition, as the thresholds are less clearly defined than for bearings.

The results show that the IB mechanical seal remains in normal condition, with a current leakage of 2 psiG. This aligns with the historical data, which show an initial leakage in the 0–1 psiG range and average lifespans of approximately four years. The current IB seal was installed in 2023 and remains within expected parameters.

The OB seal, however, is flagged as alarming, with a leakage of 8 psiG. This is not yet considered faulty, as the leakage pipe's safety valve is set to release at 15 psiG. The trend is consistent with the statistically estimated lifespan. Historical records indicate that the OB seal's service life has declined from five years to three years over recent replacements, which may warrant a further inspection of the sealing system.

4.3.3. Coupling

The coupling condition remains acceptable according to the visual inspection data. It was replaced in 2023 and has only one year of service, which is consistent with its expected service life.

4.4. Maintenance Optimisation

The results shown in Table 4 demonstrate that the optimised maintenance routine (Route 2) is significantly more cost-effective than the traditional approach (Route 1), in which each component is maintained independently based on its individual estimated lifespan and condition. Although Route 1 reflects a component-level optimisation approach, it does not consolidate maintenance activities into shared downtime windows. Consequently, this approach results in substantially higher total costs—up to six times greater than Route 2. The elevated cost arises primarily from operational losses due to frequent pump downtimes, which far exceed the cost of component replacement in the oil and gas sector.

Table 4. Optimised route cost-effectiveness comparison (Unit: USD).

	Route 1	**Route 2**
Pump IB Bearing	1,507,364	1250
Pump OB Bearing	2,261,046	2500
Motor IB Bearing	2,261,046	2500
Motor OB Bearing	2,261,046	2500
IB Mechanical Seal	3,023,763	10,000
OB Mechanical Seal	3,023,763	10,000
Coupling	1,508,267	2000
Total	1,584,6293	2,530,750
Route 1/Route 2	6.26	

In contrast, Route 2 consolidates component replacements into common shutdown periods. This reduces the total downtime by approximately 80% and leads to an overall cost reduction of the same magnitude. Notably, Route 2 includes more frequent component replacements than Route 1, but this trade-off is economically favourable due to the dominance of downtime costs. The result reflects the conservative assumption that cost savings are driven largely by uptime optimisation, rather than spare part conservation. Figure 13 highlights the increase in the number of component replacements when comparing Route 1 and Route 2 over a ten-year period.

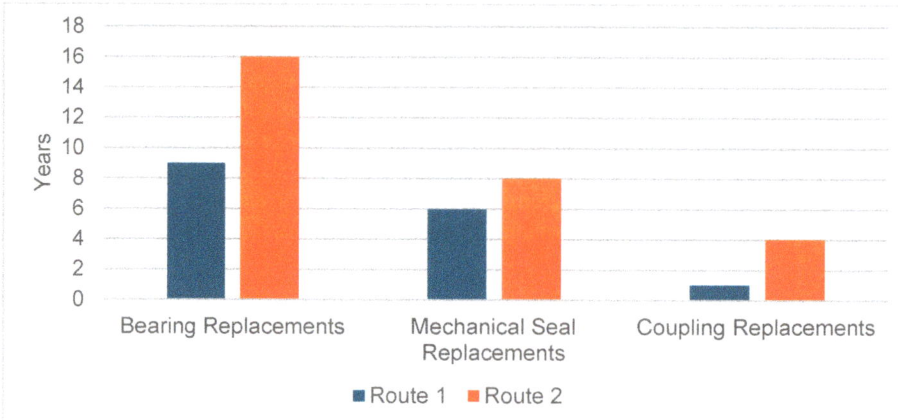

Figure 13. Route 1 and Route 2 comparison in terms of component replacements over 10 years.

The optimised schedule produced by Route 2 provides replacement dates for each component from 2024 to 2034. This long-range forecast supports maintenance planning and spare part availability. Although cost inputs such as replacement costs and daily operational losses are user-defined, the values used in this case study reflect actual NGL plant figures and are consistent with the published literature.

4.5. Diagnosis and Recommended Actions

While no actual replacements of bearings or mechanical seals occurred at the time of testing, the diagnostic and recommendation functions of the tool were validated using representative dummy datasets. These test scenarios confirmed that the tool performs as expected in identifying fault causes and recommending corrective actions. Figure 14 illustrates example outputs for bearing and mechanical seal diagnostics, respectively.

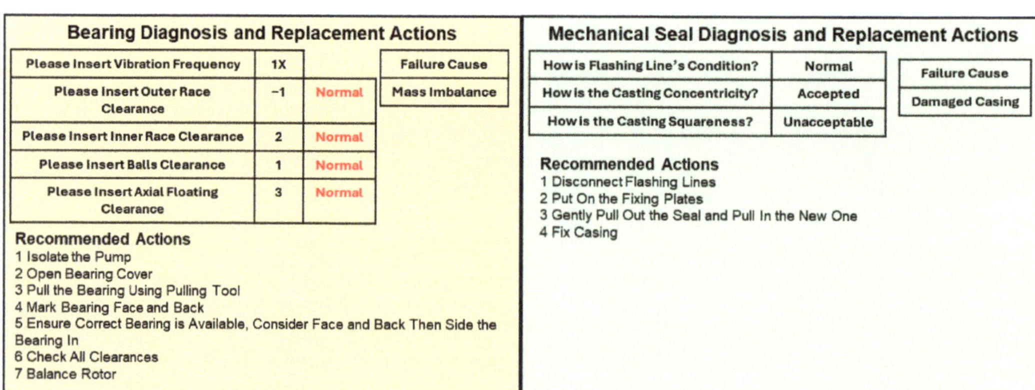

Figure 14. Diagnostic output for bearings and mechanical seals in the case study.

4.6. Comparison with Existing Studies

To further evaluate the distinctiveness and practical value of the proposed methodology, this section presents a comparative analysis against recent predictive maintenance approaches reported in the literature.

Most recent advances focus on specific aspects, such as a single component or failure prediction, and lack integrated decision support [45]. Kumar et al. [46] proposed a digital twin approach with domain adaptation to identify bearing defects. Although the method effectively handles data scarcity and adapts simulated knowledge to real systems, it remains focused on a single failure type and does not assist with maintenance execution. By comparison, our approach supports multi-component diagnostics.

Mohammed [47] developed a data-driven model using multiple linear regression to predict failures in seawater pumps. In contrast, our method not only predicts potential failures but also diagnoses fault causes and translates predictions into cost-driven, actionable maintenance schedules. Similarly, Souza et al. [48] utilised CNNs to detect faults in offshore centrifugal pumps. While the model achieved a high classification accuracy, it offered no guidance for maintenance planning and did not address how to manage operational downtime. Our method improves upon this by directly linking diagnostic results to optimised maintenance interventions, including the consolidation of shutdowns and the prioritisation of tasks based on their cost and risk. In addition, Upasane et al. [49] developed a type-2, fuzzy-based, explainable AI system to improve transparency in predictive models. The emphasis on interpretability is valuable, particularly for user trust. However, their method does not incorporate any cost modelling or multi-component planning. Our method maintains model interpretability while extending functionality to include coordinated task scheduling and cost-effective intervention strategies.

In summary, most recent studies stop short of translating prediction into specific, cost-informed actions. Our framework closes this gap by linking prediction, diagnosis, and scheduling within a single, unified maintenance tool that delivers operational benefits under constrained data conditions.

5. Validation Questionnaire Results and Discussion

Following the technical testing of the developed tool, a validation questionnaire was distributed to a selected group of 15 industry experts. The objective of this survey was to evaluate the methodology and the practical performance of the tool from an industrial perspective. This section presents the expert feedback and provides a discussion of the validation results.

The questionnaire consisted of 10 questions, each targeting a specific aspect of the tool's value and robustness. Topics included the business relevance of such a predictive maintenance system, the validity of underlying assumptions, the ease of implementation, and the effectiveness of the diagnostic functionalities. The responses were captured using a four-point Likert scale: "Strongly Agree", "Agree", "Neutral" (indicating partial familiarity with the topic), and "Disagree". This design was deliberately selected to avoid a true neutral midpoint, thereby encouraging respondents to make a clear evaluative choice. Such forced-choice formats are particularly effective when collecting expert feedback on prototype tools, as they reduce the central tendency bias and yield more actionable results.

The expert panel comprised 15 professionals from major industrial organisations in the Middle East region, as summarised in Table 5. Their areas of expertise ranged from engineering to management and technical operations, with most of the respondents being affiliated with either Saudi Aramco or SABIC. The distribution of their expertise is shown below.

Table 5. Subject matter experts' details.

Number of Experts	Expertise Field	Employer
5	Maintenance engineers	Saudi Aramco
4	Maintenance engineers	SABIC
3	Maintenance Managers	Saudi Aramco
3	Maintenance Technicians	Saudi Aramco

5.1. The Need for an AI Tool to Optimise NGL Pump Maintenance

- **Question 1: There is a business need to deploy AI for NGL pump maintenance optimisation instead of manual optimisation.**

The experts' responses to this question were highly consistent. Fourteen out of fifteen participants selected "Strongly Agree", indicating a clear and unanimous recognition of the value and necessity of AI-based maintenance optimisation in the NGL pump context. One respondent selected "Neutral", which was interpreted as reflecting limited familiarity with AI technologies, rather than disagreement with the concept.

This strong consensus aligns with the current gap in the literature, where existing approaches primarily focus on predictive modelling or fault detection in isolation, with little emphasis on integrated, AI-supported decision tools tailored to the scheduling and operational realities of NGL systems. The expert feedback, therefore, reinforces both the relevance and the timeliness of the proposed tool.

5.2. Assumption Technical Validation

- **Question 2: The assumptions regarding bearing conditions are accurate.**

All the subject matter experts strongly agreed that the assumptions used to assess bearing conditions based on vibration measurements are technically valid. This consensus aligns with the clearly defined thresholds provided in the ISO and Saudi Aramco standards.

- **Question 3: The correlation used to estimate bearing RUL based on vibration readings is applicable and provides a fair estimation.**

Figure 15 presents the results of expert opinions regarding the Questions 3 to 8. All the maintenance engineers expressed strong agreement regarding the applicability of the correlation in Equation (1) for estimating the RUL. They confirmed the appropriateness of using vibration readings and supported the use of a linear relationship between the initial vibration and the coefficient n. The maintenance managers also agreed once the theoretical basis of the correlation was explained in line with the literature. However, some of the maintenance technicians selected "Neutral", citing that the correlation is not explicitly recognised in Saudi Aramco's current standards. Nevertheless, the overall feedback supports the validity and applicability of the correlation approach used in the tool.

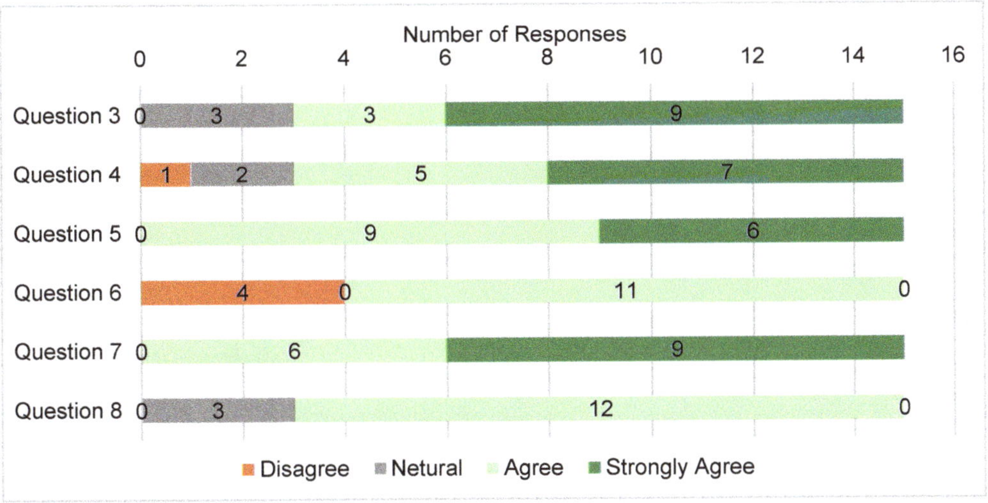

Figure 15. Questionnaire results for Questions 3 to 8.

- **Question 4: The adopted approach reflects mechanical seals' condition accurately.**

Most of the experts (12 of 15) supported the use of fuzzy logic and the defined fuzzy sets for classifying mechanical seal conditions. The Saudi Aramco engineers and two of the maintenance managers strongly endorsed the approach. In contrast, the SABIC engineers agreed in principle but noted that their organisation does not apply fixed criteria for seal condition statuses, which affects their evaluation. One maintenance technician endorsed the approach based on consultation with engineers, while another chose "Neutral" due to limited familiarity with seal condition assessment. One maintenance manager expressed disagreement, citing a belief that seals should be replaced once leakage reaches 5 psiG—an approach inconsistent with current standards. Despite a range of perspectives, the majority of the experts validated the fuzzy logic methodology used in the tool.

5.3. *Tool Practicality*

- **Question 5: The proposed methodology helps reduce the total cost significantly.**

All the field experts—including the maintenance managers and the technicians—strongly agreed with the methodology of aligning component maintenance within shared shutdown periods. Many noted that this practice is informally adopted even in the absence of formal directives from engineers. The engineering staff also agreed with the principle, recognising its potential to reduce the overall maintenance costs through improved scheduling.

- **Question 6: The tool is simple and user-friendly.**

All the Saudi Aramco experts agreed that the tool is intuitive and easy to operate. However, the SABIC engineers provided a more critical assessment, suggesting that integration with the SAP maintenance system would enhance its usability. It is worth noting that, unlike SABIC, Saudi Aramco enforces strict IT policies that prohibit third-party software from directly interfacing with SAP, which influences perceptions of tool compatibility.

- **Question 7: The methodology and developed tool address the business need for using AI to optimise NGL pump maintenance.**

Both the Saudi Aramco and the SABIC maintenance engineers strongly agreed that the methodology and tool address the core business need for AI-driven maintenance optimisation. The field technicians also acknowledged the tool's practical contribution to improving NGL pump availability and reducing costs.

Collectively, the responses to Questions 1 through 7 provide a secondary validation of the tool's effectiveness and alignment with the operational goals. This complements the primary validation obtained through the technical testing of the tool.

5.4. Tool Diagnostic Functionality

- **Question 8: The tool provides useful diagnoses and recommended actions that reduce pump downtime.**

Both the Saudi Aramco and the SABIC engineers agreed that the tool delivers effective diagnostic insights and maintenance recommendations that help reduce reliance on manual inspection processes and contribute to shorter downtimes. This feedback confirms the technical soundness of the diagnostic module.

In contrast, the maintenance managers and the technicians selected "Neutral", primarily because they are not directly involved in diagnostic tasks for NGL pumps and thus could not fully assess this aspect of the tool's functionality.

6. Conclusions

This study introduced an AI-based methodology and digital tool specifically developed to optimise the maintenance of NGL pumps. Unlike conventional approaches that focus solely on failure prediction or isolated condition monitoring, the proposed framework integrates reliability-based scheduling, real-time condition assessment, and AI-driven diagnostics within a unified system. This comprehensive approach allows for the development of optimised maintenance routines that balance component longevity with the operational cost, even under limited data availability—a common constraint in industrial settings.

The significance of the work lies in its ability to reduce unplanned downtime and operational losses by aligning component maintenance within consolidated shutdown windows. In the case study involving pump G221, the optimised routine reduced the total cost by approximately 80% compared to conventional individual replacement scheduling. This improvement was primarily achieved by minimising downtime, which was reduced by a similar margin. Furthermore, while the number of replacements increased under the optimised plan, the overall routine cost remained significantly lower due to reduced production losses.

The tool was validated through both technical application and structured expert feedback from 15 industry professionals. The survey results showed that over 90% of the experts strongly agreed with the need for such a tool and acknowledged its effectiveness in addressing key operational and diagnostic challenges.

While this study focuses on a single equipment type (NGL pumps), the proposed methodology is designed to be generalisable to other rotating assets, such as compressors and fans. Future work will involve extending the tool's application to a broader range of equipment under varied operational conditions.

Author Contributions: Conceptualisation, A.A. and M.K.; methodology, F.H., A.A. and M.K.; validation, A.A. and M.K.; formal analysis, A.A., F.H. and M.K.; investigation, A.A.; resources, A.A. and M.K.; data curation, A.A., F.H. and M.K.; writing—original draft preparation, A.A. and F.H.; writing—review and editing, F.H. and M.K.; visualisation, A.A. and F.H.; supervision, F.H. and M.K.;

project administration, F.H. and M.K.; funding acquisition, M.K. All authors have read and agreed to the published version of the manuscript.

Funding: This research received no external funding.

Data Availability Statement: The original contributions presented in this study are included in the article. Further inquiries can be directed to the corresponding authors.

Conflicts of Interest: The authors declare no conflicts of interest.

Nomenclature

AI	Artificial Intelligence
API	American Petroleum Institute
ANN	artificial neural network
CCW	Counter-Clockwise
CNN	Convolutional Neural Network
C-MAPSS	Commercial Modular Aero-Propulsion System Simulation
FMEA	Failure Modes and Effects Analysis
IMS	Inboard Mechanical Seal
ISO	International Organization for Standardization
LSTM	Long Short-Term Memory
MOB	Motor Outboard Bearing
MEP	mechanical, electrical, and plumbing
MIB	Motor Inboard Bearing
MTBF	mean time between failures
NGL	Natural Gas Liquid
OB	outboard
OEM	Original Equipment Manufacturer
OMS	Outboard Mechanical Seal
PIB	Pump Inboard Bearing
POB	Pump Outboard Bearing
RAM	Reliability, Availability, and Maintainability
RPN	risk priority number
RUL	remaining useful life
SAP	Systems, Applications, and Products (in Data Processing)
SVM	support vector machine

References

1. Average Cost of Downtime per Industry. Available online: https://www.pingdom.com/outages/average-cost-of-downtime-per-industry/ (accessed on 22 April 2025).
2. Gulati, R. *Maintenance & Reliability Best Practices*, 2nd ed.; Industrial Press: New York, NY, USA, 2012; p. 162.
3. Sharma, A.; Yadava, G.S.; Deshmukh, S.G. A Literature Review and Future Perspectives on Maintenance Optimization. *J. Qual. Maint. Eng.* **2011**, *17*, 5–25. [CrossRef]
4. Van Horenbeek, A.; Pintelon, L.; Muchiri, P. Maintenance Optimization Models and Criteria. *Int. J. Syst. Assur. Eng. Manag.* **2010**, *1*, 189–200. [CrossRef]
5. de Jonge, B.; Scarf, P.A. A Review on Maintenance Optimisation. *Eur. J. Oper. Res.* **2020**, *285*, 805–824. [CrossRef]
6. Lipol, L.S.; Haq, J. Risk Analysis Method: FMEA/FMECA in the Organizations. *Int. J. Basic Appl. Sci.-Int. J. Eng. Sci.* **2011**, *11*, 74–82.
7. Clavijo Mesa, M.V.; Patino-Rodriguez, C.E.; Guevara Carazas, F.J.; Gunawan, I.; Droguett, E.L. Asset Management Strategies Using Reliability, Availability, and Maintainability (RAM) Analysis. *J. Braz. Soc. Mech. Sci. Eng.* **2021**, *43*, 495. [CrossRef]
8. Hauser, A.; Fenski, B.; Cavalli, L. Maximise Asset Availability and Reduce Maintenance Costs—An Integrated Approach Combining Condition Assessment with Data Analytics. *CIRED Open Access Proc. J.* **2017**, *2017*, 316–319. [CrossRef]
9. Zhang, S.; Yan, Y.; Wang, P.; Xu, Z.; Yan, X. Sustainable Maintainability Management Practices for Offshore Assets: A Data-Driven Decision Strategy. *J. Clean. Prod.* **2019**, *237*, 117730. [CrossRef]

10. Errandonea, I.; Beltrán, S.; Arrizabalaga, S. Digital Twin for Maintenance: A Literature Review. *Comput. Ind.* **2020**, *123*, 103316. [CrossRef]
11. Magargle, R.; Johnson, L.; Mandloi, P.; Davoudabadi, P.; Kesarkar, O.; Krishnaswamy, S.; Batteh, J.; Pitchaikani, A. A Simulation-Based Digital Twin for Model-Driven Health Monitoring and Predictive Maintenance of an Automotive Braking System. In Proceedings of the 12th International Modelica Conference, Prague, Czech Republic, 15–17 May 2017; Volume 132.
12. Canizo, M.; Onieva, E.; Conde, A.; Charramendieta, S.; Trujillo, S. Real-Time Predictive Maintenance for Wind Turbines Using Big Data Frameworks. In Proceedings of the 2017 IEEE International Conference on Prognostics and Health Management, ICPHM 2017, Dallas, TX, USA, 19–21 June 2017.
13. Fleet, T.; Kamei, K.; He, F.; Khan, M.A.; Khan, K.A.; Starr, A. A Machine Learning Approach to Model Interdependencies between Dynamic Response and Crack Propagation. *Sensors* **2020**, *20*, 6847. [CrossRef]
14. Tiwari, R.; Bordoloi, D.J.; Dewangan, A. Blockage and Cavitation Detection in Centrifugal Pumps from Dynamic Pressure Signal Using Deep Learning Algorithm. *Measurement* **2021**, *173*, 108676. [CrossRef]
15. Prognostic Health, H.; Carvalho, A.; Aliyu, R.; Akmar Mokhtar, A.; Hussin, H. Prognostic Health Management of Pumps Using Artificial Intelligence in the Oil and Gas Sector: A Review. *Appl. Sci.* **2022**, *12*, 11691. [CrossRef]
16. Garg, A.; Vijayaraghavan, V.; Tai, K.; Singru, P.M.; Jain, V.; Krishnakumar, N. Model Development Based on Evolutionary Framework for Condition Monitoring of a Lathe Machine. *Measurement* **2015**, *73*, 95–110. [CrossRef]
17. Amruthnath, N.; Gupta, T. A Research Study on Unsupervised Machine Learning Algorithms for Early Fault Detection in Predictive Maintenance. In Proceedings of the 2018 5th International Conference on Industrial Engineering and Applications, ICIEA 2018, Singapore, 26–28 April 2018.
18. Paolanti, M.; Romeo, L.; Felicetti, A.; Mancini, A.; Frontoni, E.; Loncarski, J. Machine Learning Approach for Predictive Maintenance in Industry 4.0. In Proceedings of the 2018 14th IEEE/ASME International Conference on Mechatronic and Embedded Systems and Applications, MESA 2018, Oulu, Finland, 2–4 July 2018.
19. Cheng, J.C.P.; Chen, W.; Chen, K.; Wang, Q. Data-Driven Predictive Maintenance Planning Framework for MEP Components Based on BIM and IoT Using Machine Learning Algorithms. *Autom. Constr.* **2020**, *112*, 103087. [CrossRef]
20. Çinar, Z.M.; Nuhu, A.A.; Zeeshan, Q.; Korhan, O.; Asmael, M.; Safaei, B. Machine Learning in Predictive Maintenance towards Sustainable Smart Manufacturing in Industry 4.0. *Sustainability* **2020**, *12*, 8211. [CrossRef]
21. Prytz, R.; Nowaczyk, S.; Rögnvaldsson, T.; Byttner, S. Predicting the Need for Vehicle Compressor Repairs Using Maintenance Records and Logged Vehicle Data. *Eng. Appl. Artif. Intell.* **2015**, *41*, 139–150. [CrossRef]
22. Mechanical Shaft Seals for Pumps—Grundfos. Available online: https://api.grundfos.com/literature/Grundfosliterature-5768950.pdf (accessed on 22 April 2025).
23. Falamarzi, A.; Moridpour, S.; Nazem, M.; Cheraghi, S. Prediction of Tram Track Gauge Deviation Using Artificial Neural Network and Support Vector Regression. *Aust. J. Civ. Eng.* **2019**, *17*, 63–71. [CrossRef]
24. Susto, G.A.; Beghi, A.; De Luca, C. A Predictive Maintenance System for Epitaxy Processes Based on Filtering and Prediction Techniques. *IEEE Trans. Semicond. Manuf.* **2012**, *25*, 638–649. [CrossRef]
25. Abu-Samah, A.; Shahzad, M.K.; Zamai, E.; Ben Said, A. Failure Prediction Methodology for Improved Proactive Maintenance Using Bayesian Approach. *IFAC-PapersOnLine* **2015**, *28*, 844–851. [CrossRef]
26. Mathew, V.; Toby, T.; Singh, V.; Rao, B.M.; Kumar, M.G. Prediction of Remaining Useful Lifetime (RUL) of Turbofan Engine Using Machine Learning. In Proceedings of the IEEE International Conference on Circuits and Systems, ICCS 2017, Thiruvananthapuram, India, 20–21 December 2017; Volume 2018.
27. Lias, M.R.; Rao, T.V.V.L.N.; Awang, M.; Khan, M.A. The Stress Distribution of Gear Tooth due to Axial Misalignment Condition. *J. Appl. Sci.* **2012**, *12*, 2404–2410. [CrossRef]
28. Janssens, O.; Van De Walle, R.; Loccufier, M.; Van Hoecke, S. Deep Learning for Infrared Thermal Image Based Machine Health Monitoring. *IEEE/ASME Trans. Mechatron.* **2018**, *23*, 151–159. [CrossRef]
29. Scalabrini Sampaio, G.; Vallim Filho, A.R.d.A.; Santos da Silva, L.; Augusto da Silva, L. Prediction of Motor Failure Time Using An Artificial Neural Network. *Sensors* **2019**, *19*, 4342. [CrossRef]
30. Bekar, E.T.; Nyqvist, P.; Skoogh, A. An Intelligent Approach for Data Pre-Processing and Analysis in Predictive Maintenance with an Industrial Case Study. *Adv. Mech. Eng.* **2020**, *12*, 1687814020919207. [CrossRef]
31. Praveenkumar, T.; Saimurugan, M.; Krishnakumar, P.; Ramachandran, K.I. Fault Diagnosis of Automobile Gearbox Based on Machine Learning Techniques. *Procedia Eng.* **2014**, *97*, 2092–2098. [CrossRef]
32. Durbhaka, G.K.; Selvaraj, B. Predictive Maintenance for Wind Turbine Diagnostics Using Vibration Signal Analysis Based on Collaborative Recommendation Approach. In Proceedings of the 2016 International Conference on Advances in Computing, Communications and Informatics, ICACCI 2016, Jaipur, India, 21–24 September 2016.

33. Su, C.J.; Huang, S.F. Real-Time Big Data Analytics for Hard Disk Drive Predictive Maintenance. *Comput. Electr. Eng.* **2018**, *71*, 93–101. [CrossRef]
34. Butte, S.; Prashanth, A.R.; Patil, S. Machine Learning Based Predictive Maintenance Strategy: A Super Learning Approach with Deep Neural Networks. In Proceedings of the 2018 IEEE Workshop on Microelectronics and Electron Devices, WMED 2018, Boise, ID, USA, 20 April 2018.
35. Gosavi, A.; Le, V.K. Maintenance Optimization in a Digital Twin for Industry 4.0. *Ann. Oper. Res.* **2024**, *340*, 245–268. [CrossRef]
36. Zai, B.A.; Khan, M.A.; Khan, K.; Nisar, S.; Shahzad, M.; Shah, A. The Role of Dynamic Response Parameters in Damage Prediction. *Proc. Inst. Mech. Eng. Part C J. Mech. Eng. Sci.* **2019**, *233*, 4620–4636. [CrossRef]
37. Zai, B.A.; Khan, M.A.; Khan, K.A.; Mansoor, A. A Novel Approach for Damage Quantification Using the Dynamic Response of a Metallic Beam under Thermo-Mechanical Loads. *J. Sound Vib.* **2020**, *469*, 115134. [CrossRef]
38. Zai, B.A.; Mansoor, A.; Siddiqui, U.A.; Javed, J.; us Saqib, N. Damage Quantification of a Metallic Beam under Thermo-Mechanical Loads Using Novel Empirical Correlations and Neural Network. *Noise Vib. Worldw.* **2024**, *55*, 68–83. [CrossRef]
39. Zai, B.A.; Khan, M.A.; Khan, S.Z.; Asif, M.; Khan, K.A.; Saquib, A.N.; Mansoor, A.; Shahzad, M.; Mujtaba, A. Prediction of Crack Depth and Fatigue Life of an Acrylonitrile Butadiene Styrene Cantilever Beam Using Dynamic Response. *J. Test. Eval.* **2020**, *48*, 1520–1536. [CrossRef]
40. Kamei, K.; Khan, M.A. Current Challenges in Modelling Vibrational Fatigue and Fracture of Structures: A Review. *J. Braz. Soc. Mech. Sci. Eng.* **2021**, *43*, 77. [CrossRef]
41. *API 610/ISO 13709:2009*; Centrifugal Pumps for Petroleum, Petrochemical and Natural Gas Industries. International Organization for Standardization (ISO): Geneva, Switzerland, 2009.
42. *ISO 20816-1:2016*; Mechanical Vibration—Measurement and Evaluation of Machine Vibration. International Organization for Standardization (ISO): Geneva, Switzerland, 2016.
43. Huebner, M.; Buck, G.; Azibert, H. Advancements in Mechanical Sealing—API 682 Fourth Edition. In Proceedings of the Twenty-Ninth International Pump Users Symposium, Houston, TX, USA, 1–3 October 2013.
44. Jeffrey, D. *Principles of Machine Operation and Maintenance*; Routledge: London, UK, 2013.
45. Crespo Márquez, A.; Crespo Del Castillo, A.; Gómez Fernández, J.F. Integrating Artificial Intelligent Techniques and Continuous Time Simulation Modelling. Practical Predictive Analytics for Energy Efficiency and Failure Detection. *Comput. Ind.* **2020**, *115*, 103164. [CrossRef]
46. Kumar, A.; Kumar, R.; Xiang, J.; Qiao, Z.; Zhou, Y.; Shao, H. Digital Twin-Assisted AI Framework Based on Domain Adaptation for Bearing Defect Diagnosis in the Centrifugal Pump. *Measurement* **2024**, *235*, 115013. [CrossRef]
47. Mohammed, A. Data Driven-Based Model for Predicting Pump Failures in the Oil and Gas Industry. *Eng. Fail. Anal.* **2023**, *145*, 107019. [CrossRef]
48. e Souza, A.C.O.; de Souza, M.B.; da Silva, F.V. Development of a CNN-Based Fault Detection System for a Real Water Injection Centrifugal Pump. *Expert Syst. Appl.* **2024**, *244*, 122947. [CrossRef]
49. Upasane, S.J.; Hagras, H.; Anisi, M.H.; Savill, S.; Taylor, I.; Manousakis, K. A Type-2 Fuzzy-Based Explainable AI System for Predictive Maintenance Within the Water Pumping Industry. *IEEE Trans. Artif. Intell.* **2024**, *5*, 490–504. [CrossRef]

processes

MDPI

Article

Artificial Intelligence and Extraction of Bioactive Compounds: The Case of Rosemary and Pressurized Liquid Extraction

Martha Mantiniotou [1], Vassilis Athanasiadis [1], Konstantinos G. Liakos [2], Eleni Bozinou [1] and Stavros I. Lalas [1,*]

[1] Department of Food Science and Nutrition, University of Thessaly, Terma N. Temponera Street, 43100 Karditsa, Greece; mmantiniotou@uth.gr (M.M.); vaathanasiadis@uth.gr (V.A.); empozinou@uth.gr (E.B.)

[2] Department of Electrical and Computer Engineering, University of Thessaly, Sekeri Street, 38334 Volos, Greece; kliakos@uth.gr

[*] Correspondence: slalas@uth.gr; Tel.: +30-24410-64783

Abstract: Rosemary (*Rosmarinus officinalis* or *Salvia rosmarinus*) is an aromatic herb that possesses numerous health-promoting and antioxidant properties. Pressurized Liquid Extraction (PLE) is an efficient, environmentally friendly technique for obtaining valuable compounds from natural sources. The optimal PLE conditions were established as 25% *v/v* ethanol at 160 °C for 25 min, and a liquid-to-solid ratio of 10 mL/g. The optimal extract exhibited high polyphenol and antioxidant content through various assays. The recovered bioactive compounds possess potential applications in the food, pharmaceutical, and cosmetics sectors, in addition to serving as feed additives. This research compares two distinct optimization models: one statistical, derived from experimental data, and the other based on artificial intelligence (AI). The objective was to evaluate if AI could replicate experimental models and ultimately supplant the laborious experimental process, yielding the same results more rapidly and adaptably. To further enhance data interpretation and predictive capabilities, six machine learning models were implemented on the original dataset. Due to the limited sample size, synthetic data were generated using Random Forest (RF)-based resampling and Gaussian noise addition. The augmented dataset significantly improved the model performance. Among the models tested, the RF algorithm achieved the highest accuracy.

Keywords: *Rosmarinus officinalis*; polyphenols; antioxidants; HPLC-DAD; response surface methodology; machine learning; regression models; generative models; random forest; ensemble learning

1. Introduction

Rosemary (*Rosmarinus officinalis* L.), a perennial species of the Lamiaceae family, is distinguished for its distinctive scent, culinary use, and therapeutic properties [1]. It has been established by phylogenetic studies that rosemary is classified within the genus Salvia, specifically referred to as *Salvia rosmarinus* [2]. Rosemary originates from the Mediterranean region; however, it has been cultivated successfully in numerous other locations globally [3]. This is an aromatic plant characterized by its needle-like foliage, widely cultivated around the world [3]. Rosemary's therapeutic properties have been utilized in traditional folk medicine to address a range of ailments, such as pain relief, headaches, stomach discomfort, respiratory disorders, and others [1,4–6].

Several methodologies have been investigated for the recovery of bioactive compounds, mainly rosmarinic acid and carnosic acid, from rosemary leaves at a laboratory scale. Certain extraction processes, especially conventional methods, are often associated with various disadvantages, including the utilization of hazardous solvents, the degradation of target compounds resulting from elevated temperatures, prolonged extraction durations, challenges in implementation, and significant economic and energy expenditures. In recent years, the concepts of "Green chemistry" and "eco-extraction" have emerged [7]. Recent studies indicate that extraction processes have become more energy-efficient, safer for users, and environmentally friendly compared to previous methods, all while maintaining extraction efficiency. The intensification of extraction processes, considering these various aspects, should emerge as a new challenge for the design of such processes.

Despite extensive research on rosemary leaf extracts, gaps remain in exploring less-studied aspects and emerging opportunities driven by technological advancements. One such critical area is the use of green, non-toxic solvents for the sustainable recovery of bioactive compounds from plant materials. This study aims to bridge this gap by investigating the combined effects of green solvent mixtures, optimizing extraction conditions, and evaluating their performance. Rosemary leaves were subjected to Pressurized Liquid Extraction (PLE), which combines elevated pressure to enhance mass transfer and elevated temperatures to help facilitate the diffusion of the solvent into the sample by diminishing its viscosity [8]. The study examined the influence of eco-friendly solvent mixtures, specifically water and ethanol, alongside key process parameters such as temperature and extraction duration. Additionally, a partial least squares (PLS) model was utilized to identify the optimal extraction conditions.

In parallel, the growing integration of artificial intelligence (AI), particularly machine learning (ML) and Deep Learning (DL), has enabled more accurate modeling, prediction, and optimization across the food, beverage, pharmaceutical, and cosmetic industries. ML techniques are increasingly applied in bioactive compound prediction, formulation optimization [9], sensory analysis, and green extraction process modeling [10]. Recent studies demonstrate the effectiveness of algorithms such as Random Forest (RF), Support Vector Machines (SVMs), and Artificial Neural Networks (ANNs) in predicting antioxidant capacity, total polyphenol content (TPC), and other physicochemical properties from experimental variables [11]. In this context, the present study incorporates multiple machine learning approaches to predict the antioxidant potential of rosemary extracts under varying PLE conditions, thereby enhancing process efficiency and supporting sustainable product development in food, nutraceutical, and cosmetic applications [12].

While ML methods have been increasingly used for modeling extraction processes, most prior studies rely on relatively large experimental datasets or focus on specific extraction methods with extensive data availability. In contrast, PLE of rosemary leaves is a process with high experimental cost and limited data availability, which hinders the effective application of conventional ML techniques. This study addresses this gap by integrating ML models with data augmentation strategies to enhance model robustness under small-sample conditions. To our knowledge, this is one of the first studies applying such an approach to optimize green extraction of bioactive compounds from rosemary, offering insights that can support broader adoption of AI-assisted extraction workflows.

2. Materials and Methods

2.1. Chemicals and Reagents

A deionizing column was used to produce deionized water for all the experiments performed. The deionized column contains mixed-bed ion exchange resin, ensuring conductivity below 1 μS/cm, with a standard flow rate and operating pressure. All polyphenolic standards for the HPLC determination, along with L-ascorbic acid (99%), 2,4,6-tris(2-

pyridyl)-s-triazine (TPTZ) (≥98%), 2,2-diphenyl-1-picrylhydrazyl (DPPH•), and hydrochloric acid (37%), were bought from Sigma-Aldrich (Darmstadt, Germany) and were at least 97% purity or higher. Acetonitrile was acquired from Labkem (Barcelona, Spain). Sodium carbonate (anhydrous, 99.5%), rutin (≥94%), and formic acid (99.8%) were bought from Penta (Prague, Czech Republic). Iron (III) chloride hexahydrate (97%) was obtained from Merck (Darmstadt, Germany). Folin–Ciocalteu reagent, gallic acid (97%), and ethanol (99.8%) were acquired from Panreac Co. (Barcelona, Spain).

2.2. Rosemary Leaves' Raw Material and Sample Preparation

Rosemary leaves were obtained from a local plant shop from the Karditsa region (Central Greece). The rosemary leaves were washed carefully and manually dried with paper towels. Then, they were subjected to lyophilization through a Biobase BK-FD10 (Jinan, China) freeze-drier. The moisture content was determined as 53.2 ± 3.8%. Then, the dried material was sieved in an Analysette 3 PRO (Fritsch GmbH, Oberstein, Germany) sieving machine, and the powder, consisting of an average particle diameter of 497 μm, was obtained. The obtained powder was kept in a freezer at up to –40 °C until further analysis.

2.3. Experimental Design

A custom-designed Response Surface Methodology (RSM) with four factors at five levels was employed to optimize the extraction conditions for TPC, antioxidant activity (FRAP and DPPH assays), and ascorbic acid content (AAC) using the Pressurized Liquid Extraction (PLE) technique on rosemary powder. A Pressurized Liquid Extraction (PLE) system (Fluid Management Systems, Inc., Watertown, MA, USA) was used to facilitate all extractions. The independent variables examined included the ethanol concentration (*C*, % *v/v*) as X_1, liquid-to-solid ratio (*R*, mL/g) as X_2, extraction temperature (*T*, °C) as X_3, and extraction time (*t*, min) as X_4, each assigned five levels. To assess the method's repeatability, 17 experimental runs, including one central point, were conducted, with each run replicated three times, and the average response values were documented for subsequent analysis.

Stepwise regression was utilized to refine the model's predictive precision by reducing variance from superfluous term estimation, leading to a second-order polynomial equation that delineates the interactions between the three independent variables:

$$Y_k = \beta_0 + \sum_{i=1}^{2} \beta_i X_i + \sum_{i=1}^{2} \beta_{ii} X_i^2 + \sum_{i=1}^{2} \sum_{j=i+1}^{3} \beta_{ij} X_i X_j \quad (1)$$

where the independent variables are denoted by X_i and X_j, and the predicted response variable is defined by Y_k. In the model, the intercept and regression coefficients β_0, β_i, β_{ii}, and β_{ij} represent the linear, quadratic, and interaction terms, respectively.

2.4. Total Polyphenolic Content (TPC) Determination Through Spectrophotometric Evaluation

The Folin–Ciocalteu methodology [13] was used to evaluate TPC and express the results in milligrams of gallic acid equivalents (GAEs) per gram of dry weight (dw). A calibration curve (10–100 mg/L of gallic acid, R^2 = 0.9996) in water was used to assess the results. Briefly, after mixing 100 μL of the properly diluted extract with 100 μL of the Folin–Ciocalteu reagent for 2 min, 800 μL of a 5% *w/v* sodium carbonate solution was subsequently added. Following a 20 min incubation at 40 °C, in the absence of light exposure, the absorbance of the solution was measured at 740 nm in a Shimadzu UV-1900i UV/Vis spectrophotometer (Kyoto, Japan). Sample incubation at 40 °C was conducted utilizing an Elmasonic P70H ultrasonic bath from Elma Schmidbauer GmbH (Singen,

Germany). Each analysis was performed in triplicate and the average was used to assess the results.

2.5. Ferric-Reducing Antioxidant Power (FRAP) Evaluation of Antioxidant Activity

A previously established study provides a thorough description of the method used to test the antioxidant capacity of the extracts utilizing the common electron-transfer method [13]. This method entailed identifying the decrease in the iron oxidation state from +3 to +2. Briefly, 50 μL of the properly diluted sample was combined with 50 μL of $FeCl_3$ solution (4 mM in 0.05 M HCl). Subsequently, the samples were incubated at 37 °C for 30 min. After a 5 min interval, 900 μL of TPTZ solution (1 mM in 0.05 M HCl) was added, and the absorbance was measured at 620 nm. A calibration curve of ascorbic acid (50–500 μM in 0.05 M HCl, $R^2 = 0.9997$) was utilized, and the results were expressed as μmol of ascorbic acid equivalents (AAEs) per gram of dw. Each analysis was performed in triplicate and the average was used to evaluate the results.

2.6. Evaluation of Radical Scavenging Activity

A previously described assay [14] for $DPPH^\bullet$ scavenging was employed. The absorbance at 515 nm was initially measured immediately and 30 min later by combining 25 μL of properly diluted sample extract with 975 μL of $DPPH^\bullet$ solution (100 μmol/L in methanol). A calibration curve of the antiradical activity of ascorbic acid (100–1000 μmol/L in methanol, $R^2 = 0.9926$) was used, and the results were expressed as μmol of ascorbic acid equivalents (AAEs) per gram of dw. Each analysis was performed in triplicate and the average was used to evaluate the results.

2.7. HPLC Quantification of Polyphenolic Compounds

High-Performance Liquid Chromatography coupled with Diode Array Detector (HPLC-DAD) identification of individual polyphenols from the rosemary leaves' extracts was based on our prior research [15]. The liquid chromatograph (model CBM-20A) and diode array detector (model SPD-M20A) utilized in this investigation were supplied by Shimadzu Europa GmbH, Duisburg, Germany. The detection wavelength ranges from 200 to 800 nm. The compounds were injected at a volume of 20 μL and separated at 40 °C using a Phenomenex Luna C18(2) column (100 Å, 5 μm, 4.6 mm × 250 mm) from Phenomenex Inc. in Torrance, CA, USA. The mobile phase consisted of 0.5% formic acid in acetonitrile (B) and 0.5% formic acid in aqueous solution (A). The gradient program involved a gradual initiation from 0 and increase to 40% B, followed by 50% B for 10 min, 70% B for another 10 min, and a constant value for 10 min. The mobile phase flow rate was kept constant at 1 mL/min. By comparing the absorbance spectrum and retention time to those of purified standards, the compounds were identified and subsequently quantified using calibration curves (0–50 μg/mL).

2.8. Ascorbic Acid Content (AAC)

The ascorbic acid content of the samples was quantified as mg/g of dry weight, as previously described by Athanasiadis et al. [15]. A total of 500 μL of 10% (*v/v*) Folin–Ciocalteu reagent and 100 μL of sample extract were combined with 900 μL of 10% (*w/v*) trichloroacetic acid in an Eppendorf tube. The absorbance was promptly assessed at 760 nm following 10 min of storage in darkness.

2.9. Statistical Analysis

The RSM and distribution analysis were statistically evaluated utilizing JMP® Pro 16 software (SAS, Cary, NC, USA). The Kolmogorov–Smirnov test assessed the normality of the data. ANOVA and the Tukey HSD multiple comparison test were employed to ascertain any significant differences. The results were reported as means accompanied by measures of variability.

2.10. Initial Data Set Exploration and Visualization

The initial dataset comprised 17 experimental samples of rosemary extract that, as mentioned, were evaluated under varying PLE conditions. For the development of ML models, four features were used as inputs: ethanol concentration (% *v/v*), liquid-to-solid ratio (mL/g), extraction temperature (°C), and extraction time (min). Additionally, four features were used as the outputs: TPC, FRAP, DPPH, and AAC. The initial dataset comprised only 17 samples. Of the data, 80% was allocated for training and 20% for testing our ML models. Figure 1 presents the distribution of experimental variables and antioxidant responses, illustrating balanced sampling across extraction parameters and greater variability in antioxidant outcomes.

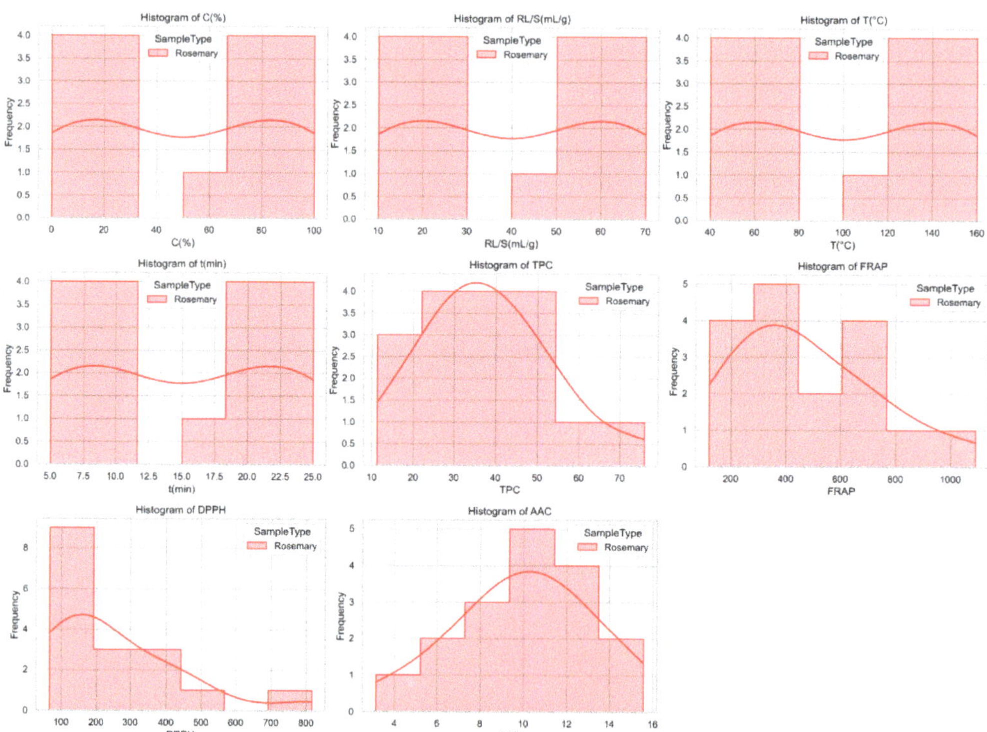

Figure 1. Histograms with kernel density estimates showing the distribution of extraction parameters and antioxidant response variables for rosemary samples. While extraction settings are uniformly distributed due to the design structure, antioxidant responses such as FRAP and DPPH exhibit skewed distributions, indicating variability in sample performance.

To assess variability and central tendencies, a combined boxplot was created (Figure 2). The extraction parameters (C, $R_{L/S}$, T, and t) showed narrow interquartile ranges and symmetry, indicating controlled conditions. In contrast, the antioxidant responses—especially FRAP and DPPH—exhibited wide variability and outliers, reflecting greater sensitivity to extraction settings. TPC showed moderate spread, while AAC remained tightly clustered. These patterns suggest that antioxidant outcomes are more affected by experimental variation than the input parameters.

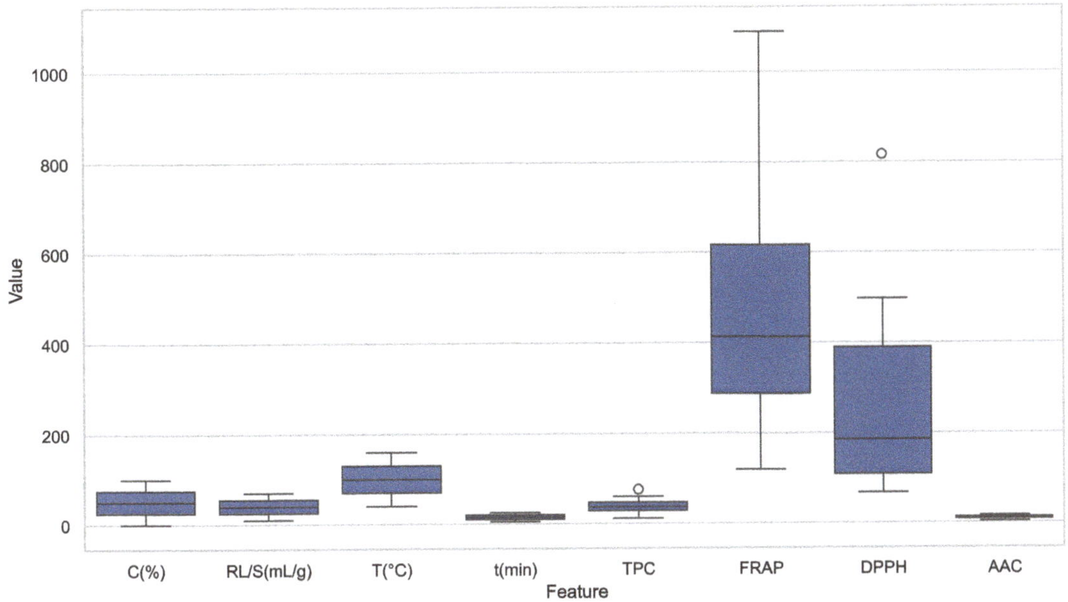

Figure 2. Combined boxplot of all extraction parameters and antioxidant response variables. Antioxidant metrics (FRAP and DPPH) exhibit larger spread and more outliers than extraction parameters, indicating greater variability and sensitivity to experimental conditions.

To assess the variation across extraction conditions and antioxidant responses, two complementary heatmaps were produced and are presented together in Figure 3. Plot (A) displays the raw value matrix, highlighting absolute differences among samples. The high-intensity region in the FRAP column (design point 13) corresponds to elevated antioxidant activity, also reflected in TPC. AAC values remained consistently low across all samples. Plot (B) shows the standardized (z-score) version of the same matrix, enabling scale-independent comparison. The same sample exhibited z-scores above +2 in FRAP and TPC, confirming its outlier status. Other samples with moderate DPPH or AAC responses became more distinct through normalization. Together, the heatmaps reveal both high-performing conditions and hidden patterns across the dataset.

To explore the variable relationships, a Pearson correlation matrix was computed (Figure 4). Strong positive correlations between TPC, FRAP, and DPPH (r = 0.81–0.91) indicate phenolics' central role in antioxidant capacity. The ethanol concentration (C, %) was negatively correlated with both TPC and FRAP, suggesting diminishing returns at higher concentrations. The temperature and solvent ratio showed moderate positive correlations with AAC, while the extraction time had minimal influence on any response. These findings align with the heatmap results and underscore the compound-specific effects of extraction parameters.

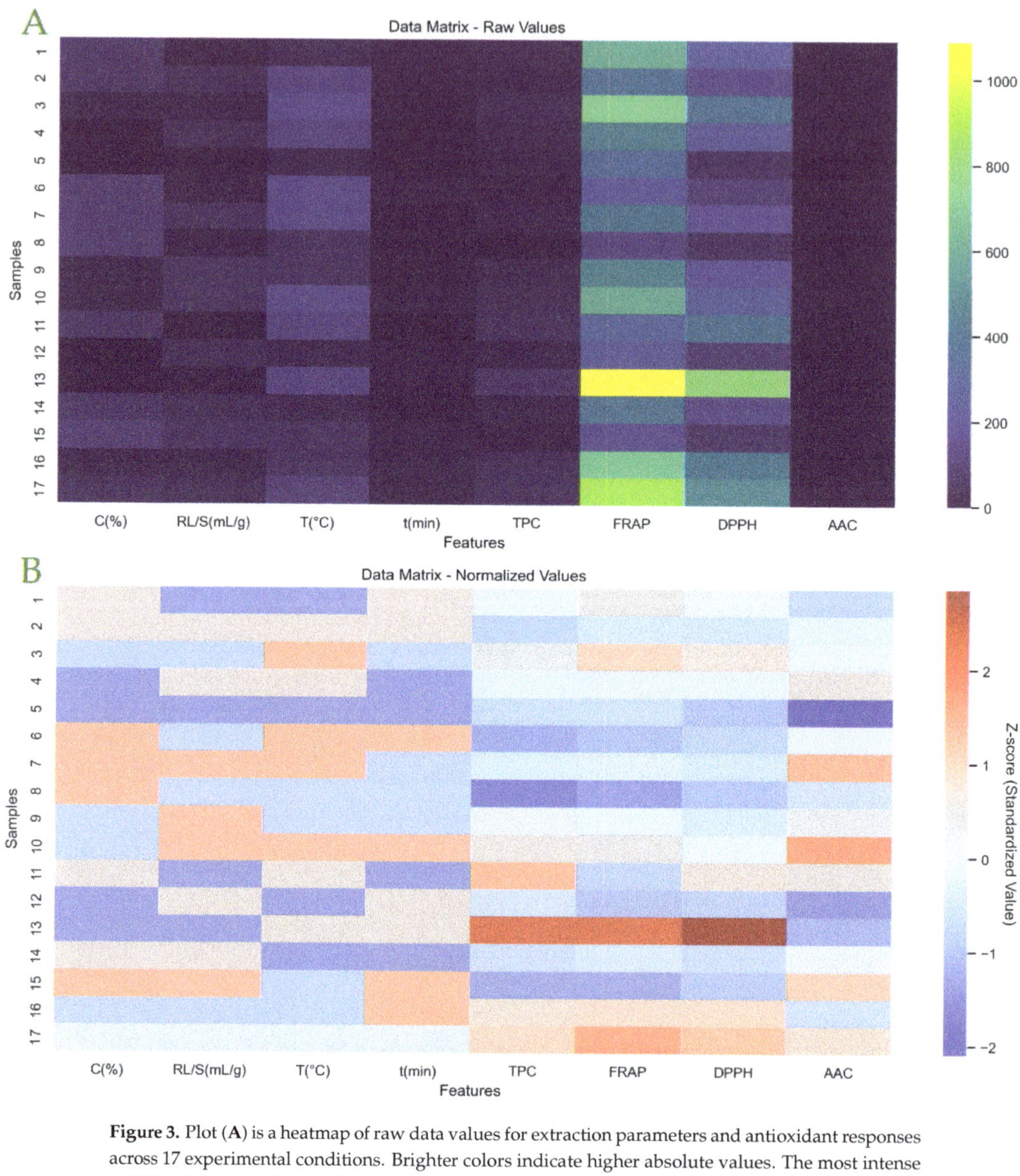

Figure 3. Plot (**A**) is a heatmap of raw data values for extraction parameters and antioxidant responses across 17 experimental conditions. Brighter colors indicate higher absolute values. The most intense FRAP activity was observed in sample #13. Plot (**B**) is a z-score normalized heatmap of the same dataset. Standardization enables direct comparison across all features. Red shades represent values above the mean, while blue shades indicate values below the mean. Strong positive deviations in FRAP and TPC are evident in sample #13.

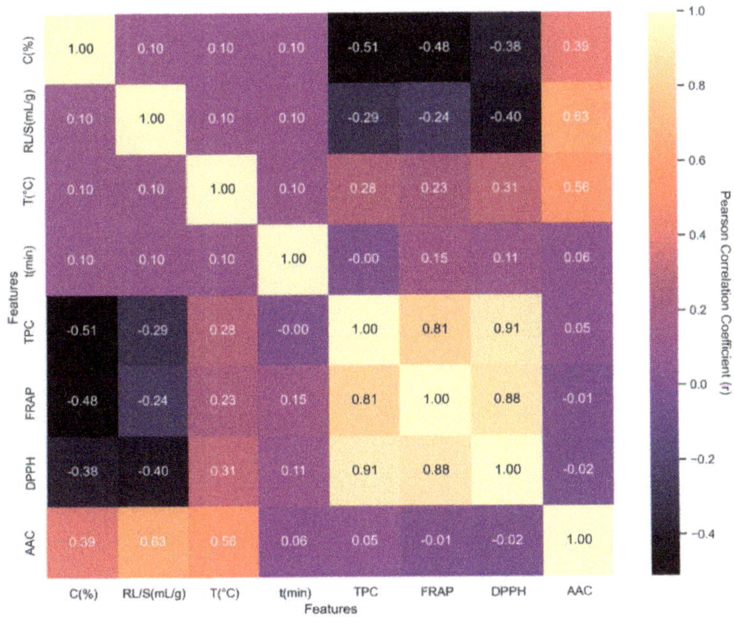

Figure 4. Pearson correlation matrix among extraction parameters and antioxidant response variables. Values range from −1 (perfect negative correlation) to +1 (perfect positive correlation). Strong internal consistency was observed among antioxidant metrics (TPC, FRAP, and DPPH), while concentration (C, %) negatively correlates with TPC and FRAP.

2.11. ML Regressor Development

To be able to develop our ML-based regressors for our initial data set, we trained six regression algorithms that were applied to model the relationships between the extraction parameters and the antioxidant responses. The models included Linear Regression [16], Ridge Regression [17], Lasso Regression [18], RF regression, Gradient Boosting (GB) Regression [19], and Adaptive Boosting (AdaBoost) Regression [20].

Each model was implemented as a multi-output regressor. Hyperparameter tuning was conducted using grid search with 5-fold cross-validation, due to the limited data. Ridge and Lasso regressors were tuned for regularization strength (α), while tree-based models were optimized for the number of estimators, maximum depth, learning rate for boosting models, and minimum samples per split. The full list of model parameters and their tested values is shown in Table 1.

Table 1. Summary of machine learning regression models and the corresponding hyperparameters tuned during grid search. Default parameters were used for Linear Regression, while regularization strengths (α) and core structural parameters (number of estimators, max depth, min samples split, tree depth, and learning rate) were varied for the other models.

Model	Tuned Parameters	Values Tested
Linear Regression	None	-
Ridge	α	[0.1, 1.0, 10.0]
Lasso	α	[0.001, 0.01, 0.1, 1.0]
RF	n_estimators, max_depth, min_samples_split	[100, 200], [None, 10], [2, 5]
GB	n_estimators, learning_rate, max_depth	[100, 200], [0.05, 0.1], [3, 5]
AdaBoost	n_estimators, learning_rate	[50, 100], [0.5, 1.0]

The selected hyperparameters were chosen for their direct impact on model complexity, generalization, and performance. For regularized linear models, Ridge and Lasso, the regularization strength "α" controls the degree of penalty applied to large coefficients—helping to reduce overfitting, especially in small datasets. Smaller values allow more flexibility, while larger values enforce stronger shrinkage of less informative predictors. To further mitigate the risk of overfitting, all models were trained and evaluated using 5-fold cross-validation. In addition, regularization techniques (Ridge and Lasso penalties) and parameter tuning were applied to control model complexity and improve generalization performance, particularly given the small size of the original dataset.

In the tree-based models RF, GB, and AdaBoost, the "number of estimators" determines how many trees are used to build the ensemble; more trees generally improve performance but increase computation. The "maximum tree depth" controls how complex each tree can be, balancing fit versus overfitting. The "minimum samples split" parameter sets the minimum number of samples required to split a node, helping to regularize the model by preventing overly deep trees. The "learning rate", used in boosting algorithms, scales how much each tree contributes to the final prediction—lower rates typically yield better generalization at the cost of longer training.

2.12. Machine Learning Regressor Evaluation

In this study, the performance of the regression models was evaluated using four standard metrics: Mean Absolute Error (MAE) [21], Mean Squared Error (MSE) [21], Root Mean Squared Error (RMSE) [22], and the Coefficient of Determination (R^2) [23]. These metrics quantify the difference between the predicted values $\hat{y}_i$ and actual experimental values y_i, based on a total of n observations.

MAE measures the average magnitude of errors in a set of predictions, without considering their direction. It is calculated as the mean of the absolute differences between actual and predicted values (2).

$$\text{MAE} = \frac{1}{n}\sum_{i=1}^{n}|y_i - \hat{y}_i| \tag{2}$$

MSE penalizes larger errors more strongly by squaring them. It is the average of the squared differences between actual and predicted values (3).

$$\text{MSE} = \frac{1}{n}\sum_{i=1}^{n}(y_i - \hat{y}_i)^2 \tag{3}$$

RMSE is the square root of the MSE and provides an error measure in the same units as the original response variable, making it more interpretable (4):

$$\text{RMSE} = \sqrt{\frac{1}{n}\sum_{i=1}^{n}(y_i - \hat{y}_i)^2} \tag{4}$$

R^2 represents the proportion of variance in the actual values that is predictable from the independent variables. A value of 1 indicates perfect prediction, while 0 means the model explains none of the variance (5):

$$R^2 = 1 - \frac{\sum_{i=1}^{n}(y_i - \hat{y}_i)^2}{\sum_{i=1}^{n}(y_i - \overline{y})^2} \tag{5}$$

These metrics together offer a robust framework for comparing the prediction accuracy, error dispersion, and explanatory power of each machine learning regressor tested in this study.

2.13. Generative Model Development

To address the limitations imposed by the small sample size (n = 17), a synthetic dataset was generated to enhance the modeling capacity and generalization of the machine learning algorithms. New input combinations were uniformly sampled within the observed range of the original extraction parameters: ethanol concentration (C, % v/v), liquid-to-solid ratio ($R_{L/S}$, mL/g), extraction temperature (T, °C), and extraction time (t, min). A total of 100 synthetic input samples were created to expand the dataset in a balanced and controlled manner.

To estimate the corresponding antioxidant responses—total polyphenol content (TPC), ferric reducing antioxidant power (FRAP), DPPH radical scavenging activity, and ascorbic acid content (AAC)—a pre-trained RF model, previously identified as the best-performing regressor, was employed. RF is an ensemble learning method based on decision trees that captures nonlinear relationships by aggregating the predictions of multiple base learners. The predicted output $\hat{y}$ for a given input vector x is computed as the average of predictions from all trees in the forest (6):

$$\hat{y} = \frac{1}{T}\sum_{t=1}^{T} f_t(x) \quad (6)$$

where T is the total number of decision trees in the ensemble, and $f_t(x)$ is the prediction from the t-th tree for input x. In this study, RF was applied both as a predictive model and, in the generative phase, to estimate antioxidant outcomes for synthetically generated feature combinations.

To simulate natural variability and reduce overfitting to deterministic predictions, Gaussian noise was added to the RF-generated outputs. This augmentation mimics experimental uncertainty and improves the realism of synthetic samples. The final synthetic response $\tilde{y}$ was computed as follows (7):

$$\tilde{y} = \hat{y} + \epsilon,\ \epsilon \sim \mathcal{N}\left(0, \sigma^2\right) \quad (7)$$

where ϵ is a noise term drawn from a normal distribution with zero mean and variance σ^2. In this study, σ was set to 5% of the standard deviation of the respective real target variable, providing a balance between stability and variability.

The resulting synthetic data points were merged with the original experimental dataset to create a mixed dataset. The rationale for this approach is that the RF model captures complex nonlinear interactions in the original data, and the controlled addition of Gaussian noise (set at 5% of the standard deviation of each real target variable) introduces realistic variability while avoiding overfitting to deterministic predictions. This method balances model fidelity with enhanced generalization potential.

The resulting synthetic data points were merged with the original experimental dataset to create a mixed dataset. Due to the limitations of our available computational resources, the size of the synthetic dataset was intentionally kept small (100 samples) to perform an initial proof-of-concept evaluation. Future work will explore more extensive data augmentation using more advanced generative techniques and more powerful hardware.

3. Results and Discussion

3.1. Optimization of PLE Parameters

The extraction procedure may be challenging due to the presence of several distinct bioactive compounds which lead to variations in solubility and polarity [24]. Moreover, various processing parameters along with the extraction technique might significantly affect both the extract yield and antioxidant capacity. Table 2 presents how the variables under investigation affect the examined responses, while in Table 3 the ANOVA applied to the RSM quadratic polynomial model is presented.

Table 2. Experimental results for the four examined independent variables and the dependent variables' responses to the PLE technique.

Design Point	Independent Variables				Actual PLE Responses *			
	C (%) (X_1)	$R_{L/S}$ (mL/g) (X_2)	T (°C) (X_3)	t (min) (X_4)	TPC	FRAP	DPPH	AAC
1	75	10	40	20	37.20	612.74	243.81	7.47
2	75	55	130	20	26.84	366.39	152.88	9.46
3	25	25	160	10	44.43	740.02	389.29	9.95
4	0	55	130	5	37.03	451.66	230.22	12.02
5	0	10	40	5	28.84	331.68	67.14	3.14
6	100	25	160	25	18.96	214.42	108.73	9.90
7	100	70	160	10	34.14	413.30	174.82	14.65
8	100	25	70	10	11.26	119.68	66.67	8.35
9	25	70	70	10	41.10	467.27	185.26	11.05
10	25	70	160	25	47.41	615.91	251.75	15.56
11	75	10	130	5	60.26	286.87	388.69	11.89
12	0	55	40	20	31.88	233.39	100.86	5.45
13	0	10	130	20	75.85	1089.81	816.36	6.38
14	75	55	40	5	30.09	393.47	150.16	10.24
15	100	70	70	25	19.17	169.93	85.11	13.34
16	25	25	70	25	49.64	696.79	427.46	8.44
17	50	40	100	15	52.69	893.17	497.11	12.45

* Values represent the mean of triplicate determinations; TPC, total polyphenol content (in mg GAE/g dw); FRAP, ferric reducing antioxidant power (in μmol AAE/g dw); DPPH, antiradical activity (in μmol AAE/g dw); AAC, ascorbic acid content (in mg/g dw).

Table 3. Analysis of variance (ANOVA) is performed for the response surface quadratic polynomial model in the context of the PLE technique.

Factor	TPC	FRAP	DPPH	AAC
Stepwise regression coefficients				
Intercept	42.88 *	715.8 *	356.2 *	11.04 *
X_1—ethanol concentration	−10.7 *	−170 *	−96.8 *	1.204 *
X_2—liquid-to-solid ratio	−5.69	−81.1	−103 *	2.283 *
X_3—temperature	7.183	96.73 *	93.54 *	1.956 *
X_4—extraction time	0.854	66.9	39.24	-
X_1X_2	3.746	166.6 *	92.01	-
X_1X_3	-	−114	−54.1	−1.02
X_1X_4	−8.19	-	−91	-
X_2X_3	−4.4	-	−52.1	-
X_2X_4	−4.51	−131 *	−82.1	-

Table 3. *Cont.*

Factor	TPC	FRAP	DPPH	AAC
X_3X_4	-	-	-	-
X_1^2	−13.8	−268 *	−103	−1.69
X_2^2	15.5	-	79.73	-
X_3^2	−8.56	-	−130	-
X_4^2	-	−131	-	-
ANOVA				
F-value	4.507	7.716	4.362	10.65
p-Value	0.0545 ns	0.0067 *	0.0833 ns	0.0006 *
R^2	0.908	0.908	0.929	0.829
Adjusted R^2	0.707	0.791	0.716	0.751
RMSE	8.747	121.9	104.7	1.633
PRESS	3913	363,375	580,653	57.61
CV	42.46	55.93	77.03	32.76
DF (total)	16	16	16	16

* The values significantly affected responses at a probability level of 95% ($p < 0.05$). TPC, total polyphenol content; FRAP, ferric reducing antioxidant power; DPPH, antiradical activity; AAC, ascorbic acid content; ns, non-significant; *F*-value, test for comparing model variance with residual (error) variance; *p*-Value, probability of seeing the observed *F*-value if the null hypothesis is true; RMSE, root mean square error; PRESS, predicted residual error sum of squares; CV, coefficient of variation; DF, degree of freedom.

3.1.1. Model Analysis

The following Equations (8)–(11) represent regression models related to the extraction process, predicting key response variables: total phenolic content (TPC), ferric reducing antioxidant power (FRAP), DPPH radical scavenging capacity, and ascorbic acid content (AAC). Each equation includes linear, quadratic, and interaction terms, highlighting the complex relationships between experimental factors. The models contain only significant terms. The regression models highlight the impact of solvent composition, temperature, and duration on extraction efficiency. Notably, the linear and quadratic terms suggest nonlinear relationships between variables, indicating optimal conditions for maximizing antioxidant yield. The FRAP and DPPH equations show a strong dependency on the extraction conditions, particularly the extraction time (X_4). The presence of interaction terms suggests that the combined effects of multiple variables influence antioxidant potential, emphasizing the need for precise parameter optimization. Longer extraction times allow more bioactive compounds, including antioxidants, to dissolve into the solvent. The presence of quadratic terms (X_4^2) and interactions (X_2X_4 and X_1X_4) suggests that the extraction time has an optimal range—too short may limit compound release, while excessive duration could lead to degradation or reduced efficiency. The interaction terms imply that the extraction time does not act alone. For example, X_2X_4 in DPPH suggests that time interacts with another variable, the liquid-to-solid ratio, to influence antioxidant capacity.

$$TPC = 11.74 + 0.49X_1 - 1.22X_2 + 0.69X_3 + 1.51X_4 - 0.006X_1^2 + 0.017X_2^2 - 0.002X_3^2 + 0.002X_1X_2 - 0.016X_1X_4 - 0.002X_2X_3 - 0.015X_2X_4 \quad (8)$$

$$FRAP = -58.84 + 6.67X_1 - 1.73X_2 + 3.51X_3 + 63.39X_4 - 0.107X_1^2 - 1.31X_4^2 + 0.11X_1X_2 - 0.038X_1X_3 - 0.44X_2X_4 \quad (9)$$

$$DPPH = -330.79 + 4.24X_1 - 6.60X_2 + 10.86X_3 + 23.97X_4 - 0.041X_1^2 + 0.089X_2^2 - 0.036X_3^2 + 0.061X_1X_2 - 0.018X_1X_3 - 0.182X_1X_4 - 0.029X_2X_3 - 0.274X_2X_4 \quad (10)$$

$$AAC = 0.14 + 0.13X_1 + 0.08X_2 + 0.05X_3 - 0.0007X_1^2 - 0.0003X_1X_3 \quad (11)$$

Figure 5 shows how each parameter and their combinations affect the responses of the parameters under study. The predicted optimal values of PLE parameters along with the predicted TPC and FRAP, DPPH, and AAC values, along with the desirability of the model, are presented in Table 4.

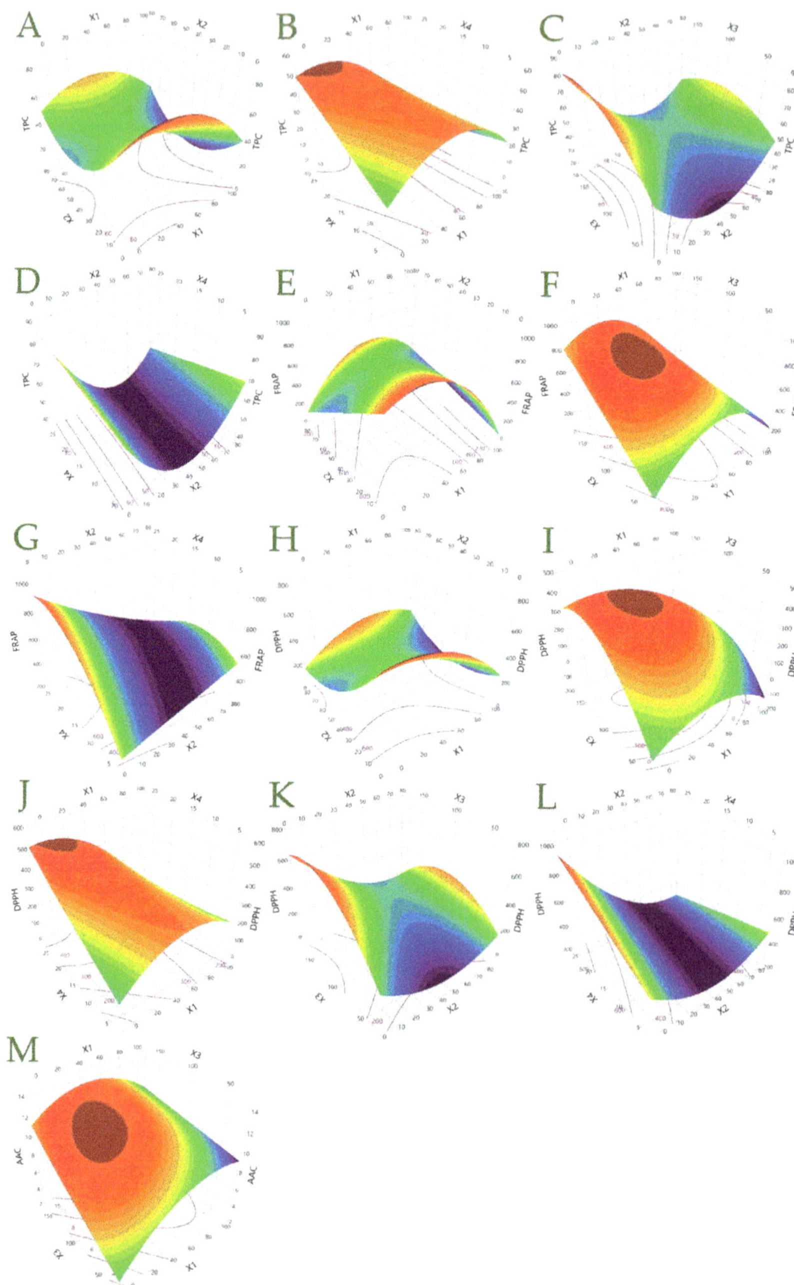

Figure 5. TPC, showing the (**A**) covariation of X_1 (ethanol concentration, *C*, % *v/v*) and X_2 (liquid-to-solid ratio, *R*, mL/g); (**B**) covariation of X_1 and X_4 (extraction time, *t*, min); (**C**) covariation of X_2

and X_3 (extraction temperature, *T*, °C); and (**D**) covariation of X_2 and X_4. FRAP, showing the (**E**) covariation of X_1 and X_2; (**F**) covariation of X_1 and X_3; and (**G**) covariation of X_2 and X_4. DPPH, showing the (**H**) covariation of X_1 and X_2; (**I**) covariation of X_1 and X_3; (**J**) covariation of X_1 and X_4; (**K**) covariation of X_2 and X_3; and (**L**) covariation of X_2 and X_4. AAC, showing the (**M**) covariation of X_1 and X_3.

Table 4. Maximum predicted responses and optimum extraction conditions for the dependent variables.

Parameters	Independent Variables				Desirability	Stepwise Regression
	C (%) (X_1)	$R_{L/S}$ (mL/g) (X_2)	*T* (°C) (X_3)	*t* (min) (X_4)		
TPC (mg GAE/g dw)	25	10	130	20	0.9292	76.19 ± 17.92
FRAP (μmol AAE/g dw)	25	10	160	20	0.9907	1117.68 ± 212.88
DPPH (μmol AAE/g dw)	0	10	130	20	0.8728	799.03 ± 271.68
AAC (mg/g dw)	50	70	160	-	0.9318	15.28 ± 2.23

3.1.2. Impact of Extraction Parameters on Assays Through Pareto Plot Analysis

In a Pareto plot (Figure 6), the orthogonal estimate typically refers to a statistical method used to estimate the effects of different factors while minimizing the correlation between them. This approach helps in identifying the most significant contributors to a given outcome by ensuring that the estimates are independent of each other.

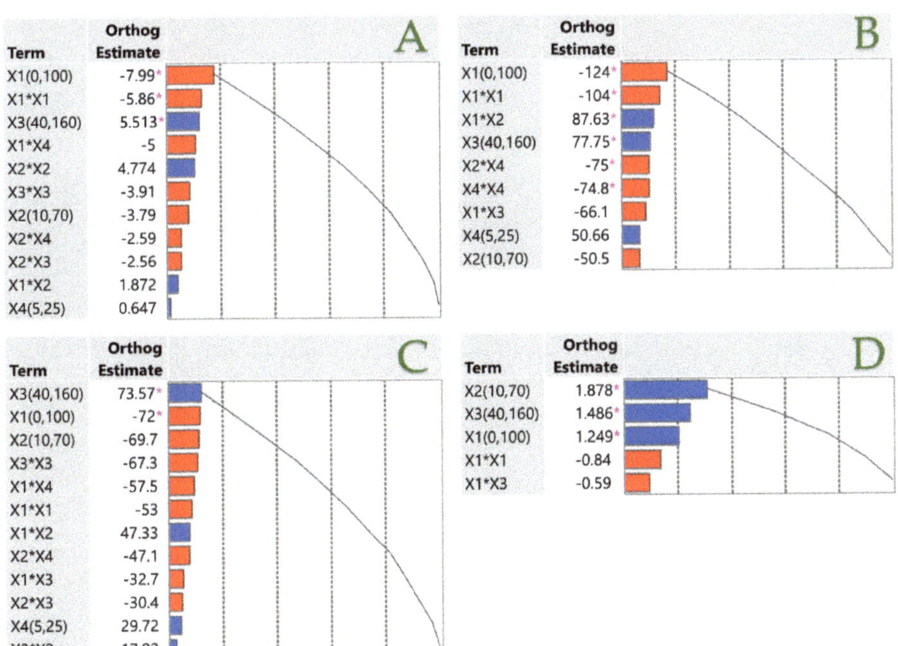

Figure 6. Pareto plots illustrating the significance of parameter estimates for the PLE technique across TPC (**A**), FRAP (**B**), DPPH (**C**), and AAC (**D**), with a pink asterisk marking significant values ($p < 0.05$). Positive estimates are shown in blue, while negative ones are represented in red.

It seems that temperature (parameter X_3) has a significantly positive effect on all responses. Another factor that seems to be very important is the solvent composition (X_1), where increasing the percentage of ethanol has a negative effect on all responses except ascorbic acid, where it has a positive effect. It is worth noting that the extraction duration (X_4) does not significantly affect any of the responses, but there is a trend where increased extraction times positively affect all responses.

3.2. Principal Component Analysis (PCA) and Multivariate Component Analysis (MCA)

The interactions between assays and extraction conditions were investigated through correlation analyses, which included PCA and MCA, as illustrated in Figure 7 and described in Table 5, respectively. The correlation analyses were conducted to ascertain the relationships between the variables and TPC, FRAP, DPPH, and AAC within the context of PCA. The chart demonstrates that PC1 and PC2 each contributed 67.6% and 24.9% of the variance, respectively, accounting for 92.5% of the variance. The analysis was deemed to be significantly influenced by the independent variables. The graph demonstrated that TPC, FRAP, DPPH, and extraction temperature and duration (X_3 and X_4) were positively correlated within both components and were represented in close proximity. AAC was considerably improved by the increased concentration of ethanol (X_1) and liquid-to-solid ratio (X_2), which explains their strong correlation. Their combined impact on extraction parameters was comparable. Conversely, the favorable placement of AAC in PC2, which is situated at a significant distance from the other variables, may indicate a diminished relationship between them. Previous research has suggested a positive correlation between an increase in ethanol concentration and AAC recovery [25].

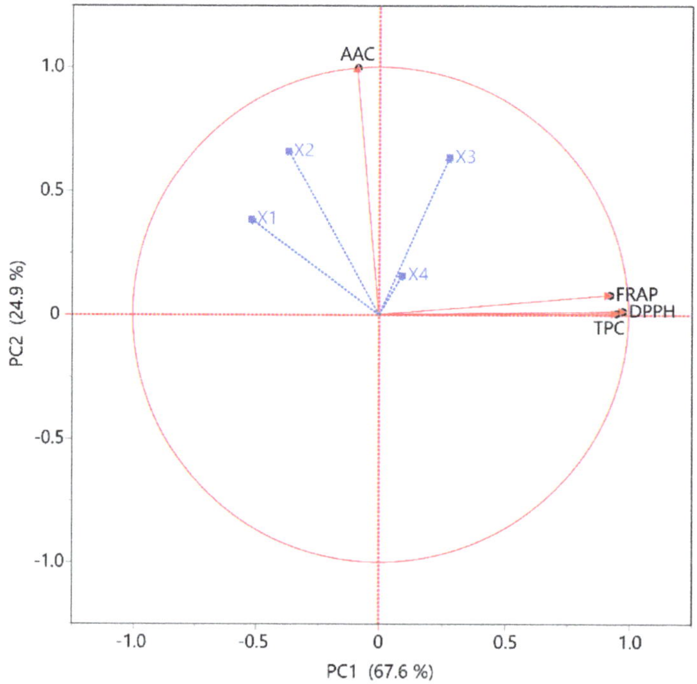

Figure 7. PCA for the measured variables. Each X variable is presented with a blue color.

Table 5. Multivariate correlation analysis of measured variables.

Responses	TPC	FRAP	DPPH	AAC
TPC	–	0.7796	0.9129	−0.0738
FRAP		–	0.8549	−0.0139
DPPH			–	−0.0722
AAC				–

In addition, the MCA provides further insights into the interrelationships between variables. The primary benefit of this approach is its ability to determine the degree of positive or negative correlation between the variables under investigation. Table 5 delineates the results of this investigation. The pattern of robust positive correlations (>0.77) between antioxidant assays and total phenolic content (TPC) was previously substantiated [26]. Ultimately, the negative correlation between ascorbic acid (AAC) and all other responses (TPC, FRAP, and DPPH) is highly emphasized. Nevertheless, it is particularly noteworthy that molecules exhibiting considerable antioxidant activity demonstrate a negative correlation with antioxidant assays.

3.3. Partial Least Squares (PLS) Analysis

The PLS model was employed to assess the influence of the extraction condition parameters (X_1, X_2, X_3, and X_4). Figure 8 illustrates the prediction profiler alongside a desirability function that features extrapolation control and includes a variable importance plot (VIP). The extraction of bioactive compounds is significantly influenced by various factors, with temperature, solvent composition, and extraction duration being the most critical [27]. Initially, it is important to note that the extraction process can be complicated by the differing solubility and polarity of polyphenols [28]. Concerning the PLE technique, it is evident that the X_1 parameter exhibited the most statistically significant impact ($p < 0.05$) compared to other parameters in the extraction process, as demonstrated by the Variance Importance Plot (VIP) presented in Figure 8B. The observations previously noted from the 3D models of the response surface were corroborated in Figure 8A, indicating that the optimal concentration was 25% *v/v* aqueous ethanol, a liquid-to-solid ratio of 10 mL/g, and the optimal temperature was 160 °C. The high efficiency observed at 160 °C is likely due to the enhanced solubility of polyphenols and increased solvent penetration into plant material. Elevated temperatures reduce surface tension, improving mass transfer and extraction yield. Concerning the duration of extraction, it appeared to exert the least significant influence on the process; consequently, the longest duration was favored, as it favored AAC recovery. The extraction process was not significantly influenced by the temperature or extraction duration; nevertheless, elevated temperatures coupled with long extraction times were favored. The solute–matrix interaction can be significantly reduced by the PLE technique, which is primarily due to the influence of van der Waals forces or hydrogen bonds, particularly in the presence of elevated temperature and pressure. This reduces energy demands, improves the efficacy of solute molecular extraction, and decreases the viscosity of the solvent. This reduces the solvent's resistance to the matrix, thereby facilitating its diffusion into the sample [29]. The model exhibited a prolonged extraction duration, as prior research has substantiated the effectiveness of both brief [30] and prolonged [28] intervals. While elevated temperatures facilitate the extraction of bioactive compounds by enhancing their solubility in other techniques, like stirring [31], it is important to note that many thermolabile compounds may experience degradation under these conditions [32].

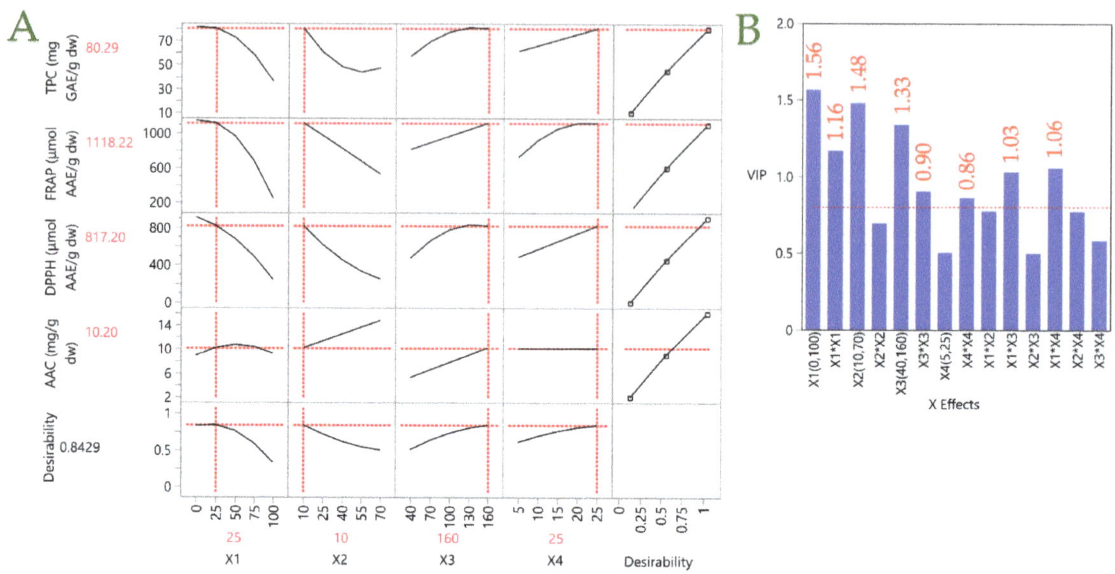

Figure 8. Plot (**A**) shows the optimization of the PLE technique for rosemary extracts through the partial least squares (PLS) prediction profiler and a desirability function with extrapolation control. Plot (**B**) shows the Variable Importance Plot (VIP) graph, showing the VIP values for each predictor variable in the PLE technique, with a red dashed line marking the 0.8 significance level.

Table 6 shows the values of TPC, FRAP, DPPH, and AAC of the optimal extract. The results of the present study are worth comparing with those of our previous work, where four different extraction techniques from rosemary leaves were studied, namely stirring, pulsed electric field (PEF)-assisted extraction, and ultrasound probe- and ultrasound bath-assisted extraction [31]. It is noteworthy that in that work, the highest TPC was given by stirring, and yet in the present work, PLE gave ~320% higher yield. A similar pattern was observed for FRAP, where PLE resulted in a ~455% higher yield. The highest value of DPPH was observed in ultrasound bath-assisted extraction; however, PLE gave a ~516% greater result. Regarding AAC, ultrasound probe-assisted extraction was the best value, and PLE only gave a ~10% greater performance than PEF-assisted extraction. Unlike conventional methods, PLE offers improved recovery of bioactive compounds in a shorter time frame, reducing energy consumption and solvent waste. A comparison with traditional extraction methods, such as ultrasound-assisted extraction (UAE), exhibits lower polyphenol recovery compared to PLE. Some other researchers also studied the TPC and antioxidant capacity of rosemary leaves. More specifically, Hashem Hashempur et al. [33] utilized a deep eutectic solvent, consisting of ammonium acetate and lactic acid, along with ultrasound, and their TPC was 334% lower than our result. Kabubii et al. [34] also determined a TPC in crude extracts, which was ~52% lower than ours. In general, PLE treatment of rosemary leaves seems to lead to higher yields than other extraction techniques.

Table 6. The partial least squares (PLS) prediction profiler determined the maximum desirability for all variables under optimal extraction condition for PLE technique.

Parameters	Independent Variables				Desirability	Partial Least Squares (PLS) Regression	Experimental Values
	C (%) (X_1)	$R_{L/S}$ (mL/g) (X_2)	T (°C) (X_3)	t (min) (X_4)			
TPC (mg GAE/g dw)	25	10	160	25	0.8429	80.29	78.23 ± 0.63
FRAP (μmol AAE/g dw)						1118.22	914.82 ± 1.53
DPPH (μmol AAE/g dw)						817.20	878.7 ± 6.34
AAC (mg/g dw)						10.20	17.83 ± 0.25

The experimental results and PLS model predictions exhibit outstanding concordance, as evidenced by the high correlation coefficient of 0.981 and substantial R^2 value of 0.962. Furthermore, the p-value being less than 0.0001 indicates that the deviations between the actual and predicted values are statistically insignificant.

Table 7 presents a list of the individual polyphenols identified in the optimal extract by HPLC-DAD, while Table 8 provides information on the equations of the standard compounds. The compound with the highest concentration is hesperidin, followed by rosmarinic acid and Quercetin 3-*D*-galactoside. In our previous work, the compound with the highest concentration was rosmarinic acid in all cases, and here it is worth noting how the parameters applied by each different technique during the extraction process greatly affect the profile of the final extracts obtained. However, the same compounds were also identified in this work. Other researchers, like Xie et al. [35], Sammer and Samarrai [36], Baptista et al. [37], and Miljanović et al. [38], determined compounds like hesperidin, apigenin and its derivatives, rosmarinic acid, and carnosic acid in rosemary leaves.

Table 7. Optimal extraction conditions for polyphenolic compounds using the PLE technique of rosemary extraction.

Polyphenolic Compound	Concentration (μg/g dw)
Catechin	239 ± 11
Quercetin 3-*D*-galactoside	1114 ± 45
Luteolin-7-glucoside	236 ± 13
Kaempferol-3-glucoside	442 ± 19
Hesperidin	3711 ± 96
Rosmarinic acid	1570 ± 58
Apigenin	245 ± 7
Kaempferol	72 ± 2
Rosmanol	731 ± 27
Carnosic acid	889 ± 32
Total identified	9250 ± 311

Values represent the mean of triplicate determinations ± standard deviation.

Table 8. Equation of calibration curves for each compound identified through HPLC-DAD.

Polyphenolic Compounds (Standards)	Equation (Linear)	R^2	Retention Time (min)	UVmax (nm)
Catechin	y = 11,920.78x − 128.19	0.997	20.933	278
Quercetin 3-D-galactoside	y = 41,489.69x − 35,577.55	0.993	34.598	257
Luteolin-7-glucoside	y = 34,875.94x − 16,827.36	0.999	35.949	347
Kaempferol-3-glucoside	y = 50,916.85x − 42,398.83	0.996	38.724	265

Table 8. *Cont.*

Polyphenolic Compounds (Standards)	Equation (Linear)	R^2	Retention Time (min)	UVmax (nm)
Hesperidin	y = −30,502.75x − 30,502.75	0.995	40.249	283
Rosmarinic acid	y = 50,281.27x − 113,633.31	0.995	41.644	329
Apigenin	y = 95,483.52x − 5214.26	0.998	55.860	227
Kaempferol	y = 93,385.02x − 18,613.03	0.999	56.883	265
Rosmanol	y = 5509.45x − 10,899.23	0.994	65.924	288
Carnosic acid	y = 8883.45x + 101,483.30	0.992	77.870	284

3.4. *Performance of Machine Learning Regressors on the Original Data*

In the following sections, we focus on reporting the key findings regarding the performance of the ML regressors, with an emphasis on practical insights relevant to extraction optimization. The detailed technical analysis is intentionally limited, in line with the overall scope of this experimental study.

From Figure 9, it can be observed that all regression models achieved relatively good performance on the training dataset. In particular, RF, GB, and AdaBoost demonstrated very high training accuracy, with GB achieving an R^2 of 1.00 and RF and AdaBoost closely following with R^2 values of 0.87 and 0.99, respectively. These results indicate that ensemble-based models fit the training data extremely well, though they may risk overfitting.

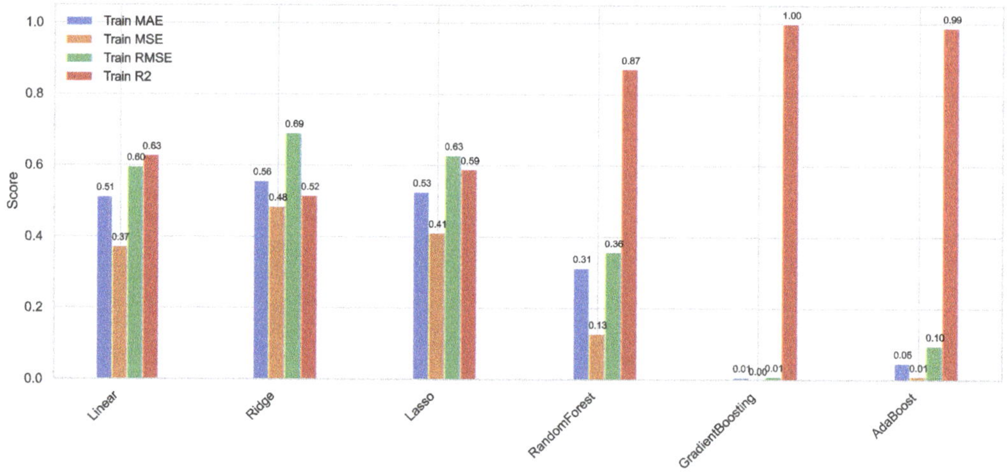

Figure 9. Histograms of the performance of our six ML models on our original training set.

In contrast, Figure 10 reveals substantial performance degradation across all models when evaluated on the testing dataset. The Linear and Ridge regressors showed moderate predictive ability with test R^2 scores around −3.29 and −1.31, respectively. Among all models, RF achieved the best test performance, with lower error scores, 0.81 MAE, 0.91 MSE, 0.91 RMSE, and a test R^2 of −1.66, outperforming the other regressors under the given constraints.

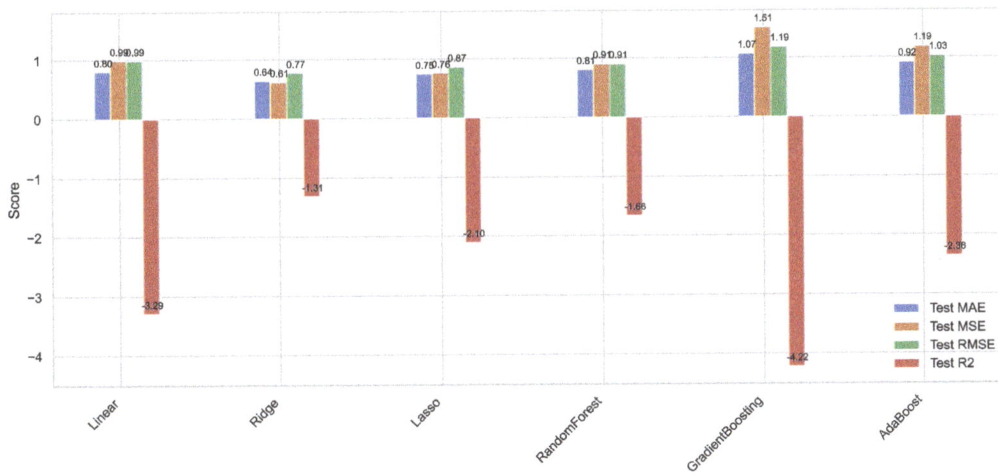

Figure 10. Histograms of the performance of our six ML models on our original test set.

Despite the overall lower performance on test data, RF was selected as the most efficient regressor due to its balance between training fit and test error, as well as its suitability for data generation in the augmentation phase that followed. Based on this model, synthetic data were produced and evaluated in combination with the original dataset.

3.5. Performance of Machine Learning Regressors on the Synthetic Dataset

Our next step was to compare our regressors based on our synthetic data. A synthetic dataset, consisting of 100 samples, was generated using RF-based predictions with Gaussian noise added to introduce controlled variability. This synthetic dataset maintained the same feature distribution as the original dataset to ensure comparability.

Ensemble-based models demonstrated high performance compared to linear models. Specifically, the GB regressor achieved the highest training accuracy with an R^2 of 1.00, followed by the RF regressor with an R^2 of 0.98 and AdaBoost with an R^2 of 0.95. These models also recorded notably low error metrics, indicating an almost perfect fit to the training data. In contrast, Linear Regression, Ridge Regression, and Lasso Regression yielded identical training performance, each reaching an R^2 of 0.71, which indicates a moderate capacity to model the underlying relationships within the synthetic data. Their MAE and RMSE values were also consistently higher than those of the ensemble models (Figure 11).

For the test set, the GB-based and RF-based regressors again outperformed the other models, with R^2 values of 0.93 and 0.91 respectively, along with the lowest RMSE scores, suggesting strong predictive ability on unseen data. The AdaBoost regressor also performed well with an R^2 of 0.88, although with slightly higher error values. Linear models showed consistent but comparatively limited predictive performance on the test data, with all three achieving an R^2 of 0.70 and similar error magnitudes (Figure 12).

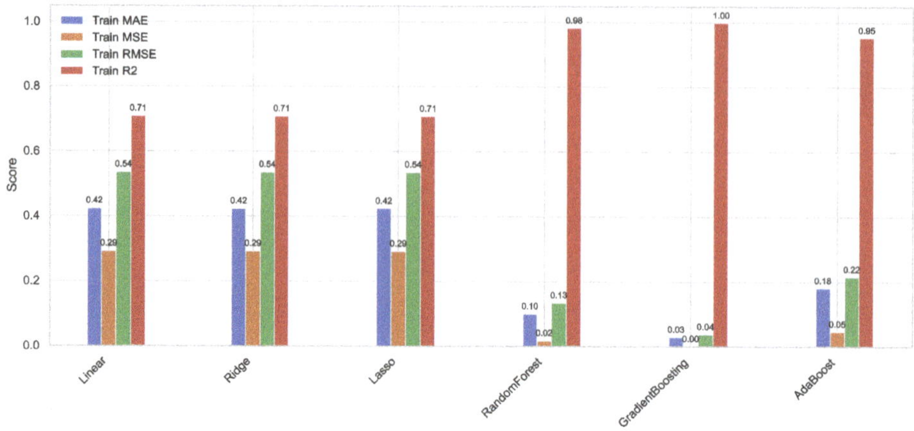

Figure 11. Histograms of the performance of our six ML models on our synthetic training set.

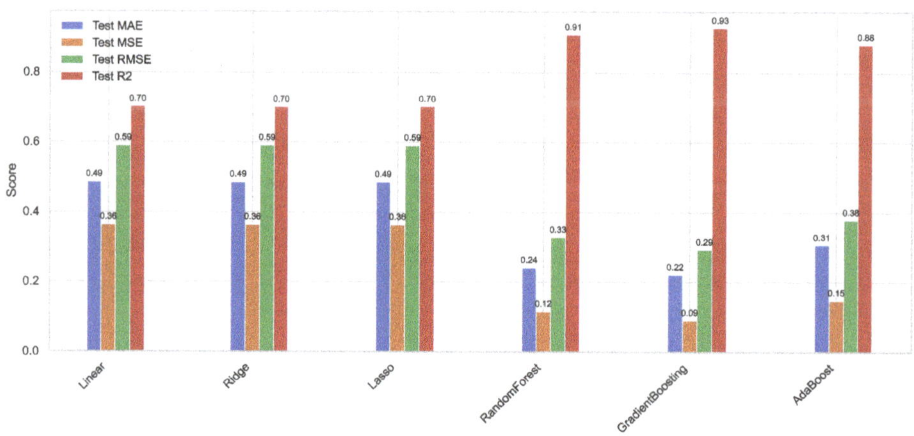

Figure 12. Histograms of the performance of our six ML models on our synthetic test set.

3.6. Performance of Machine Learning Regressors on the Mixed Dataset

To further evaluate the model performance in a data-rich scenario, we trained and tested our regressors using a mixed dataset consisting of both real experimental samples from our initial dataset and synthetically generated data.

The model training results showed clear performance differentiation among algorithm families. Ensemble models again demonstrated high accuracy. The GB-based regressor achieved the highest training performance, with an R^2 of 1.00 and near-zero error values across all metrics: MAE = 0.03, MSE = 0.001, and RMSE = 0.04. The RF-based regressor followed with an R^2 of 0.95 and comparatively low error values: MAE = 0.13, MSE = 0.05, and RMSE = 0.22. The AdaBoost regressor also performed well during training, with an R^2 of 0.93. In contrast, the linear models—Linear Regression, Ridge Regression, and Lasso Regression—yielded nearly identical training outcomes, each with an R^2 of 0.61 and RMSE of approximately 0.39 to 0.40. These results indicate that the linear models were only moderately successful in capturing the increased variance introduced through the mixed data (Figure 13).

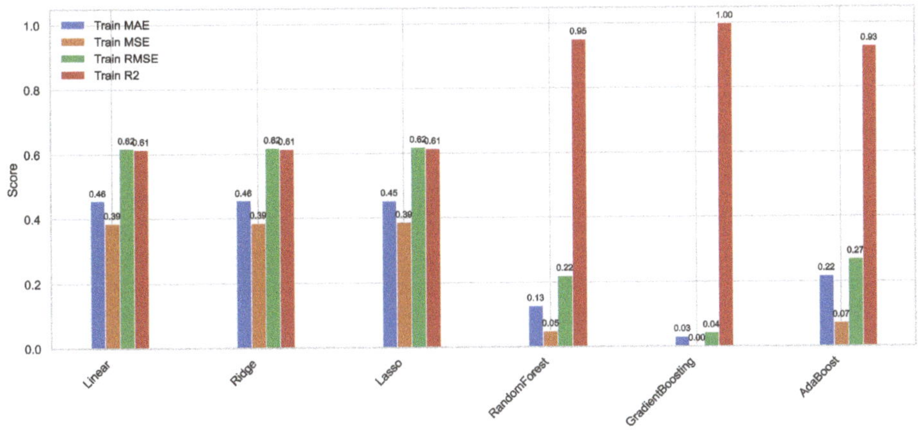

Figure 13. Histograms of the performance of our six ML models on our mixed training set.

The testing performance followed a similar trend but revealed more nuanced differences in generalization capability. The GB-based regressor achieved a test R^2 of 0.76, while the RF-based regressor reached 0.84, indicating that both models retained strong generalization on unseen data. The AdaBoost regressor also maintained respectable performance with a test R^2 of 0.74. Notably, the GB and RF regressors both achieved low RMSE values on the test set of 0.16 and 0.21, respectively, underscoring their effectiveness in handling diverse and noise-augmented data distributions. The linear models again exhibited limited predictive strength on the test set, with all three achieving an R^2 from 0.59 to 0.60 and RMSE values ranging from 0.28 to 0.29 (Figure 14).

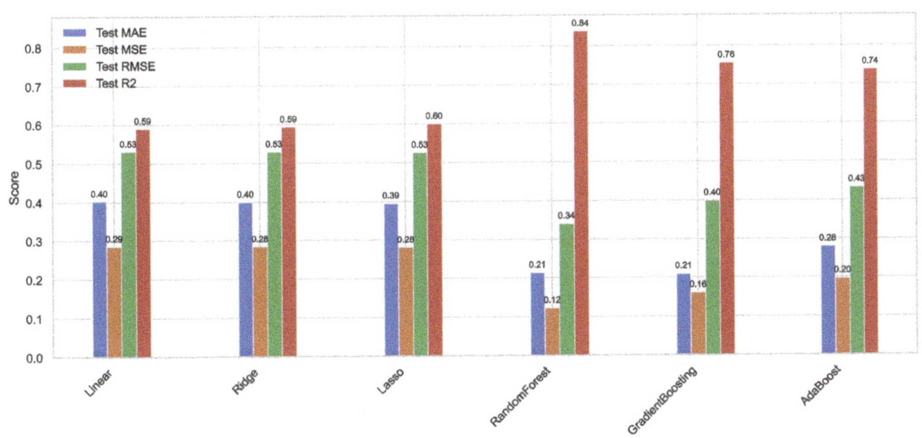

Figure 14. Histograms of the performance of our six ML models on our mixed test set.

3.7. Cross-Evaluation of RF Models on Real, Synthetic, and Mixed Data

Based on the previous results, our best regressor is the RF mixed-based regressor which was developed based on the original and synthetic data. To further examine the robustness and generalization capabilities of the RF mixture-based model, we conducted a cross-dataset evaluation in which models trained on one dataset were tested on different datasets. The training datasets included the original experimental data, a synthetically generated dataset, and a mixed dataset combining both sources. Each model's predictive accuracy was then assessed across all three datasets using standard performance metrics.

The variability introduced by the Gaussian noise in synthetic data was controlled and systematically evaluated, as described in Section 2.13, to ensure that the augmented dataset preserved meaningful variance while remaining consistent with the statistical characteristics of the original data.

When the model trained on the original dataset was tested on the synthetic dataset, it produced a test MAE of 0.60 and a RMSE of 0.80, with an R^2 of 0.48. This indicated moderate generalization capacity to the artificial data. Slightly better performance was observed when the same model was evaluated on the mixed dataset, yielding a lower test RMSE of 0.62 and a marginally improved R^2 of 0.43.

In contrast, the model trained solely on the synthetic data demonstrated poor performance on the original data, with an R^2 of only 0.04 and an RMSE of 0.44, suggesting a substantial gap in representational fidelity between the synthetic and real data distributions. However, when the same synthetically trained model was tested on the mixed dataset, performance improved drastically, achieving an R^2 of 0.86 and RMSE of 0.31. This highlights that the synthetic model generalized well within synthetic-heavy contexts but struggled with real experimental variability.

The model trained on the mixed dataset exhibited the strongest overall generalization. It achieved a low RMSE of 0.27 and a high R^2 of 0.63 when tested on the original data. Most notably, it yielded the best cross-dataset performance when tested on the synthetic dataset, with an RMSE of 0.38 and an R^2 of 0.88. With our new RF mixed regressor we had a 20% increase in the performance, which is satisfactory due to the lack of samples (Figure 15).

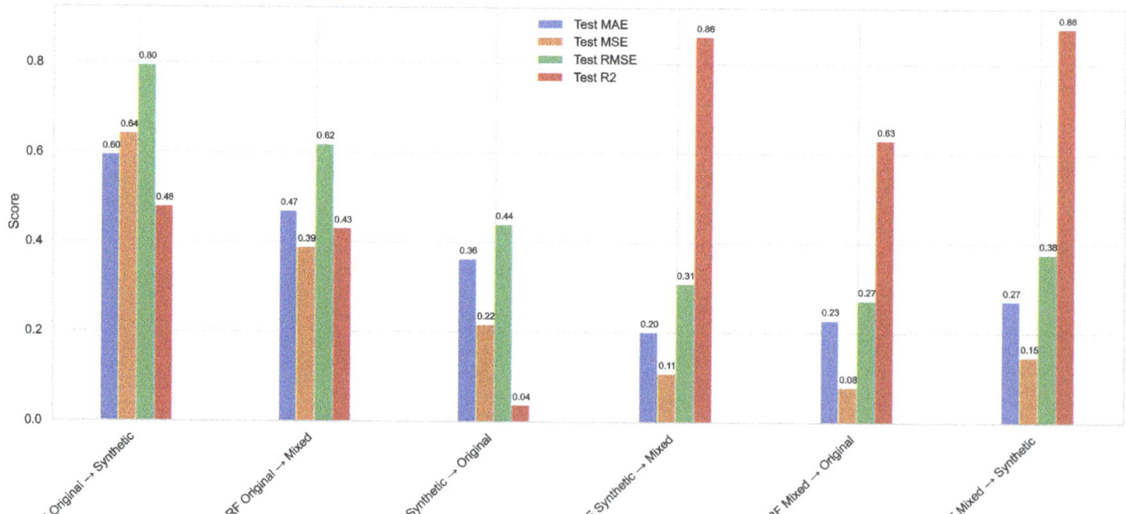

Figure 15. Cross-dataset evaluation of the RF regression model trained on the original, synthetic, and mixed datasets. Each group of bars represents the performance metrics MAE, MSE, RMSE, and R^2 obtained when the model trained on one dataset was tested on another. Results highlight the generalization ability of each training regime across data domains. Models trained on the mixed dataset showed superior cross-domain performance, particularly when evaluated on both the original and synthetic test sets.

These results validate our objective of generating new samples, demonstrating that synthetic data can enhance the development of robust machine learning regressors for accurately predicting total phenolic content.

3.8. Feature Importance Analysis Across RF-Based Models

To investigate how the model training data influences the learned relationships between extraction parameters and antioxidant responses, feature importance scores were extracted from each RF model trained on the original, synthetic, and mixed datasets. These scores were derived from the individual estimators trained for each target (TPC, FRAP, DPPH, and AAC), and visualized side by side (Figure 16).

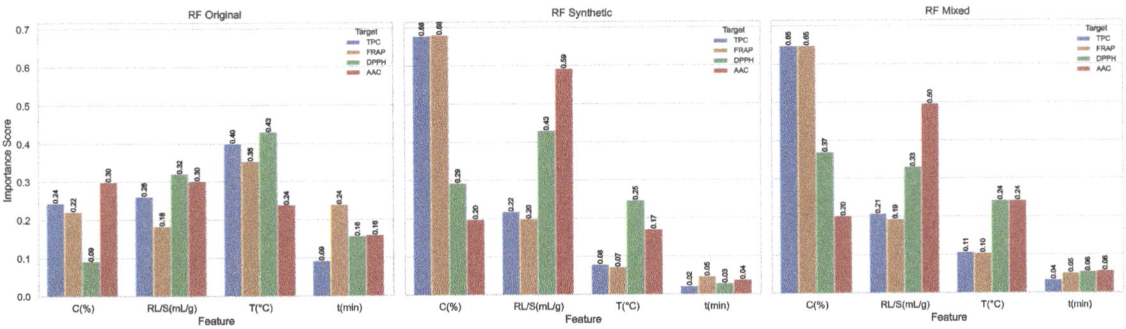

Figure 16. Feature importance scores from RF models trained on the (**left**) original, (**middle**) synthetic, and (**right**) mixed datasets. Importance scores reflect each feature's contribution to predicting antioxidant targets (TPC, FRAP, DPPH, and AAC). Feature influence varies significantly depending on the dataset used for training.

In the model trained on the original dataset, the most influential feature overall was temperature (*T*, °C), particularly for FRAP and TPC, where it accounted for over 35–40% of total importance. This aligns with experimental expectations, as thermal energy often enhances compound release. $R_{L/S}$ and *C* (%) also showed moderate contributions, while extraction time had the lowest influence across all targets.

In contrast, the model trained exclusively on synthetic data placed much greater emphasis on *C* (%), especially for FRAP (0.78) and TPC (0.72). This shift likely reflects the statistical bias introduced during synthetic generation, where concentration appeared as a dominant predictor due to its nonlinear interactions captured by RF. Meanwhile, *T* (°C) and *t* (min) showed very low influence (<0.10) across all targets.

In the mixed model, a balanced importance distribution emerged. *C* (%) again held strong predictive power for TPC and FRAP (~0.66), but now $R_{L/S}$ and *T* (°C) also gained relevance, particularly for AAC (0.57) and DPPH (0.39), indicating a more nuanced learning of underlying relationships. Time remained the least influential, consistent across all models.

This comparison reveals how the training dataset affects not only model accuracy but also which experimental parameters are deemed most critical. Mixed training produces models that are both accurate and biologically plausible, while synthetic-only training can exaggerate the significance of specific variables.

However, it should be noted that the feature importance results presented here are influenced by the synthetic component of the mixed dataset. As such, there is a potential risk of bias in the interpretation of variable importance, particularly for features that may exhibit amplified or diminished effects in synthetic samples. This limitation highlights the need for cautious interpretation and suggests that future studies should validate these findings using larger experimental datasets.

3.9. Actual vs. Predicted Performance Across RF-Based Models

To visually assess prediction accuracy and generalization, scatter plots comparing actual vs. predicted values were generated for all four antioxidant targets, TPC, FRAP, DPPH, and AAC, using the RF-based models trained on the original, synthetic, and mixed datasets (Figure 17). All values were plotted in standardized, z-score space, and each subplot includes the R^2 as a quantitative measure of fit.

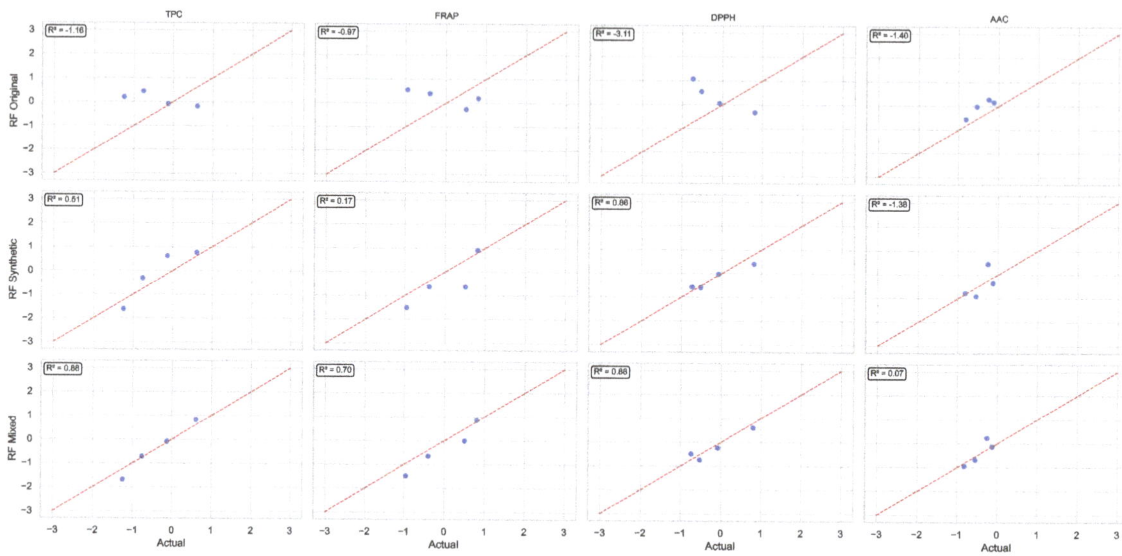

Figure 17. Predicted vs. actual standardized values for antioxidant targets using RF-based models trained on the original (**top row**), synthetic (**middle row**), and mixed (**bottom row**) datasets. Diagonal red dashed lines represent the ideal 1:1 relationship. R^2 values quantify model fit for each case.

The model trained exclusively on the original dataset exhibited poor predictive performance across all targets. R^2 values were consistently negative, indicating substantial overfitting and a lack of generalization to unseen data. DPPH with $R^2 = -3.11$ and AAC with $R^2 = -1.40$ showed a complete breakdown of predictive capacity, while even the best-performing targets, TPC with $R^2 = -1.16$ and FRAP with $R^2 = -0.97$, failed to demonstrate any meaningful alignment between predicted and actual values.

In contrast, the model trained on synthetic data produced improved results, with three of the four targets yielding positive R^2 values. DPPH was best predicted with $R^2 = 0.86$, followed by TPC with $R^2 = 0.51$ and FRAP with $R^2 = 0.17$. Nevertheless, AAC continued to exhibit poor predictability, with an R^2 of -1.38. These results suggest that while synthetic data can partially capture the data structure of some antioxidant properties, it remains insufficient for accurately modeling targets like AAC without real data inputs.

The RF-based model trained on the mixed dataset, which combined both the original and synthetic samples, delivered the most balanced and reliable performance. TPC and DPPH both achieved strong fits with R^2 values of 0.88, while FRAP also performed well with $R^2 = 0.70$. AAC, however, remained a challenging target, with only a marginally positive R^2 of 0.07. The mixed data approach thus demonstrated the strongest generalization overall, effectively integrating the empirical variability of real measurements with the expanded coverage of synthetically generated patterns.

These results highlight that while the RF-based mixed model improved the predictive alignment for TPC and DPPH, challenges remain in accurately modeling AAC, which may reflect inherent biological variability or limited representation in the training data.

3.10. Partial Dependence Analysis of RF-Based Models

To better interpret how individual input features influenced the model predictions, partial dependence plots (PDPs) were generated for each antioxidant response (TPC, FRAP, DPPH, and AAC) across the four predictors: concentration of solvent (C, %), solvent ratio ($R_{L/S}$, mL/g), temperature (T, °C), and extraction time (t, min). The PDPs visualize the marginal effect of each predictor after averaging out the influence of other variables, offering insight into the modeled relationships learned by RF models trained on different datasets (original, synthetic, and mixed).

For the model trained on the original dataset, PDPs (Figure 18) revealed generally smooth but shallow response curves, indicating low model sensitivity to input variation. The results of C (%) and $R_{L/S}$ showed modest negative slopes for most targets, particularly TPC and FRAP, suggesting that higher solvent concentrations and solvent ratios tended to reduce predicted antioxidant values. Temperature exhibited a more pronounced positive relationship with TPC and FRAP, while time (t) influenced predictions positively in nearly all cases, though effects on AAC appeared erratic. These modest trends are consistent with the limited representational power of the model trained on the small original dataset, which likely restricted the learned functional relationships to low-variance approximations.

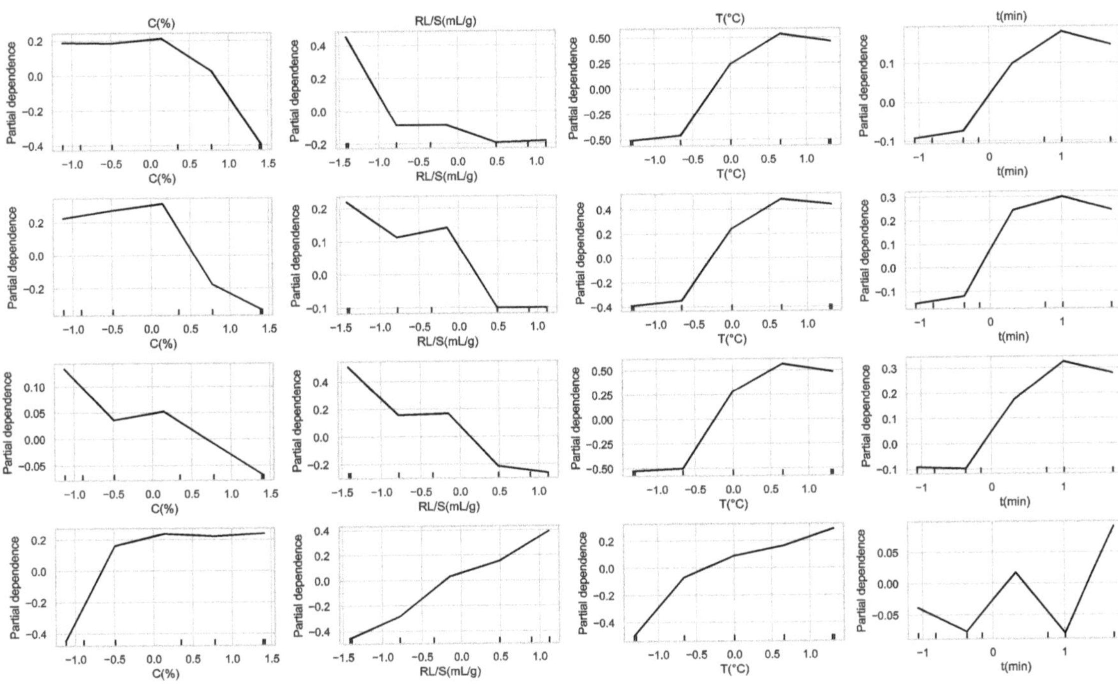

Figure 18. Partial dependence plots showing the marginal effect of each feature on antioxidant predictions from the RF original model.

In contrast, the synthetic-trained model (Figure 19) exhibited steeper and more structured response patterns across nearly all features. For example, C (%) and $R_{L/S}$ displayed strong nonlinear declines for TPC, FRAP, and DPPH, whereas temperature exhibited a clear sigmoidal increase, particularly evident for DPPH. The time variable contributed positively to predictions, with increasing slopes across most plots. These sharper transitions suggest that the synthetic model was able to capture more defined patterns between variables, although some over-smoothing and artifacts were apparent in less reliable targets such as AAC, where PDP curves were more erratic.

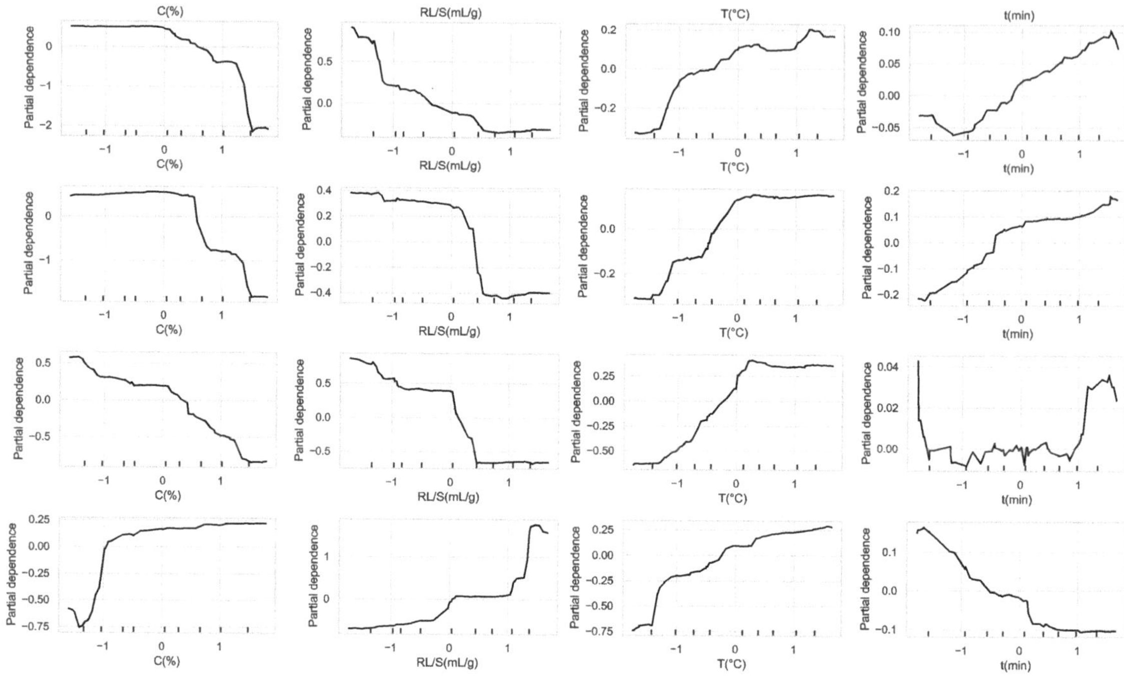

Figure 19. Partial dependence plots showing the marginal effect of each feature on antioxidant predictions from the RF synthetic model.

The model trained on the mixed dataset (Figure 20) presented the most coherent and biologically plausible trends. The PDPs across targets demonstrated well-defined, monotonic relationships. *C* (%) consistently showed negative associations with antioxidant capacity, while $R_{L/S}$ exhibited declining effects, particularly for FRAP and TPC. Temperature maintained a strong positive relationship, especially for DPPH and FRAP. Extraction time (*t*) showed clear and mostly monotonic increases in partial dependence, indicating its significant influence on yield-related outcomes. Compared to the other models, the mixed-trained model exhibited more stable and interpretable PDPs, which is consistent with its higher predictive accuracy and generalization capacity.

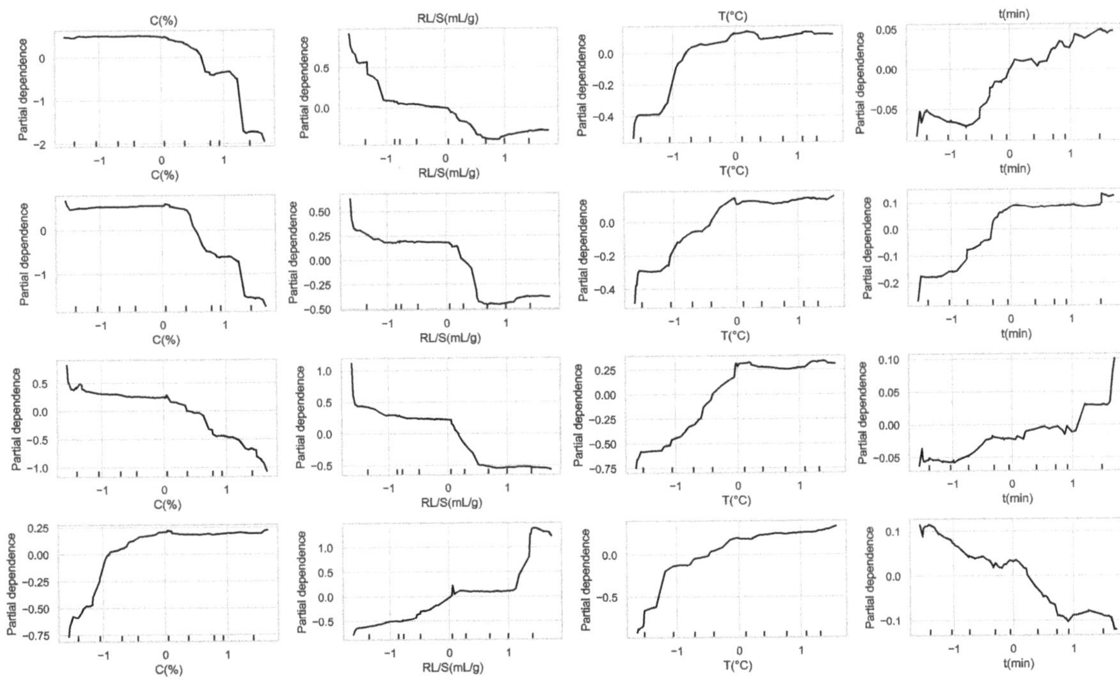

Figure 20. Partial dependence plots showing the marginal effect of each feature on antioxidant predictions from the RF mixed model.

Overall, the partial dependence analysis reinforces the conclusion that models trained on a mixture of real and synthetic data achieve superior learning of underlying relationships between process variables and antioxidant outcomes. The synthetic model was able to capture sharp feature effects, but only the mixed model exhibited smooth, consistent trends aligned with expected extraction behavior. These findings support the inclusion of controlled synthetic data to augment and stabilize learning in low-sample experimental contexts.

Despite these encouraging results, several limitations remain. The relatively small size of the original dataset constrains the ability of the models to fully capture the underlying variability of the extraction process. The current synthetic data generation approach, while effective, is based on RF predictions with Gaussian noise, which may not fully reflect the true complexity of the system. Future work should explore more advanced generative modeling techniques, such as Variational Autoencoders or Generative Adversarial Networks, to enhance the diversity and realism of synthetic data. Additionally, expanding the experimental dataset and incorporating additional physicochemical or spectral variables could further improve model accuracy and generalization, supporting more robust and transferable AI-assisted extraction models.

3.11. Model Prediction Accuracy at Optimal Conditions

To assess how accurately RF models predicted antioxidant outcomes under optimal extraction conditions, we compared model predictions against experimentally reported values for four antioxidant metrics: TPC, FRAP, DPPH, and AAC. Figure 21 presents a comparative bar plot showing the predicted values from each model along with their absolute errors, visualized as value $\pm$ error for each target.

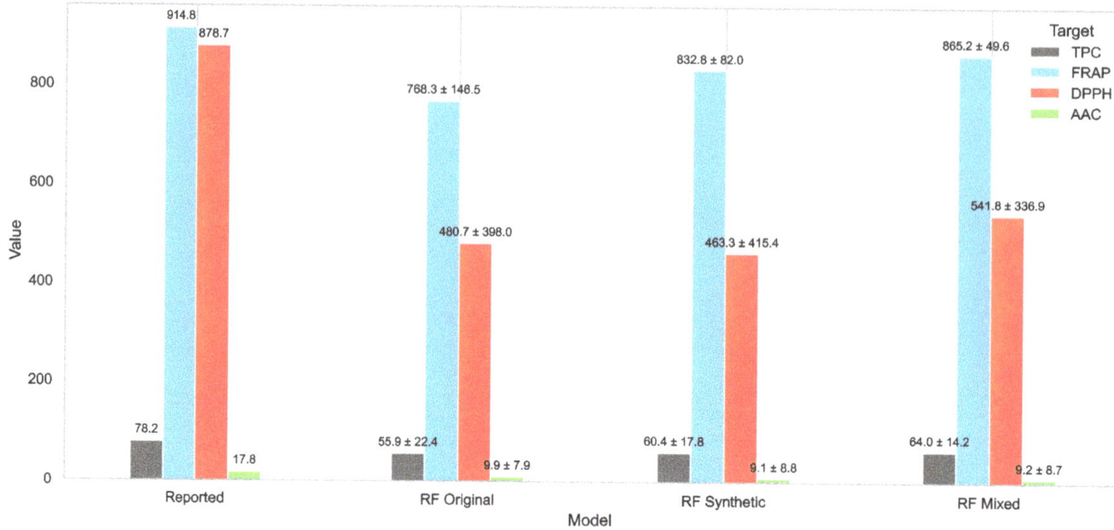

Figure 21. Comparison of predicted versus reported antioxidant values under optimal extraction conditions using RF-based models trained on the original, synthetic, and mixed datasets. Bars represent predicted values for TPC, FRAP, DPPH, and AAC, with numeric labels showing predicted value ± absolute error relative to the experimentally reported values. The mixed model produced the most accurate predictions overall, with reduced errors across most targets.

The model trained solely on the original data exhibited the highest prediction error across all targets. For TPC, the model predicted 55.9 compared to the reported 78.2, yielding an absolute error of 22.4. Similarly, FRAP and DPPH were underestimated by 768.3 and 480.7, corresponding to absolute errors of 146.5 and 398.0, respectively. AAC was also significantly underestimated with 9.9, with an error of 7.9. These results highlight the limitations of training exclusively on small, original datasets, especially when attempting to extrapolate to optimal regions.

The model trained on the synthetic data demonstrated improved performance across most targets. TPC was predicted at 60.4 with error = 17.8, while FRAP reached 832.8 with error = 82.0, and DPPH was predicted at 463.3 with error = 415.4. Although the DPPH prediction remained notably poor, AAC predictions were marginally closer to the reported value, with a predicted value of 9.1 with an error = 8.8.

The model trained on the mixed dataset yielded the most accurate and consistent predictions across all targets. TPC was predicted at 64.0 with error = 14.2, and FRAP at 865.2 with error = 49.6, showing strong alignment with the experimental values. DPPH prediction also improved, with a value of 541.8 and an error of 336.9. AAC was predicted at 9.2 with an error of 8.7. While some targets, particularly DPPH and AAC, remained challenging to predict accurately, the mixed model consistently outperformed the other models in terms of proximity to the experimental data.

In summary, these results confirm that training on a mixed dataset comprising both real and synthetic samples enhances the model's ability to generalize and make reliable predictions under optimal conditions. The mixed model showed the lowest aggregate absolute error and the closest alignment to the reported values across all antioxidant targets. Models trained solely on synthetic data exhibited poor generalization to real samples, underscoring the need for empirical grounding. In addition, predictions for specific targets such as AAC and DPPH remained less accurate, likely due to high intrinsic variability or limited representation within the training data.

4. Conclusions

This study successfully optimized the extraction conditions for rosemary leaves using PLE, demonstrating its potential as an efficient and environmentally friendly technique for recovering bioactive compounds. The optimized PLE parameters yielded extracts rich in antioxidants, polyphenols, and ascorbic acid, highlighting the suitability of PLE for such applications.

In parallel, ML approaches were applied to model and predict antioxidant responses based on extraction parameters. While the RF-based mixed model showed improved generalization compared to models trained solely on experimental or synthetic data, the small sample size and reliance on data augmentation introduce limitations to the robustness of the conclusions. In particular, feature importance results may be influenced by synthetic data, and test set performance indicates that further model refinement is needed to ensure reliable predictions in real-world scenarios.

Future research should focus on expanding experimental datasets to improve model training and validation, applying advanced generative methods for more realistic data augmentation, and conducting real-world testing of model predictions. Additionally, evaluating model transferability across different plant matrices and extraction systems, as well as validating the process at industrial scale, will be important steps toward broader practical implementation of AI-assisted extraction optimization.

Author Contributions: Conceptualization, V.A. and S.I.L.; methodology, V.A., software, V.A.; validation, V.A.; formal analysis, M.M. and V.A.; investigation, M.M. and E.B.; resources, S.I.L.; data curation, M.M. and K.G.L.; writing—original draft preparation, M.M. and K.G.L.; writing—review and editing, V.A., M.M., K.G.L., E.B. and S.I.L.; visualization, M.M. and K.G.L.; supervision, V.A. and S.I.L.; project administration, S.I.L.; funding acquisition, S.I.L. All authors have read and agreed to the published version of the manuscript.

Funding: This research received no external funding.

Data Availability Statement: The original contributions presented in this study are included in the article. Further inquiries can be directed to the corresponding author.

Conflicts of Interest: The authors declare no conflicts of interest.

References

1. Ghasemzadeh Rahbardar, M.; Hosseinzadeh, H. Toxicity and Safety of Rosemary (*Rosmarinus officinalis*): A Comprehensive Review. *Naunyn. Schmiedebergs Arch. Pharmacol.* **2025**, *398*, 9–23. [CrossRef] [PubMed]
2. Aamer, H.A.; Al-Askar, A.A.; Gaber, M.A.; El-Tanbouly, R.; Abdelkhalek, A.; Behiry, S.; Elsharkawy, M.M.; Kowalczewski, P.Ł.; El-Messeiry, S. Extraction, Phytochemical Characterization, and Antifungal Activity of Salvia Rosmarinus Extract. *Open Chem.* **2023**, *21*, 20230124. [CrossRef]
3. de Macedo, L.M.; Santos, É.M.d.; Militão, L.; Tundisi, L.L.; Ataide, J.A.; Souto, E.B.; Mazzola, P.G. Rosemary (*Rosmarinus officinalis* L., Syn *Salvia rosmarinus* Spenn.) and Its Topical Applications: A Review. *Plants* **2020**, *9*, 651. [CrossRef]
4. Ahmed, H.M.; Babakir-Mina, M. Investigation of Rosemary Herbal Extracts (*Rosmarinus officinalis*) and Their Potential Effects on Immunity. *Phytother. Res.* **2020**, *34*, 1829–1837. [CrossRef]
5. González-Minero, F.J.; Bravo-Díaz, L.; Ayala-Gómez, A. *Rosmarinus officinalis* L. (Rosemary): An Ancient Plant with Uses in Personal Healthcare and Cosmetics. *Cosmetics* **2020**, *7*, 77. [CrossRef]
6. Aziz, E.; Batool, R.; Akhtar, W.; Shahzad, T.; Malik, A.; Shah, M.A.; Iqbal, S.; Rauf, A.; Zengin, G.; Bouyahya, A.; et al. Rosemary Species: A Review of Phytochemicals, Bioactivities and Industrial Applications. *S. Afr. J. Bot.* **2022**, *151*, 3–18. [CrossRef]
7. Dhenge, R.; Rinaldi, M.; Ganino, T.; Lacey, K. Recent and Novel Technology Used for the Extraction and Recovery of Bioactive Compounds from Fruit and Vegetable Waste. In *Wealth out of Food Processing Waste*; CRC Press: Boca Raton, FL, USA, 2024; ISBN 978-1-00-326919-9.

8. Christoforidis, A.; Mantiniotou, M.; Athanasiadis, V.; Lalas, S.I. Caffeine and Polyphenolic Compound Recovery Optimization from Spent Coffee Grounds Utilizing Pressurized Liquid Extraction. *Beverages* **2025**, *11*, 74. [CrossRef]
9. Galanakis, C.M.; Aldawoud, T.M.S.; Rizou, M.; Rowan, N.J.; Ibrahim, S.A. Food Ingredients and Active Compounds against the Coronavirus Disease (COVID-19) Pandemic: A Comprehensive Review. *Foods* **2020**, *9*, 1701. [CrossRef]
10. Martins, R.; Barbosa, A.; Advinha, B.; Sales, H.; Pontes, R.; Nunes, J. Green Extraction Techniques of Bioactive Compounds: A State-of-the-Art Review. *Processes* **2023**, *11*, 2255. [CrossRef]
11. Kim, H.C.; Ha, S.Y.; Yang, J.-K. Antioxidant Activity of Ultrasonic Assisted Ethanol Extract of *Ainsliaea acerifolia* and Prediction of Antioxidant Activity with Machine Learning. *BioResours.* **2024**, *19*, 7637–7652. [CrossRef]
12. Kunjiappan, S.; Ramasamy, L.K.; Kannan, S.; Pavadai, P.; Theivendren, P.; Palanisamy, P. Optimization of Ultrasound-Aided Extraction of Bioactive Ingredients from Vitis Vinifera Seeds Using RSM and ANFIS Modeling with Machine Learning Algorithm. *Sci. Rep.* **2024**, *14*, 1219. [CrossRef] [PubMed]
13. Kalompatsios, D.; Athanasiadis, V.; Mantiniotou, M.; Lalas, S.I. Optimization of Ultrasonication Probe-Assisted Extraction Parameters for Bioactive Compounds from *Opuntia macrorhiza* Using Taguchi Design and Assessment of Antioxidant Properties. *Appl. Sci.* **2024**, *14*, 10460. [CrossRef]
14. Shehata, E.; Grigorakis, S.; Loupassaki, S.; Makris, D.P. Extraction Optimisation Using Water/Glycerol for the Efficient Recovery of Polyphenolic Antioxidants from Two Artemisia Species. *Sep. Purif. Technol.* **2015**, *149*, 462–469. [CrossRef]
15. Athanasiadis, V.; Chatzimitakos, T.; Mantiniotou, M.; Kalompatsios, D.; Bozinou, E.; Lalas, S.I. Investigation of the Polyphenol Recovery of Overripe Banana Peel Extract Utilizing Cloud Point Extraction. *Eng* **2023**, *4*, 3026–3038. [CrossRef]
16. Freedman, D.A. *Statistical Models: Theory and Practice*, 2nd ed.; Cambridge University Press: Cambridge, UK, 2009; ISBN 978-0-511-81586-7.
17. Hoerl, A.E.; Kennard, R.W. Ridge Regression: Biased Estimation for Nonorthogonal Problems. *Technometrics* **1970**, *12*, 55–67. [CrossRef]
18. Tibshirani, R. Regression Shrinkage and Selection via the Lasso. *J. R. Stat. Soc. Ser. B Methodol.* **1996**, *58*, 267–288. [CrossRef]
19. Friedman, J.H. Greedy Function Approximation: A Gradient Boosting Machine. *Ann. Stat.* **2001**, *29*, 1189–1232. [CrossRef]
20. Freund, Y.; Schapire, R.E. A Decision-Theoretic Generalization of On-Line Learning and an Application to Boosting. *J. Comput. Syst. Sci.* **1997**, *55*, 119–139. [CrossRef]
21. Willmott, C.J.; Matsuura, K. Advantages of the Mean Absolute Error (MAE) over the Root Mean Square Error (RMSE) in Assessing Average Model Performance. *Clim. Res.* **2005**, *30*, 79–82. [CrossRef]
22. Chai, T.; Draxler, R.R. Root Mean Square Error (RMSE) or Mean Absolute Error (MAE)?—Arguments against Avoiding RMSE in the Literature. *Geosci. Model Dev.* **2014**, *7*, 1247–1250. [CrossRef]
23. Kvålseth, T.O. Cautionary Note about R^2. *Am. Stat.* **1985**, *39*, 279–285. [CrossRef]
24. Thoo, Y.; Ng, S.Y.; Khoo, M.; Mustapha, W.; Ho, C. A Binary Solvent Extraction System for Phenolic Antioxidants and Its Application to the Estimation of Antioxidant Capacity in *Andrographis paniculata* Extracts. *Int. Food Res. J.* **2013**, *20*, 1103–1111.
25. Mantiniotou, M.; Athanasiadis, V.; Kalompatsios, D.; Lalas, S.I. Optimization of Carotenoids and Other Antioxidant Compounds Extraction from Carrot Peels Using Response Surface Methodology. *Biomass* **2025**, *5*, 3. [CrossRef]
26. Segovia, F.J.; Luengo, E.; Corral-Pérez, J.J.; Raso, J.; Almajano, M.P. Improvements in the Aqueous Extraction of Polyphenols from Borage (*Borago officinalis* L.) Leaves by Pulsed Electric Fields: Pulsed Electric Fields (PEF) Applications. *Ind. Crops Prod.* **2015**, *65*, 390–396. [CrossRef]
27. More, P.R.; Jambrak, A.R.; Arya, S.S. Green, Environment-Friendly and Sustainable Techniques for Extraction of Food Bioactive Compounds and Waste Valorization. *Trends Food Sci. Technol.* **2022**, *128*, 296–315. [CrossRef]
28. Athanasiadis, V.; Mantiniotou, M.; Kalompatsios, D.; Makrygiannis, I.; Alibade, A.; Lalas, S.I. Evaluation of Antioxidant Properties of Residual Hemp Leaves Following Optimized Pressurized Liquid Extraction. *AgriEngineering* **2025**, *7*, 1. [CrossRef]
29. Zhou, J.; Wang, M.; Carrillo, C.; Zhu, Z.; Brncic, M.; Berrada, H.; Barba, F.J. Impact of Pressurized Liquid Extraction and pH on Protein Yield, Changes in Molecular Size Distribution and Antioxidant Compounds Recovery from Spirulina. *Foods* **2021**, *10*, 2153. [CrossRef] [PubMed]
30. Anticona, M.; Blesa, J.; Lopez-Malo, D.; Frigola, A.; Esteve, M.J. Effects of Ultrasound-Assisted Extraction on Physicochemical Properties, Bioactive Compounds, and Antioxidant Capacity for the Valorization of Hybrid Mandarin Peels. *Food Biosci.* **2021**, *42*, 101185. [CrossRef]
31. Athanasiadis, V.; Chatzimitakos, T.; Mantiniotou, M.; Kalompatsios, D.; Kotsou, K.; Makrygiannis, I.; Bozinou, E.; Lalas, S.I. Optimization of Four Different Rosemary Extraction Techniques Using Plackett–Burman Design and Comparison of Their Antioxidant Compounds. *Int. J. Mol. Sci.* **2024**, *25*, 7708. [CrossRef]
32. Antony, A.; Farid, M. Effect of Temperatures on Polyphenols during Extraction. *Appl. Sci.* **2022**, *12*, 2107. [CrossRef]
33. Hashem Hashempur, M.; Zareshahrabadi, Z.; Shenavari, S.; Zomorodian, K.; Rastegari, B.; Karami, F. Deep Eutectic Solvent-Based Extraction of Rosemary Leaves: Optimization Using Central Composite Design and Evaluation of Antioxidant and Antimicrobial Activities. *New J. Chem.* **2025**, *49*, 4495–4505. [CrossRef]

34. Kabubii, Z.N.; Mbaria, J.M.; Mathiu, P.M.; Wanjohi, J.M.; Nyaboga, E.N. Diet Supplementation with Rosemary (*Rosmarinus officinalis* L.) Leaf Powder Exhibits an Antidiabetic Property in Streptozotocin-Induced Diabetic Male Wistar Rats. *Diabetology* **2024**, *5*, 12–25. [CrossRef]
35. Xie, L.; Li, Z.; Li, H.; Sun, J.; Liu, X.; Tang, J.; Lin, X.; Xu, L.; Zhu, Y.; Liu, Z.; et al. Fast Quantitative Determination of Principal Phenolic Anti-Oxidants in Rosemary Using Ultrasound-Assisted Extraction and Chemometrics-Enhanced HPLC–DAD Method. *Food Anal. Methods* **2023**, *16*, 386–400. [CrossRef]
36. Samer, A.J.A.; Samarrai, O.R.A. Phytochemical Screening, Antioxidant Power and Quantitative Analysis by HPLC of Isolated Flavonoids from Rosemary. *Samarra J. Pure Appl. Sci.* **2025**, *6*, 15–29. Available online: https://sjpas.com/index.php/sjpas/article/view/861 (accessed on 10 June 2025).
37. Baptista, A.; Menicucci, F.; Brunetti, C.; dos Santos Nascimento, L.B.; Pasquini, D.; Alderotti, F.; Detti, C.; Ferrini, F.; Gori, A. Unlocking the Hidden Potential of Rosemary (*Salvia rosmarinus* Spenn.): New Insights into Phenolics, Terpenes, and Antioxidants of Mediterranean Cultivars. *Plants* **2024**, *13*, 3395. [CrossRef]
38. Miljanović, A.; Dent, M.; Grbin, D.; Pedisić, S.; Zorić, Z.; Marijanović, Z.; Jerković, I.; Bielen, A. Sage, Rosemary, and Bay Laurel Hydrodistillation By-Products as a Source of Bioactive Compounds. *Plants* **2023**, *12*, 2394. [CrossRef]

processes

MDPI

Article

One-Pot Improvement of Stretchable PEDOT/PSS Alginate Conductivity for Soft Sensing Biomedical Processes

Somayeh Zanganeh [1], Alberto Ranier Escobar [2], Hung Cao [1,2,*] and Peter Tseng [1,2]

[1] Department of Electrical Engineering and Computer Science, University of California, Irvine, Irvine, CA 92697, USA
[2] Department of Biomedical Engineering, University of California, Irvine, Irvine, CA 92697, USA
* Correspondence: hungcao@uci.edu

Abstract: Hydrogels have immense potential in soft electronics due to their similarity to biological tissues. However, for applications in fields like tissue engineering and wearable electronics, hydrogels must obtain electrical conductivity, stretchability, and implantability. This article explores recent advancements in the development of electrically conductive hydrogel composites with high conductivity, low Young's modulus, and remarkable stretchability. By incorporating conductive particles into hydrogels, such as poly(3,4-ethylenedioxythiophene)/poly (styrenesulfonate) (PEDOT/PSS) researchers have enhanced their conductivity. This study presents a one-pot synthesis method for creating electrically conductive hydrogel composites by combining PEDOT/PSS with alginate. The hydrogel reveals changes in chemical composition upon treatment with dimethyl sulfoxide (DMSO). Additionally, surface morphology analysis via Field Emission Scanning Electron Microscopy (FESEM) and Atomic Force Microscopy (AFM) demonstrate the impact of DMSO treatment on PEDOT/PSS/alginate films. Furthermore, electrical conductivity measurements highlighted the effectiveness of the conductive hydrogels in Electromyography (EMG) and human motion detection. This study offers insights into the fabrication and characterization of stretchable, conductive hydrogels, advancing their potential for various soft sensing biomedical applications. The optimized PDOT/PSS/alginate composite under dry condition shows a conductivity of 0.098 S/cm and can be stretched without significant loss in conductivity or mechanical stability. This one-pot method provides a simple and effective way to improve the properties of conductive hydrogel-based sensors.

Keywords: biomaterials; conductive hydrogel; PEDOT/PSS; sensors and electronics

1. Introduction

Advancements in wearable bioelectronics are facilitating the shift towards patient-centric, personalized healthcare [1,2]. Traditional electronic systems are composed of rigid materials, such as metals and silicon, in a two-dimensional (2D) plane and are not suitable for interfacing with the human body [1,3–8]. Wearable bioelectronics adapt to the soft, curvilinear surfaces of the body in order to provide noninvasive, real-time monitoring of a patient's physiological state, including their heart rate, respiration, and blood oxygen levels, for in situ clinical monitoring and personal health management [2,9,10].

There has been significant progress in the development of stretchable conductive materials for point-of-care health monitoring through creative, structural organization and/or novel material selection [11–14]. The first strategy utilizes deterministic geometrics (e.g., wave/wrinkle, kirigami tessellations, cracks) to allow otherwise rigid materials to deform out of plane in response to stress [15–17]. The second strategy swaps rigid substrate materials for stretchable, conductive materials whose conformal properties are independent of their geometry [18–20]. In combination, these strategies provide options that significantly expand the interfacing capabilities of wearable bioelectronics with the human body [18,21,22].

Poly(3,4-ethylenedioxythiophene)/poly(styrene sulfonate) (PEDOT/PSS) is a highly promising conductive polymer for use in wearable sensing systems [23–25]. PEDOT/PSS boasts tunable conductivity, good transmittance, and excellent thermal stability, in addition to being compatible with a wide range of production processes [24–28]. Furthermore, the inherent biocompatibility of the polymer allows it to readily interface with the human body and sense signals such as body temperature, humidity, and strain [27,29–31]. Despite these advantages, native PEDOT/PSS has very limited stretchability due to its rigid conjugated backbone and strong interchain interactions that can lead to crack formation under strain [25,27]. To address this key limitation, it is necessary to incorporate other materials, such as elastomers, plasticizers, and hydrogels, that provide free volume for chain stretching while also enhancing the crystallinity of PEDOT regions within PEDOT/PSS substrates to compensate for potential reductions in conductivity [31].

Here, we present a one-pot method for improving the stretchability and conductivity of hydrogel films composed of PEDOT/PSS and sodium alginate. Sodium alginate is a naturally derived polysaccharide that forms a highly tunable, stretchable hydrogel in the presence of multivalent cations—most commonly calcium. When combined with PEDOT/PSS, the resulting hydrogel films retain the inherent stretchability of alginate while also gaining the conductive properties native to PEDOT/PSS substrates. Further modification of conductive hydrogel films via exposure to dimethyl sulfoxide (DMSO) removes insulative PSS groups, strengthening interchain interactions in PEDOT-rich regions of the substrate and significantly improving the conductivity of the overall films [25]. This simple method allows for the fabrication of highly conductive and stretchable films that can be readily incorporated into wearable bioelectronic platforms.

2. Materials and Methods

2.1. Materials

PEDOT/PSS aqueous solution (PH1000, Heraeus Clevios) was purchased from Hanau, Hessen, Germany. Dimethyl sulfoxide (DMSO) (purity $\geq$ 98%), D-(+)-gluconic acid δ-lactone (GDL), and calcium carbonate ($CaCO_3$) were obtained from Sigma Aldrich (Burlington, MA, USA). Sodium alginate (Na-ALG, viscosity 80–120 cp) was purchased from FUJIFILM Wako Pure Chemical Corporation (Osaka, Japan). All chemicals were purchased and used without further purification. All aqueous solutions were prepared using deionized water (DI) unless otherwise stated.

2.2. PEDOT/ALG Hydrogel Film Preparation

In order to make hydrogels with various mechanical and electrical properties, three different PEDOT/PSS/ALG precursor solutions were prepared at room temperature. First, alginate (ALG) 10% was made by vigorously mixing 4 g of sodium alginate in 40 mL deionized water using a magnetic stirrer (Corning™, Corning, NY, USA) until it completely dissolved. This alginate was used in all samples, and the remaining amount was kept in the fridge at 4 degrees Celsius. To prepare PEDOT/PSS films, we added the pristine aqueous PEDOT/PSS solution into the alginate precursor solution. To investigate the properties of the whole range of PEDOT/PSS/ALG compositions, we prepared three different concentrations of alginate. We made PEDOT/PSS/ALG 1:1 by mixing equal volumes of 1.3 wt% PEDOT/PSS and 1.3 wt% alginate, PEDOT/PSS/ALG 1:3 by mixing equal volumes of 1.3 wt% PEDOT/PSS and 3.9 wt% alginate, and PEDOT/PSS/ALG 3:1

by mixing three times the volume of 1.3 wt% PEDOT/PSS with respect to that of 1.3 wt% alginate. We then added 16 mg/mL gluconic acid and 4.5 mg/mL calcium carbonate to accelerate the gelation of films. The solutions were then strongly mixed using vortex. Stirring the solutions between each step assured proper bonding. The precursor solutions were then transferred into a substrate, usually petri dishes, and were kept under fume hood for minimum 12 h at room temperature. The petri dishes were left open to accelerate the drying sequence. Hydrogel films made with this method were referred to as 3:1, 1:1, or 1:3 PEDOT/ALG. Figure 1a depicts this procedure.

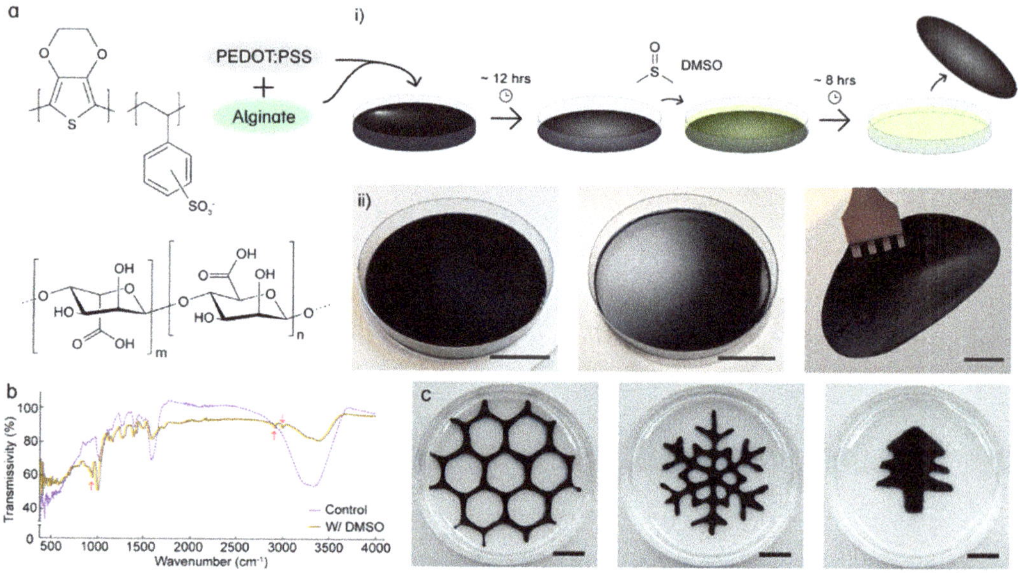

Figure 1. (**a**) A schematic illustration of preparing PEDOT/ALG films. (**b**) FTIR spectra of 1:1 PEDOT/ALG films. Characteristic bands are shown by arrows. (**c**) Simple patterning of 1:1 PEDOT/ALG films in various shapes. Scale bar: 1 cm.

2.3. DMSO Post-Treatment

In order to form a conductive and mechanically stable film using secondary doping method, DMSO > 99% was poured on the dried 1:3, 1:1 and 3:1 PEDOT/PSS, and the thoroughly submerged films, together with the petri dishes, were left at room temperature for 8 h after sealing the cap using parafilm to avoid the solution from evaporating. The films then easily detached from the substrate and could be kept in DMSO for over two months without any change in their mechanical and/or electrical properties for further characterization and analysis.

2.4. Resistance Measurements

PEDOT/ALG films were cut into 1×2 cm^2 rectangles and left to dry completely under ambient conditions (~10 min). Copper wires were attached to each end of the rectangles using conductive silver epoxy adhesive (8331D, MG Chemicals (Ontario, Canada)). A low resistivity meter (Loresta-GP MCP-T600, Mitsubishi (Tokyo, Japan)) was used to measure the electrical conductivity of the films with the probes attached to the copper wires.

2.5. Dynamic Mechanical Analysis (DMA)

Uniform rectangular sections of 1:3, 1:1, and 3:1 PEDOT/ALG films were cut from larger films for tensile testing. For each PEDOT/ALG ratio, two conditions were tested: a control condition without DMSO treatment and an experimental condition with DMSO treatment. Since the control films would disintegrate when in contact with water, they were strained as dry films. The experimental samples were lightly dried using a delicate task wipe (Kimwipes, Kimtech™ (Busan, Republic of Korea)) before straining. Stress/strain curves for each film were obtained using a DMA Q800 (TA Instruments (Dallas, TX, USA)) equipped with a tensile clamp set to perform a force ramp at 0.1 N/min until sample failure.

3. Results

3.1. Fourier-Transform Infrared Spectroscopy (FTIR)

Several methods, such as conductivity measurements, spectroscopy, microscopy, and mechanical testing, may be used to examine the obtained films. Figure 1b shows the Fourier Transform InfraRed (FTIR) spectra of 1:1 PEDOT/ALG as the control and 1:1 PEDOT/ALG with DMSO treatment using a FTIR spectrometer (JASCO Inc., Jeddah, Saudi Arabia). As well as the absorption bands at 1564 cm^{-1} for the C=C stretching in the thiophene ring, at 1270 and 1122 cm^{-1} for the vibrations of the fused dioxane ring, and at 862 cm^{-1} for the stretching of the C–S bond in the thiophene ring [8] associated with PEDOT/PSS in both spectra, the absorption bands at 980, 2900 and 3000 confirm the presence of DMSO in films after treatment with DMSO, while in the control sample, before adding DMSO, those peaks disappeared. Based on the FTIR spectra, adding DMSO did not damage the PEDOT/ALG chains, and it allowed the re-orientation of the carboxylate groups and maintained the mechanical stability of the film [9].

3.2. PEDOT/ALG Film Patterning

The patterning of conductive hydrogels is an important step in making bioelectronics. Due to their water content, it is challenging to pattern hydrogel-based films. Figure 1c shows different pattering of our stretchable and conductive hydrogel films in different shapes, including honeycomb, snowflake, and Christmas tree, to confirm the stability, patternability, and homogeneity of the films, making them perfect options for use in a wide spectrum of applications in flexible sensors and actuators [10]. These films are stable in different media, and no change in their physical condition has been noticed in DI water, PBS, and ethanol after 6 months.

3.3. Scanning Electron Microscopy (SEM)

To examine the surface morphology of the PEDOT/ALG films, Field Emission Scanning Electron Microscope (FESEM, FEI Magellan 400 XHR Scanning Electron Microscope (Hillsboro, Oregon)) images are used. Figure 2a shows the surface morphologies of all films with different ratios of PEDOT to alginate and before and after being emerged in DMSO. Calcium chloride ($CaCl_2$) is used in control samples before tests to maintain consistent ionic conditions and ensure that any observed effects in experimental samples are due to the variable being tested, not the presence or absence of calcium, plus it helps with structural stability. The control samples (before adding DMSO) show a homogeneous structure with an irregular structure and variable pore size. On the other hand, FESEM pictures of the films that were submerged in DMSO exhibit a uniform interconnected structure due to the creation of long chains on the surface across the films. These long chains accelerate electron transfer and, hence, increase the conductivity, which also have improves the mechanical responses of these films [11]. Moving from one side to the other side for electrons via these fibrous chains is easier if there are more PEDOT/PSS nanoparticles in the material rather than insulating polymers like hydrogels [12]. This belief is clearly validated in Figure 2a through increasing the percentage of alginate in the film. Moreover, using 1:3 PEDOT/ALG in control films demonstrates more nanoparticles than 1:1 and more than 3:1. This also has been shown for the same films after treatment with DMSO overnight.

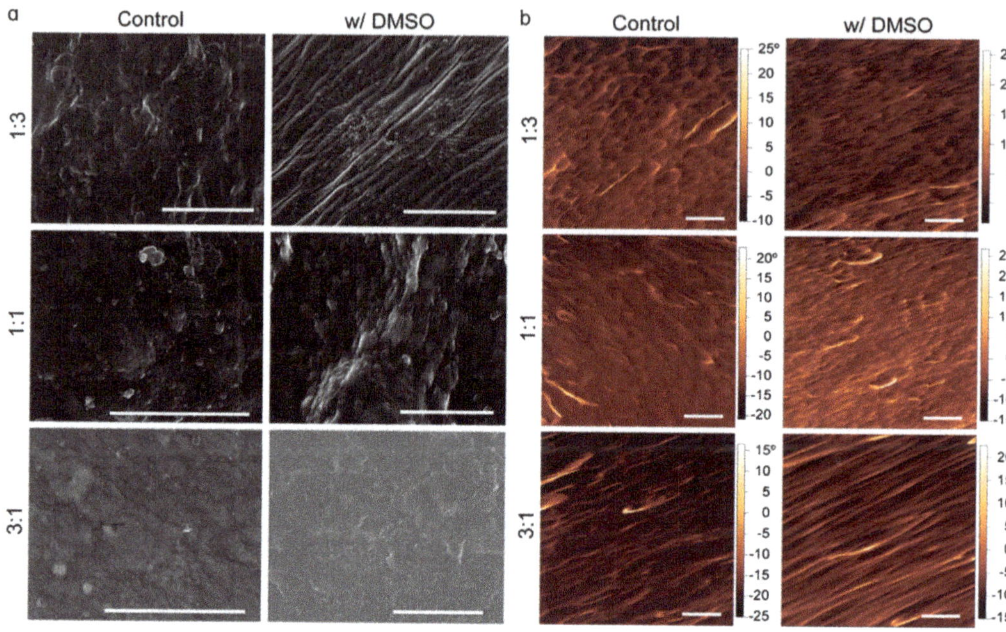

Figure 2. (**a**) SEM images of 1:3, 1:1, and 3:1 PEDOT/ALG films in $CaCl_2$ (control) and after they had been emerged in DMSO. Samples were dried before we took the images. (**b**) Corresponding AFM topography images. Scale bar: 10 µm.

3.4. Atomic Force Microscopy (AFM)

Additionally, Figure 2b provides the corresponding AFM topography images of PEDOT/PSS/ALG 1:3, 1:1, and 3:1 before and after submerging them in DMSO. The roughness and height of the 1:3 PEDOT/PSS/ALG samples clearly show that the compositions of the samples are different, and exposing the samples to DMSO creates stronger chains and, therefore, non-uniform and more stable films [13]. Upon adding more PEDOT/PSS, the roughness increases, as does the average height.

3.5. Resistance Measurements

The addition of PEDOT/PSS into alginate-based hydrogel films imparts the films with conductive properties, enabling them to function as substrates for biosensing applications [14]. Upon the addition of PEDOT/PSS, a notable decrease in resistance was observed in the control samples, corresponding to an increase in the film's conductivity (Figure 3a). Furthermore, it was observed that films with a higher ratio of PEDOT/PSS exhibited increased conductivity, implying a correlation between the concentration of PEDOT/PSS and the enhancement of electrical conductivity in the films.

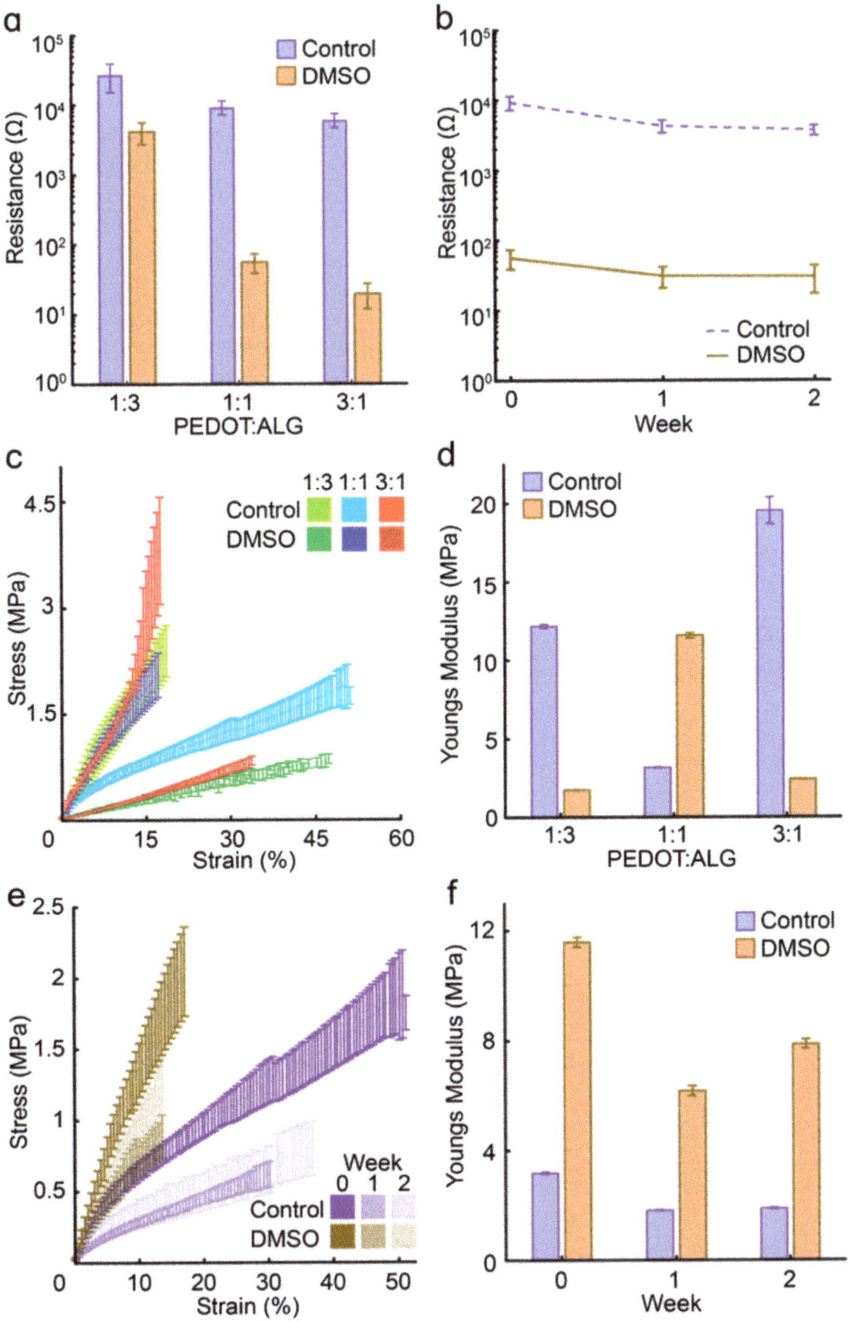

Figure 3. (**a**) Electrical conductivities of 1:3, 1:1, and 3:1 for PEDOT/ALG films in $CaCl_2$ (control) and DMSO. (**b**) Resistance of 1:1 for PEDOT/ALG films in $CaCl_2$ (control) and DMSO over time. (**c**,**d**) Mechanical properties of 1:3, 1:1, and 3:1 for PEDOT/ALG films in $CaCl_2$ (control) and DMSO. (**e**,**f**) Mechanical properties of 1:1 for PEDOT/ALG films in $CaCl_2$ (control) and DMSO over time.

Exposure to DMSO further increases the conductivity of the PEDOT/ALG films due to improved chain alignment [15]. The conductivity of the films after DMSO exposure increases by a single order of magnitude for 1:3 PEDOT/ALG, two orders of magnitude for 1:1 PEDOT/ALG, and almost three orders of magnitude for 3:1 PEDOT/ALG. DMSO-treated PEDOT/ALG films retain this improvement over the course of two weeks and exhibit a similar slight increase in conductivity, corresponding to the decrease in resistance, as shown in Figure 3b, as that for the non-treated films. The conductivity values were measured as 0.055 S/cm for the 1:3 film, 0.098 S/cm for the 1:1 film, and 0.124 S/cm for the 3:1 film.

The observed decrease in resistance could be attributed to various factors such as increased intermolecular interactions, improved polymer chain alignment, or enhanced charge carrier mobility within the film structures [16]. The enhancement of electrical conductivity in PEDOT/PSS films through the incorporation of dimethyl sulfoxide (DMSO) is attributed to significant morphological and structural modifications within the polymer matrix. DMSO functions as a secondary dopant, inducing phase separation between the conductive PEDOT domains and the insulating PSS component. This redistribution leads to the formation of a PEDOT-enriched network, which facilitates more efficient charge transport. Furthermore, DMSO promotes a conformational transition of PEDOT chains from a coiled to a more linear and expanded structure, thereby increasing carrier mobility. Concurrently, the partial removal or rearrangement of PSS reduces the density of insulating regions, further enhancing the film's electrical conductivity. These findings suggest that while DMSO treatment initially impacts conductivity adversely, the long-term trend reflects an overall enhancement in electrical properties for both control and treated samples. These films can be stored at room temperature for up to two months without significant change in their conductivity. Further analysis is warranted to elucidate the underlying mechanisms driving this phenomenon and to optimize the fabrication process for desired conductivity outcomes.

3.6. Dynamic Mechanical Analysis (DMA)

Tensile tests were performed on rectangular sections of PEDOT/ALG films with varying PEDOT/PSS-to-alginate ratios (1:3, 1:1, 3:1) to assess the impact of DMSO on their mechanical properties. The DMSO-treated samples exhibited an increase in stretchability, as shown by their increased strain-to-break relative to the dry control samples, in exchange for a decrease in stiffness (Figure 3c,d). This is likely attributed to the lateral association of alginate chains caused by exposure to DMSO and additional solvent-induced gelation within the alginate network [17]. The solvent acts as a plasticizer, reducing intermolecular rigidity and increasing the film's ability to deform under strain. Enhanced chain alignment and increased crystallinity, facilitated by DMSO, result in a more mechanically robust and stretchable network. These combined effects render DMSO-treated PEDOT/PSS films highly suitable for application in next-generation stretchable and wearable electronic devices. The underlying mechanisms have been substantiated by multiple studies, including the detailed structural and electrical characterizations reported in the recent literature. Long term, the PEDOT/ALG films show minimal mechanical degradation within a week of synthesis, followed by consistent mechanical behavior afterwards (Figure 3e,f). Stretchability is found to be 45% strain for 1:3, 68% for 1:1, and 53% for 3:1. The corresponding Young modulus values were 0.46 MPa (1:3), 0.33 MPa (1:1), and 0.52 MPa (3:1).

DMSO treatment decreased the Young modulus for the 1:3 and 3:1 films, indicating a softening effect, whereas for the 1:1 film, the modulus increased, suggesting stiffening. This contrasting result may arise from the unique balance between PEDOT/PSS and alginate content in the 1:1 formulation, which allows for a different interaction mechanism upon DMSO exposure.

In the 1:3 film, the high alginate concentration results in a dominant hydrogel matrix, and DMSO treatment likely disrupts hydrogen bonding and slightly swells the gel, leading to a softer and more elastic network. In the 3:1 film, the PEDOT/PSS is more dominant, and DMSO, known to enhance chain mobility and plasticization, reduces the rigidity of the film by weakening intermolecular interactions among PEDOT/PSS chains.

In contrast, the 1:1 film presents a more balanced composition where both PEDOT/PSS and alginate are sufficiently present to form an interpenetrating network. In this formulation, DMSO treatment may enhance interactions at the interface of PEDOT/PSS and alginate, possibly through improved chain alignment or partial densification, leading to a more tightly packed and crosslinked structure. This can result in an increase in Young's modulus and a more rigid stress/strain response.

Thus, the mechanical outcome of DMSO treatment appears to be composition-dependent, and the 1:1 film is a unique case where intermediate ratios allow DMSO to enhance structural integrity rather than induce softening.

Among the three PEDOT/ALG formulations investigated (1:3, 1:1, and 3:1), the ratio of PEDOT/PSS to alginate played a critical role in determining the electrical and mechanical properties of the hydrogel films. As the concentration of PEDOT/PSS increased, the electrical conductivity of the films improved significantly, with the 3:1 formulation achieving the highest conductivity. However, this enhancement came at the cost of reduced stretchability and increased stiffness, as evidenced by a higher Young modulus and diminished mechanical compliance. Conversely, increasing the alginate content (as in the 1:3 formulation) improved the stretchability and softness of the hydrogel but resulted in lower electrical conductivity and reduced performance stability under repeated mechanical deformation. The 1:1 formulation demonstrated a favorable balance, achieving a conductivity of 0.098 S/cm while maintaining excellent mechanical flexibility, moderate Young's modulus, and good stability during cyclic strain tests. This combination of properties makes the 1:1 PEDOT/ALG hydrogel the most promising candidate for bioelectronic applications, particularly for use in flexible and stretchable EMG sensors where both conductivity and mechanical conformity are essential.

3.7. Electromyography (EMG)

One of the main applications of conductive films is in electronic circuits and EMG. Firstly, a simplified circuit model was utilized to characterize the RC time constant of the material, as illustrated in Figure 1a. The schematic of the parallel plate capacitor, featuring planar copper plates separated by a PEDOT/PSS conductive film, was employed to understand the electrical behavior of the material. The fabricated conductive films were placed in between two copper layers to make a capacitor (Figure 4a) and then connected to a voltage source.

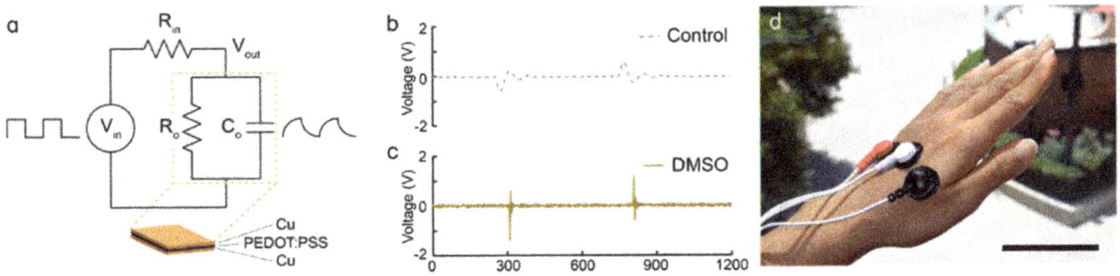

Figure 4. *Cont.*

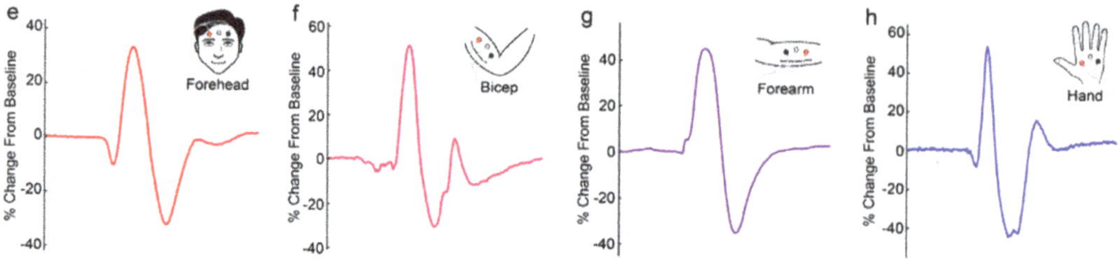

Figure 4. The applications of conductive stretchable PEDOT/ALG films (**a**) A simplified circuit model for characterizing the RC time constant of the material (**top**). A schematic of the parallel plate capacitor, wherein planar copper plates are interceded by a PEDOT/PSS conductive film (**bottom**). (**b**,**c**) Output signals for (**d**) EMG measurement using the PEDOT/PSS conductive electrodes from the hand. The electrodes were prepared and then commercial three leads were connected to a portable Arduino microcontroller to collect the data. Scale bar (5 cm). (**e**) The EMG signal for facial expression collected from the forehead. (**f–h**) The EMG signals collected from the biceps, forearm, and hand during exercise.

Figure 4b,c show real-time recording voltage signals for control and DMSO-treated samples, respectively. The PEDOT/PSS conductive electrodes were prepared and then commercial three leads were connected to a portable Arduino microcontroller to collect the data indicating the potential of PEDOT/ALG films in wearable healthcare devices. In electrochemical property tests for skin-conformable bioelectronic devices, a three-electrode system is commonly used to accurately characterize the electrochemical behavior of the working material: (a) Working electrode (WE)—This is the electrode made of the material being tested (in this case, the PEDOT/ALG hydrogel). This is where the electrochemical reactions of interest occur. The performance of this electrode reflects the material's charge storage, conductivity, and electrochemical stability. (b) Reference electrode (RE)—This provides a stable and known potential against which the working electrode's potential can be measured. It does not pass current and ensures accurate and reproducible voltage control. Common examples include Ag/AgCl or saturated calomel electrodes. (c) Counter electrode (CE)—Also known as the auxiliary electrode, it completes the circuit by allowing the current to flow through the system. It balances the current at the working electrode so that potential control and measurement are not influenced by the electrode's resistance. The electrodes are placed on hand (Figure 4d), and upon closing or opening, a spike will happen on the output signal. By comparing these two diagrams, we show that the PEDOT/PSS films that are treated with DMSO are more sensitive to motion than those without DMSO post-treatment, and they are considered better motion sensors, benefiting many applications.

Figure 4e showcases EMG signals for facial expression collected from the forehead, while Figure 4f–h display EMG signals acquired from the biceps, forearm, and hand during exercise, highlighting the adaptability and reliability of PEDOT/ALG films in physiological monitoring applications. Overall, these findings underscore the promising prospects of PEDOT/ALG films in the realm of flexible electronics and biomedical engineering.

4. Conclusions

In this study, a one-pot synthesis strategy was employed to enhance the electrical conductivity of stretchable PEDOT/PSS/alginate hydrogels without compromising their mechanical compliance or biocompatibility. The proposed formulation demonstrated significantly improved conductivity, rendering it highly suitable for applications in soft, skin-conformable bioelectronic devices. This work addresses a critical challenge in the development of stretchable electronics by presenting a facile and scalable fabrication method that does not require complex post-processing or chemical doping.

Future work will involve evaluating the long-term mechanical stability and electrical performance of the hydrogel under cyclic loading and varied environmental conditions to establish its reliability in real-world applications. In vivo biocompatibility studies and degradation assessments will also be conducted to ensure clinical safety and efficacy. Furthermore, integration with wireless modules for data transmission and power supply will be explored to facilitate the development of autonomous wearable or implantable devices. Expanding the functionality of the hydrogel to include the multiplexed sensing of biochemical or biomechanical signals is also envisioned to broaden its application scope in personalized healthcare and soft robotics.

Author Contributions: Conceptualization, S.Z. and P.T.; methodology, S.Z. and A.R.E.; validation, S.Z. and A.R.E.; formal analysis, S.Z. and A.R.E.; resources, H.C. and P.T.; data curation, S.Z. and A.R.E.; writing—original draft preparation, S.Z. and A.R.E.; writing—review and editing, S.Z. and A.R.E.; H.C. and P.T.; visualization, S.Z.; supervision, H.C. and P.T.; project administration, H.C. and P.T. All authors have read and agreed to the published version of the manuscript.

Funding: This research received no external funding.

Data Availability Statement: The original contributions presented in this study are included in the article. Further inquiries can be directed to the corresponding author.

Conflicts of Interest: The authors declare no conflicts of interest.

References

1. Libanori, A.; Chen, G.; Zhao, X.; Zhou, Y.; Chen, J. Smart textiles for personalized healthcare. *Nat. Electron.* **2022**, *5*, 142–156. [CrossRef]
2. Xu, X.; Zhao, Y.; Liu, Y. Wearable electronics based on stretchable organic semiconductors. *Small* **2023**, *19*, 2206309. [CrossRef] [PubMed]
3. Gong, S.; Cheng, W. One-dimensional nanomaterials for soft electronics. *Adv. Electron. Mater.* **2017**, *3*, 1600314. [CrossRef]
4. Wang, C.; Wang, C.; Huang, Z.; Xu, S. Materials and structures toward soft electronics. *Adv. Mater.* **2018**, *30*, 1801368. [CrossRef]
5. Jang, K.-I.; Li, K.; Chung, H.U.; Xu, S.; Jung, H.N.; Yang, Y.; Kwak, J.W.; Jung, H.H.; Song, J.; Yang, C.; et al. Self-assembled three-dimensional network designs for soft electronics. *Nat. Commun.* **2017**, *8*, 15894. [CrossRef]
6. Keplinger, C.; Sun, J.-Y.; Chiang Foo, C.; Rothemund, P.; Whitesides, G.M.; Suo, Z. Stretchable, transparent, ionic conductors. *Science* **2013**, *341*, 984–987. [CrossRef]
7. Byun, S.-H.; Sim, J.Y.; Zhou, Z.; Lee, J.; Qazi, R.; Walicki, M.C.; Parker, K.E.; Haney, M.P.; Choi, S.H.; Shon, A.; et al. Mechanically transformative electronics, sensors, and implantable devices. *Sci. Adv.* **2019**, *5*, eaay0418. [CrossRef]
8. Zhao, C.; Ma, Z.; Zhu, X.X. Rational design of thermoresponsive polymers in aqueous solutions: A thermodynamics map. *Prog. Polym. Sci.* **2019**, *90*, 269–291. [CrossRef]
9. Wang, L.; Xie, S.; Wang, Z.; Liu, F.; Yang, Y.; Tang, C.; Wu, X.; Liu, P.; Li, Y.; Saiyin, H.; et al. Functionalized helical fibre bundles of carbon nanotubes as electrochemical sensors for long-term in vivo monitoring of multiple disease biomarkers. *Nat. Biomed. Eng.* **2019**, *4*, 159–171. [CrossRef]
10. Meng, K.; Zhao, S.; Zhou, Y.; Wu, Y.; Zhang, S.; He, Q.; Wang, X.; Zhou, Z.; Fan, W.; Tan, X.; et al. A wireless textile-based sensor system for self-powered personalized health care. *Matter* **2020**, *2*, 896–907. [CrossRef]
11. Choi, C.; Choi, M.K.; Hyeon, T.; Kim, D.-H. Nanomaterial-based soft electronics for healthcare applications. *ChemNanoMat* **2016**, *2*, 1006–1017. [CrossRef]
12. Dickey, M.D. Stretchable and soft electronics using liquid metals. *Adv. Mater.* **2017**, *29*, 1606425. [CrossRef] [PubMed]
13. Jo, M.; Bae, S.; Oh, I.; Jeong, J.; Kang, B.; Hwang, S.; Lee, S.; Son, H.; Moon, B.; Ko, M.; et al. 3D printer-based encapsulated origami electronics for extreme system stretchability and high areal coverage. *ACS Nano* **2017**, *11*, 2714–2721. [CrossRef]
14. Morikawa, Y.; Yamagiwa, S.; Sawahata, H.; Numano, R.; Koida, K.; Ishida, M.; Kawano, K. Ultrastretchable kirigami bioprobes. *Adv. Healthc. Mater.* **2018**, *7*, 1800130. [CrossRef]
15. Kang, D.; Pikhitsa, P.V.; Choi, Y.W.; Lee, C.; Shin, S.S.; Piao, L.; Park, B.; Suh, K.-Y.; Kim, T.-i.; Choi, M. Ultrasensitive mechanical crack-based sensor inspired by the spider sensory system. *Nature* **2014**, *516*, 222–226. [CrossRef]

16. Zhou, Y.; Wan, C.; Yang, Y.; Yang, H.; Wang, S.; Dai, Z.; Ji, K.; Jiang, H.; Chen, X.; Long, Y. Highly stretchable, elastic, and ionic conductive hydrogel for artificial soft electronics. *Adv. Funct. Mater.* **2019**, *29*, 1806220. [CrossRef]
17. Stoyanov, H.; Kollosche, M.; Risse, S.; Waché, R.; Kofod, G. Soft conductive elastomer materials for stretchable electronics and voltage controlled artificial muscles. *Adv. Mater.* **2012**, *5*, 578–583. [CrossRef]
18. Jang, K.-I.; Chung, H.U.; Xu, S.; Lee, C.-H.; Luan, H.; Jeong, J.; Cheng, H.; Kim, G.-T.; Han, S.-Y.; Lee, J.-W.; et al. Soft network composite materials with deterministic and bio-inspired designs. *Nat. Commun.* **2015**, *6*, 6566. [CrossRef]
19. Byun, J.; Oh, E.; Lee, B.; Kim, S.; Lee, S.; Hong, Y. A single droplet-printed double-side universal soft electronic platform for highly integrated stretchable hybrid electronics. *Adv. Funct. Mater.* **2017**, *27*, 1701912. [CrossRef]
20. Zhang, X.; Yang, W.; Zhang, H.; Xie, M.; Duana, X. PEDOT:PSS: From conductive polymers to sensors. *Nanotechnol. Precis. Eng.* **2021**, *4*, 045004. [CrossRef]
21. Wang, Y.-F.; Sekine, T.; Takeda, Y.; Yokosawa, K.; Matsui, H.; Kumaki, D.; Shiba, T.; Nishikawa, T.; Tokito, S. Fully printed PEDOT:PSS-based temperature sensor with high humidity stability for wireless healthcare monitoring. *Sci. Rep.* **2020**, *10*, 2467. [CrossRef] [PubMed]
22. Hou, S.; Chen, H.; Lv, D.; Li, W.; Liu, X.; Zhang, Q.; Yu, X.; Han, Y. Highly Conductive Inkjet-Printed PEDOT:PSS Film under Cyclic Stretching. *ACS Appl. Mater. Interfaces* **2023**, *15*, 28503–28515. [CrossRef] [PubMed]
23. He, H.; Zhang, L.; Yue, S.; Yu, S.; Wei, J.; Ouyang, J. Enhancement in the mechanical stretchability of PEDOT:PSS films by compounds of multiple hydroxyl groups for their application as transparent stretchable conductors. *Macromolecules* **2013**, *54*, 1234–1242. [CrossRef]
24. Yu, Y.; Peng, S.; Blanloeuil, P.; Wu, S.; Wang, C.H. Wearable temperature sensors with enhanced sensitivity by engineering microcrack morphology in PEDOT:PSS–PDMS sensors. *ACS Appl. Mater. Interfaces* **2020**, *12*, 36578–36588. [CrossRef]
25. Stříteský, S.; Marková, A.; Víteček, J.; Šafaříková, E.; Hrabal, M.; Kubáč, L.; Kubala, L.; Weiter, M.; Vala, M. Printing inks of electroactive polymer PEDOT:PSS: The study of biocompatibility, stability, and electrical properties. *J. Biomed. Mater. Res. Part A* **2018**, *106*, 1121–1128. [CrossRef]
26. Li, Y.; Cao, W.; Liu, Z.; Zhang, Y.; Chen, Z.; Zheng, X. A personalized electronic textile for ultrasensitive pressure sensing enabled by biocompatible MXene/PEDOT:PSS composite. *Carbon Energy* **2024**, *6*, e530. [CrossRef]
27. Soni, M.; Bhattacharjee, M.; Ntagios, M.; Dahiya, R.S. Printed temperature sensor based on PEDOT:PSS-graphene oxide composite. *IEEE Sens. J.* **2020**, *20*, 7525–7531. [CrossRef]
28. Sasaki, R.; Katsuhara, M.; Yoshifuji, K.; Komoriya, Y. Novel dry EEG electrode with composite filler of PEDOT:PSS and carbon particles. In Proceedings of the 45th Annual International Conference of the IEEE Engineering in Medicine & Biology Society (EMBC), Sydney, Australia, 24–27 July 2023; pp. 1–4. [CrossRef]
29. Lo, L.-W.; Zhao, J.; Wan, H.; Wang, Y.; Chakrabartty, S.; Wang, C. An inkjet-printed PEDOT:PSS-based stretchable conductor for wearable health monitoring device applications. *ACS Appl. Mater. Interfaces* **2021**, *13*, 5768–5776. [CrossRef]
30. Silva, R.; Singh, R.; Sarker, B.; Papageorgiou, D.G.; Juhasz, J.A.; Roether, J.A.; Cicha, I.; Kaschta, J.; Schubert, D.W.; Chrissafis, K.; et al. Soft-matrices based on silk fibroin and alginate for tissue engineering. *Int. J. Biol. Macromol.* **2016**, *89*, 161–171. [CrossRef]
31. Lee, K.Y.; Mooney, D.J. Alginate: Properties and biomedical applications. *Prog. Polym. Sci.* **2012**, *37*, 106–126. [CrossRef]

MDPI AG
Grosspeteranlage 5
4052 Basel
Switzerland
Tel.: +41 61 683 77 34

Processes Editorial Office
E-mail: processes@mdpi.com
www.mdpi.com/journal/processes

www.ingramcontent.com/pod-product-compliance
Lightning Source LLC
Chambersburg PA
CBHW061815210926
51850CB00006B/706
* 9 7 8 3 7 2 5 8 4 8 0 1 0 *